T0310175

High-Throughput Analysis for Food Safety

CHEMICAL ANALYSIS

A SERIES OF MONOGRAPHS ON ANALYTICAL CHEMISTRY AND ITS APPLICATIONS

Series Editor
MARK F. VITHA

Volume 179

A complete list of the titles in this series appears at the end of this volume.

High-Throughput Analysis for Food Safety

Edited by

PERRY G. WANG
MARK F. VITHA
JACK F. KAY

Published by John Wiley & Sons, Inc., Hoboken, New Jersey
Published simultaneously in Canada

***Library of Congress Cataloging-in-Publication Data*:**

High-throughput analysis for food safety / edited by Perry G. Wang, Mark F. Vitha, Jack F. Kay.
pages cm. – (Chemical analysis)
Includes index.
ISBN 978-1-118-39630-8 (cloth)
1. Food–Safety measures. 2. Food–Safety measures–Government policy. 3. Food adulteration and inspection. I. Wang, Perry G. II. Vitha, Mark F. III. Kay, Jack F.
RA601.H54 2014
363.19′26–dc23

2013051268

Printed in the United States of America

ISBN: 9781118396308

10 9 8 7 6 5 4 3 2 1

CONTENTS

PREFACE xi

CONTRIBUTORS xiii

CHAPTER 1 INTRODUCTION: BASIC PRINCIPLES OF ASSAYS TO BE COVERED, SAMPLE HANDLING, AND SAMPLE PROCESSING 1

Wanlong Zhou, Eugene Y. Chang, and Perry G. Wang

1.1 Introduction 1
1.1.1 Current Situation and Challenges of Food Safety and Regulations 1
1.1.2 Residues and Matrices of Food Analysis and High-Throughput Analysis 2
1.1.3 Food Safety Classifications 3
1.1.4 "High Throughput" Definition 3
1.1.5 Scope of the Book 4
1.2 Advanced Sample Preparation Techniques 5
1.2.1 Automation of Weighing and Preparing Standard Solutions 5
1.2.2 QuEChERS 6
1.2.3 Swedish Extraction Technique (SweEt) and Other Fast Sample Preparation Methods 6
1.2.4 Turbulent Flow Chromatography 7
1.2.5 Pressurized Liquid Extraction 7
1.2.6 Automated 96- and 384-Well Formatted Sample Preparation as well as Automated SPE Workstations 8
1.2.7 Solid-Phase Microextraction 8
1.2.8 Microextraction by Packed Sorbent 9
1.2.9 Liquid Extraction Surface Analysis 9
1.2.10 Headspace GC 10
1.2.11 Summary 10

1.3 Future Perspectives 10
Acknowledgment 11
References 11

CHAPTER 2 **SURVEY OF MASS SPECTROMETRY-BASED HIGH-THROUGHPUT METHODS IN FOOD ANALYSIS** **15**
Lukas Vaclavik, Tomas Cajka, Wanlong Zhou, and Perry G. Wang
2.1 Introduction 15
2.2 Techniques Employing Chromatographic Separation 15
2.2.1 Gas Chromatography–Mass Spectrometry 15
2.2.2 Liquid Chromatography–Mass Spectrometry 21
2.3 Direct Techniques 30
2.3.1 Matrix-Assisted Laser Desorption/Ionization-Mass Spectrometry 30
2.3.2 Headspace (Solid-Phase Microextraction)-Mass Spectrometry E-Nose 37
2.3.3 Ambient Desorption/Ionization-Mass Spectrometry 38
2.4 Concluding Remarks 62
Acknowledgments 62
References 63

CHAPTER 3 **QUALITY SYSTEMS, QUALITY CONTROL GUIDELINES AND STANDARDS, METHOD VALIDATION, AND ONGOING ANALYTICAL QUALITY CONTROL** **73**
David Galsworthy and Stewart Reynolds
3.1 Introduction 73
3.1.1 Quality System Design 73
3.1.2 Procedures 74
3.1.3 Roles and Responsibilities 74
3.1.4 Quality Manual 74
3.1.5 Document Control 74
3.1.6 Control of Records 75
3.1.7 Audits 75
3.1.8 Validation of Methodology 75
3.1.9 Staff Competency 75

3.1.10 Internal Quality Control 76
3.1.11 Method Performance Criteria 76
3.2 Qualitative Screening Methods 76
3.2.1 Selectivity of Mass Spectrometry-Based Methods 78
3.2.2 Confirmatory Methods 78
3.2.3 Validation of Qualitative Screening Multiresidue Methods for Pesticide Residues in Foods 79
3.3 Elements of the Analytical Workflow 80
3.3.1 Sample Preparation 80
3.3.2 Effects of Sample Processing 81
3.3.3 Extraction Efficiency 81
3.4 Initial Method Validation 81
3.5 Ongoing Analytical Quality Control 86
3.5.1 Internal Quality Control 86
3.5.2 Proficiency Testing 86
3.6 Validation of Qualitative Screening Multiresidue Methods for Veterinary Drug Residues in Foods 87
3.6.1 EU Legislation Covering Method Validation for Veterinary Drug Screening 87
3.6.2 Determination of Specificity/Selectivity and Detection Capability (CCβ) Using the Classical Approach 88
3.6.3 Establishment of a Cutoff Level and Calculation of CCβ 88
3.6.4 Determination of the Applicability 89
3.7 Conclusions 90
References 90

CHAPTER 4 DELIBERATE CHEMICAL CONTAMINATION AND PROCESSING CONTAMINATION 93
Stephen Lock
4.1 Introduction 93
4.2 Heat-Induced Food Processing Contaminants 97
4.3 Packaging Migrants 101
4.4 Malicious Contamination of Food 105
References 111

CHAPTER 5 MULTIRESIDUAL DETERMINATION OF 295 PESTICIDES AND CHEMICAL POLLUTANTS IN ANIMAL FAT BY GEL PERMEATION CHROMATOGRAPHY (GPC) CLEANUP COUPLED WITH GC–MS/MS, GC–NCI-MS, AND LC–MS/MS 117

Yan-Zhong Cao, Yong-Ming Liu, Na Wang, Xin-Xin Ji, Cui-Cui Yao, Xiang Li, Li-Li Shi, Qiao-Ying Chang, Chun-Lin Fan, and Guo-Fang Pang

5.1 Introduction 117
- 5.1.1 Persistent Organic Pollutants 118
- 5.1.2 Polycyclic Aromatic Hydrocarbons 119
- 5.1.3 Polychlorinated Biphenyls 119
- 5.1.4 Phthalate Esters 120
- 5.1.5 Multiclass and Multiresidue Analyses 120

5.2 Experiment 122
- 5.2.1 Instruments 122
- 5.2.2 Reagents 122
- 5.2.3 Preparation of Standard Solutions 122
- 5.2.4 Sample Preparation 123
- 5.2.5 Analytical Methods 124
- 5.2.6 Qualitative and Quantitative Determination 136

5.3 Results and Discussion 136
- 5.3.1 Selection of GPC Cleanup Conditions 136
- 5.3.2 Selection of Extraction Solvent 138
- 5.3.3 Comparison of Sample Extraction Methods 150
- 5.3.4 Comparison of Sample Cleanup 151
- 5.3.5 Linear Range, LOD, and LOQ 152
- 5.3.6 Recoveries and Precisions 152
- 5.3.7 Actual Sample Analysis 157

5.4 Conclusions 161

References 162

CHAPTER 6 ULTRAHIGH-PERFORMANCE LIQUID CHROMATOGRAPHY COUPLED WITH HIGH-RESOLUTION MASS SPECTROMETRY: A RELIABLE TOOL FOR ANALYSIS OF VETERINARY DRUGS IN FOOD 167

María del Mar Aguilera-Luiz, Roberto Romero-González, Patricia Plaza-Bolaños, José Luis Martínez Vidal, and Antonia Garrido Frenich

6.1 Introduction 167

6.2 Veterinary Drug Legislation 168
6.3 Analytical Techniques for VD Residue Analysis 172
6.3.1 Chromatographic Separation 174
6.3.2 High-Resolution Mass Spectrometers 175
6.4 Food Control Applications 181
6.4.1 Screening Applications 181
6.4.2 Confirmation and Quantification Methods 191
6.4.3 Comparison Studies 195
6.5 Conclusions and Future Trends 201
Acknowledgments 202
References 203

CHAPTER 7 A ROLE FOR HIGH-RESOLUTION MASS SPECTROMETRY IN THE HIGH-THROUGHPUT ANALYSIS AND IDENTIFICATION OF VETERINARY MEDICINAL PRODUCT RESIDUES AND OF THEIR METABOLITES IN FOODS OF ANIMAL ORIGIN 213

Eric Verdon, Dominique Hurtaud-Pessel, and Jagadeshwar-Reddy Thota

7.1 Introduction 213
7.2 Issues Associated with Veterinary Drug Residues and European Regulations 215
7.3 Choosing a Strategy: Targeted or Nontargeted Analysis? 216
7.3.1 Targeted Analysis Using HRMS 218
7.3.2 Nontargeted Analysis Using HRMS: Screening for Unknown Compounds 219
7.4 Application Number 1: Identification of Brilliant Green and its Metabolites in Fish under High-Resolution Mass Spectral Conditions (Targeted and Nontargeted Approaches) 220
7.5 Application Number 2: Targeted and Nontargeted Screening Approaches for the Identification of Antimicrobial Residues in Meat 223
7.6 Conclusions 227
References 227

CHAPTER 8 HIGH-THROUGHPUT ANALYSIS OF MYCOTOXINS 231

Marta Vaclavikova, Lukas Vaclavik, and Tomas Cajka

8.1 Introduction 231
8.1.1 Legislation and Regulatory Limits 231
8.1.2 Emerging Mycotoxins 237

8.1.3 Analysis of Mycotoxins in the High-Throughput Environment 238
8.2 Sample Preparation 239
8.2.1 Sampling 240
8.2.2 Matrices of Interest 240
8.2.3 Extraction of Mycotoxins 241
8.2.4 Purification of Sample Extracts 246
8.3 Separation and Detection of Mycotoxins 247
8.3.1 Liquid Chromatography–Mass Spectrometry-Based Methods 248
8.3.2 High-Resolution Mass Spectrometry in Mycotoxins Analysis 250
8.4 No-Separation Mass Spectrometry-Based Methods 252
8.4.1 Matrix-Assisted Laser Desorption Ionization–Mass Spectrometry 252
8.4.2 Ambient Ionization Mass Spectrometry 253
8.4.3 Ion Mobility Spectrometry 254
8.4.4 Immunochemical Methods 256
8.5 Conclusions 259
Acknowledgments 259
References 259
INDEX **267**

PREFACE

The "high throughput" concept has become popular in the pharmaceutical industry after combinatorial chemistry was introduced for drug discovery, such as "high-throughput screening" and "high-throughput drug analysis." However, this concept has drawn significant attention in the global food industry after a number of highly publicized incidents. These incidents include bovine spongiform encephalopathy (BSE) in beef and benzene in carbonated drinks in the United Kingdom, dioxins in pork and milk products in Belgium, pesticides in contaminated foods in Japan, tainted Coca-Cola in Belgium and France, melamine in milk products and pet foods in China, salmonella in peanuts and pistachios in the United States, and phthalates in drinks and foods in Taiwan. Therefore, governments all over the world have taken many measures to tighten control and ensure food safety. Moreover, an exponentially growing population also requires rapid screening assays to ensure the safety of the international food supply. To reflect the international nature of the issues, authors from across the world were invited to contribute to this book. Their chapters thus reflect the global regulatory environment and describe in detail the latest advances in high-throughput screening and confirmatory analysis of food products.

Food safety analysis can be broadly classified based on (i) the residues or analytes and (ii) the food matrices, with some crossover between groups. High-throughput analysis for food safety is aimed at rapidly analyzing and screening food samples to detect the presence of individual or multiple unwanted chemicals, even though there is no numeric definition of "high throughput." These include veterinary drugs, hormones, metals, proteins, environmental contaminants, and pesticides found in food products that could harm consumers, jeopardize the safety of the food supply, and/or disrupt the international trade. This book focuses on high-throughput analyses for food safety using advanced technologies, with many authors discussing the use of tandem mass spectrometry and high-resolution mass spectrometry (HRMS) for rapid, multiple-analyte screening and for confirmatory analyses. Chapters 1–3 provide an overview of the methods used in food analysis and the related regulatory and quality control issues. Chapters 4–8 are "application chapters" and describe the analyses of specific classes of chemicals in a variety of matrices. The contents of each chapter are described in more detail below.

Chapter 1 provides an overview of the current state of food safety analysis and the challenges involved. The common analytical techniques and the rapid sample preparation and extraction methods are also highlighted. Importantly, this chapter also introduces the Codex Alimentarius Commission as it relates to international coordination and standardization efforts. The Codex is discussed repeatedly in subsequent chapters as it pertains to specific residues and analytes. The chapter also provides a way to quantify the throughput of "high-throughput" analyses.

Chapter 2 is a survey of mass spectrometry-based methods. It includes discussions of several ambient MS techniques, including, but not limited to, desorption electrospray ionization (DESI) and direct analysis in real time (DART). It also describes mass spectrometry methods that use a front-end separation technique such as gas chromatography (GC), reverse-phase liquid chromatography (RPLC), hydrophilic interaction chromatography (HILIC), or ultrahigh-performance liquid chromatography (UHPLC). The techniques described in this chapter are routinely used in the subsequent "application" chapters.

Chapter 3 presents quality control guidelines and systems, method validation, regulatory compliance issues, and specific discussions of the Codex and EU legislation.

Chapters 4 deals with testing for deliberate contamination of food and contamination arising from food processing and packaging. Examples include the addition of carcinogenic Sudan dyes to enhance the color of chili powder and the addition of melamine to food products to enhance the apparent protein levels. The heat-induced contamination such as that produced by the Maillard reaction and the migration of molecules from packaging (most famously bisphenol A (BPA)) and inks into food products are also described in this chapter.

Chapter 5 details the ambitious analysis of 295 pesticides and persistent organic pollutants such as polyaromatic hydrocarbons (PAHs) and polychlorinated biphenyls (PCBs) in animal fat using multiple chromatographic techniques coupled with mass spectrometry. Technical aspects of the study are described in detail.

Chapters 6 and 7 deal with the analyses of veterinary drugs (VDs) or veterinary medicinal products (VMPs). Both discuss the use of HRMS coupled with chromatographic separations. They also discuss the regulatory environments, highlighting the EU, U.S., Canadian, Australian, and Japanese regulations, as well as a discussion of the Codex Commission. Specific examples such as the analysis of brilliant green in fish and antimicrobials in meats are described.

Chapter 8 relates to the analysis of mycotoxins and covers aspects such as international regulations, as well as the technical aspects of sampling, extraction, separation, and detection of mycotoxins using both mass spectrometry and biological immunoassays.

The editors hope that this book is a valuable reference as it comprehensively describes how advanced technologies are applied to strengthen food safety. We are fortunate to have a collection by the dedicated contributing authors from across the world. Their persistent efforts and sincere scientific drive have made this book possible.

Perry G. Wang
U.S. Food and Drug Administration, College Park, MD

Mark Vitha
Drake University, Des Moines, IA

Jack Kay
University of Strathclyde, Glasgow, Scotland

CONTRIBUTORS

María del Mar Aguilera-Luiz, Department of Chemistry and Physics, University of Almería, Almería, Spain

Tomas Cajka, UC Davis Genome Center—Metabolomics, University of California, Davis, CA, USA

Yan-Zhong Cao, Qinhuangdao Entry–Exit Inspection and Quarantine Bureau, Qinhuangdao, China

Eugene Y. Chang, Pacific Regional Lab Southwest, U.S. Food and Drug Administration, Irvine, CA, USA

Qiao-Ying Chang, Chinese Academy of Inspection and Quarantine, Beijing, China

Chun-Lin Fan, Chinese Academy of Inspection and Quarantine, Beijing, China

David Galsworthy, Quality Systems Team, The Food and Environment Research Agency (FERA), York, UK

Antonia Garrido Frenich, Department of Chemistry and Physics, University of Almería, Almería, Spain

Dominique Hurtaud-Pessel, French Agency for Food, Environmental and Occupational Health Safety; National Reference Laboratory for Residues of Veterinary Medicinal Products; E.U. Reference Laboratory for Antimicrobial and Dye Residues in Food from Animal Origin, Fougeres Cedex, France

Xin-Xin Ji, Qinhuangdao Entry–Exit Inspection and Quarantine Bureau, Qinhuangdao, China

Xiang Li, Qinhuangdao Entry–Exit Inspection and Quarantine Bureau, Qinhuangdao, China

Yong-Ming Liu, Qinhuangdao Entry–Exit Inspection and Quarantine Bureau, Qinhuangdao, China

Stephen Lock, ABSCIEX, Warrington, UK

José Luis Martínez Vidal, Department of Chemistry and Physics, University of Almería, Almería, Spain

Guo-Fang Pang, Qinhuangdao Entry–Exit Inspection and Quarantine Bureau, Qinhuangdao, China; Chinese Academy of Inspection and Quarantine, Beijing, China

Patricia Plaza-Bolaños, Department of Chemistry and Physics, University of Almería, Almería, Spain

Stewart Reynolds, Food Quality and Safety Programme, The Food and Environment Research Agency (FERA), York, UK

Roberto Romero-González, Department of Chemistry and Physics, University of Almería, Almería, Spain

Li-Li Shi, Nanjing Institute of Environmental Science, Ministry of Environmental Protection of China, Nanjing, China

Jagadeshwar-Reddy Thota, French Agency for Food, Environmental and Occupational Health Safety; National Reference Laboratory for Residues of Veterinary Medicinal Products; E.U. Reference Laboratory for Antimicrobial and Dye Residues in Food from Animal Origin, Fougeres Cedex, France

Lukas Vaclavik, Office of Regulatory Science, Center for Food Safety and Applied Nutrition, U.S. Food and Drug Administration, College Park, MD, USA

Marta Vaclavikova, Office of Regulatory Science, Center for Food Safety and Applied Nutrition, U.S. Food and Drug Administration, College Park, MD, USA

Eric Verdon, French Agency for Food, Environmental and Occupational Health Safety; National Reference Laboratory for Residues of Veterinary Medicinal Products; E.U. Reference Laboratory for Antimicrobial and Dye Residues in Food from Animal Origin, Fougeres Cedex, France

Na Wang, Nanjing Institute of Environmental Science, Ministry of Environmental Protection of China, Nanjing, China

Perry G. Wang, Office of Regulatory Science, Center for Food Safety and Applied Nutrition, U.S. Food and Drug Administration, College Park, MD, USA

Cui-Cui Yao, Qinhuangdao Entry–Exit Inspection and Quarantine Bureau, Qinhuangdao, China

Wanlong Zhou, Office of Regulatory Science, Center for Food Safety and Applied Nutrition, U.S. Food and Drug Administration, College Park, MD, USA

CHAPTER

1

INTRODUCTION: BASIC PRINCIPLES OF ASSAYS TO BE COVERED, SAMPLE HANDLING, AND SAMPLE PROCESSING

WANLONG ZHOU, EUGENE Y. CHANG, and PERRY G. WANG

1.1 INTRODUCTION

1.1.1 Current Situation and Challenges of Food Safety and Regulations

Food can never be entirely safe. In recent years, food safety concern has grown significantly following a number of highly publicized incidents worldwide. These incidents include bovine spongiform encephalopathy in beef and benzene in carbonated drinks in the United Kingdom, dioxins in pork and milk products in Belgium, pesticides in contaminated foods in Japan, tainted coca-cola in Belgium and France, melamine in milk products in China, salmonella in peanuts and pistachios in the U.S. [1], and phthalates in drinks and foods in Taiwan [2]. Governments all over the world have taken many measures to increase food safety, resulting in a marked increase in the number of regulated compounds.

The European Union (EU) made a considerable effort to centralize food regulatory powers. The European Food Safety Authority (EFSA) and the national competent authorities are networks for food safety. The European Commission has designated food safety as a top priority, and published a white paper on food safety [3]. Legislative documents, such as 657/2002/EC, which sets out performance criteria for veterinary drug residue methods, are published as European Commission Decisions [4].

The Japanese government implemented a "positive list" to regulate the use of pesticides, veterinary drugs, and other chemicals in 2006, which replaced the old "negative list" regulations [5]. Over 700 compounds have to be monitored and reported. A certified safety report is now a requirement for both importing and exporting countries. The new regulations are listed as addendums to the positive list. In Japan, strengthening regulations for industrial use of perfluorooctane sulfonate (PFOS) and perfluorooctanoate (PFOA), additives, and residual pharmaceutical and personal care products (PPCPs) in the environment is progressing, which in turn creates a demand for instrumentation that provides reliable trace determination.

High-Throughput Analysis for Food Safety, First Edition.
Edited by Perry G. Wang, Mark F. Vitha, and Jack F. Kay.

In the United States, federal laws are the primary source of food safety regulations, for example, related codes under CFR Title 7, 9, 21, and 40. The law enforcement network comprises state government agencies and federal government agencies, including the U.S. Department of Agriculture (USDA), Food and Drug Administration (FDA), Centers for Disease Control and Prevention (CDC), and National Oceanic and Atmospheric Administration (NOAA). The Food Safety Modernization Act (H.R. 2751) is a federal statute signed into law by President Barack Obama on January 4, 2011. The law grants FDA authority to order recalls of contaminated food, increase inspections of domestic food facilities, and enhance detection of food-borne illness outbreaks.

As a result of regulation change and globalization, most nations around the world have now increased regulations on food safety for their domestic and export markets. International coordination and standardization are mainly conducted by the Codex Alimentarius Commission (CAC). The CAC is an intergovernmental body established in 1961 by the Food and Agriculture Organization of the United Nations (FAO), and joined by the World Health Organization (WHO) in 1962 to implement the Joint FAO/WHO Food Standards Program. There are 185 member countries and one organization member (EC) in the Codex now. The Codex standards are recommendations for voluntary application by members. However, in many cases, these standards are the basis for national legislation. The Codex covers processed, semiprocessed, and raw foods. The Codex also has general standards covering (but not limited to) food hygiene, food additives, food labeling, and pesticide residues [6].

1.1.2 Residues and Matrices of Food Analysis and High-Throughput Analysis

From the examples listed above, it is simply impossible to test every single item for every imaginable food-borne pathogen, including bacteria, viruses, and parasites; food allergens such as milk, eggs, shellfish, and soybean; naturally occurring toxins and mycotoxins; residues of pesticides and veterinary drugs; environmental contaminants; processing and packaging contaminants; spoilage markers [7]; food authenticity; and labeling accuracy [8].

Fortunately, modern analytical techniques, especially mass spectrometry-based techniques, such as gas chromatography–mass spectrometry (GC–MS) and liquid chromatography–mass spectrometry (LC–MS), can help speed up the processes. In the past decade, LC–MS, including tandem LC–MS techniques, or LC–MS/MS, has been applied in pesticide residue analysis and other food safety issues. The use of LC–MS has increased exponentially in recent years [9]. For example, an LC–MS/MS method using a scheduled selected reaction monitoring (sSRM) algorithm was developed and applied to analyze 242 multiclass pesticides for fruits and vegetables [10]. The high selectivity of LC–MS can effectively reduce interference from matrices, which significantly simplifies the process of sample preparation.

In addition, other high-throughput methods, including bioactivity-based methods, have also been widely applied today and will continue to be applied at least for the foreseeable future, although false-positive results were found in a high number of cases for these methods [11]. A striking example is the rapid microbiological assays used routinely by dairies to screen milk inexpensively and rapidly for residues of antimicrobial drugs. In the United Kingdom alone, dairy companies run millions of such assays per year, with a test duration of only minutes from sampling to result. These tests are widely used internationally by dairies for completeness.

1.1.3 Food Safety Classifications

Food safety analysis can be broadly classified and grouped based on the residues or analytes and food matrices, accepting that there will be some degree of crossover between groups. Based on the analytes, it can be classified to pesticide residues, drug residues, mycotoxins and environment pollutants, and other industrial chemicals. Based on food matrices, the most accepted classification of groups consists of high-moisture foods, low-moisture foods, and fatty foods. Examples of such matrices are fruits and vegetables, dry grains (wheat, rice, bean, etc.), and tissues, including fish and meat.

Food safety analysis methods can be further divided into two categories: screening methods and confirmation methods. The regulatory agencies and international standard organizations have clear guidelines for screening methods and confirmation methods. The requirements are slightly different for both, depending on the residues to be analyzed, matrix, risk factor, and techniques available. A screening method is qualitative or semiquantitative in nature, comprises establishment of those residues likely to be present based on an interpretation of the raw data, and tries to avoid false negatives as much as possible. A false negative rate of 5% is accepted for both the EU and the US FDA [12,13]. A confirmation method can provide unequivocal confirmation of the identity of the residue and may also confirm the quantity present on residues found in screening. Therefore, an analyst has to use appropriate guidelines to develop a new method based on the regulation, residue category, and matrices and to provide expert advice on the findings to those commissioning the analysis.

1.1.4 "High Throughput" Definition

The "high throughput" concept has become popular in the pharmaceutical industry after combinatorial chemistry was introduced for drug discovery [14], such as in "high-throughput screening" and "high-throughput drug analysis." However, "high-throughput analysis for food safety" has only recently drawn more attention, especially after China's melamine milk crisis and Taiwan's phthalates scandal.

Although there is no numeric definition of "high-throughput screening" in the pharmaceutical industry, the standardized sample plate of 96-, 384-, or even 1536-well plates can indicate how quickly many analyses can be completed. Compared with single digits of targets in drug screening, food analysis often involves multiclass compounds ranging from a few dozens to a few hundred targets. All these kinds of

GC–MS or LC–MS methods can be considered as high-throughput analyses because one way to calculate sample throughput is to use the following equation [15]:

$$\text{sample throughput} = \frac{\text{screening capacity} \times \text{number of samples}}{\text{total analysis time}} \qquad (1.1)$$

where screening capacity or analysis capacity = number of target analytes that can be screened or analyzed by the method; total analysis time = time for sample preparation + instrument data acquisition + data analysis (data process) + documentation. Given this definition, analyses using GC–MS and LC–MS as already discussed can qualify as "high throughput" because their screening capacities can be, in some instances, quite high. High screening capacities eliminate the need for many analyses on the same sample that simply screen for just one or two analytes at a time. Practically, as long as the sample throughput of a new method is significantly higher than that obtained using the current prevailing method, the new method should be considered as a high-throughput method.

1.1.5 Scope of the Book

Food safety analysis usually involves the simultaneous measurement of multiple analytes from a complex matrix. Separation of the analytes from matrices is often crucial for mass spectrometry-based analyses. Although separations can be achieved electrophoretically on one- and two-dimensional gels, by capillary electrophoresis and by GC and LC, both LC and GC are still the most applied separation methods due to their good reproducibility, recovery, sensitivity, dynamic range, and quantifiability [8,16].

GC–MS has been widely used for food safety analysis for a long time. However, the use of LC–MS for food safety analysis is among the fastest developing fields in science and industry [17]. Currently, both LC–MS and GC–MS are widely used for every food safety issue, as already mentioned. There are many modern approaches in LC–MS- and GC–MS-based methods that enable the reduction of "analytical" time and increase the sample throughput.

The book is divided into eight chapters: Chapters 1–3 discuss technology background, statistical background, industrial standards, and governments' regulations. Chapters 4–8 discuss specific fields of method development, applications of new technologies, and practice of analytical work to compile industrial standards and government regulations. The topics include pesticide residues analysis, veterinary drug residue analysis, mycotoxins analysis, and industrial chemical analysis. The discussions will show not only the current dynamic interaction between technology development and laboratory practice but also the trends of food safety analysis. Advanced sample preparation techniques and future perspectives will be discussed in the following sections, with an emphasis on an evaluation of or improvements in the throughput of the methods.

1.2 ADVANCED SAMPLE PREPARATION TECHNIQUES

Food safety analysis is a difficult task because of the complexity of food matrices and the low concentrations at which target compounds are usually present. Thus, despite the advances in the development of highly efficient analytical instrumentation for their final determination, sample pretreatment remains a bottleneck and an important part of obtaining accurate quantitative results. A past survey has shown that an average chromatography separation accounts for about 15% of the total analysis time, sample preparation for about 60%, and data analysis and reporting for 25% [18,19]. However, some new technologies and automation have significantly accelerated the sample preparation process.

Sample preparation can involve a number of steps, including collection, drying, grinding, filtration, centrifugation, precipitation, dilution, and various forms of extraction. The most conventional sample preparation methods are protein precipitation (PPT), liquid–liquid extraction (LLE), and solid-phase extraction (SPE). In addition to these traditional methods, many advanced approaches have been proposed for pretreatment and/or extraction of food samples. These approaches include salting out LLE (SALLE) such as QuEChERS (quick, easy, cheap, effective, rugged, and safe) and SweEt (Swedish extraction technique), supercritical fluid extraction (SFE), pressurized liquid extraction (PLE), microwave-assisted extraction (MAE), matrix solid-phase dispersion (MSPD), solid-phase microextraction (SPME), stir bar sorptive extraction (SBSE), turbulent flow chromatography (TFC), and others [8,20–23]. To avoid overlap with other chapters, only automation of weighing and preparing standard solutions, QuEChERS, SWEET, TFC, PLE, automated 96- and 384-well formatted sample preparation, headspace, SPME, MEPS, and liquid extraction surface analysis (LESA™) are discussed in the following sections.

1.2.1 Automation of Weighing and Preparing Standard Solutions

The first step of an analysis is to weigh standards for calibration solutions. With an automatic dosing balance, a tablet, paste, or powder sample can be easily weighed into a volume flask. Combined with liquid dosing, a specified target concentration can be obtained by adding the exact amount of solvent automatically.

Many routine sample preparations, such as calibration curve generation, sample dilution, aliquoting, reconstitution, internal standard addition, or sample derivatization are often time consuming. The technology development of liquid handlers has provided full automation or semiautomation solutions. Basically, there are two approaches: one is the multiple pipette liquid handler; another is the multifunction autosampler. For example, a sample preparation workbench was applied to determine eicosapentaenoic acid (EPA) and docosahexaenoic acid (DHA) in marine oils found in today's supplement market [24]. The workbench was programmed to methylate the analytes (derivatization) for each analytical run, to avoid sample exposure to oxygen in a closed system, and to transfer the top layer of sample to a final GC vial for injection. The workbench not only gave results comparable to three widely applied methods (AOAC 991.39, AOCS Ce 1i-07, and the GOED voluntary monograph for EPA and DHA) but also reduced analysts' time and solvent consumption.

1.2.2 QuEChERS

Anastassiades et al. developed an analytical methodology combining the extraction/isolation of pesticides from food matrices with extract cleanup [25]. The traditional method was LLE followed by salting out of water and cartridge cleaning up. Their new method used dispersive SPE sorbent (d-SPE) together with salting out in a centrifugation tube, which simplifies the whole procedure and reduces solvent consumption and dilution error. They coined the acronym QuEChERS for it. Since its inception, QuEChERS has been gaining significant popularity and has achieved official method status from international organizations (AOAC Official Method 2007.01 and European Standard Method EN 15662) for pesticide analysis.

Besides pesticide residue analysis in food samples, QuEChERS has also been used for the analysis of other industrial chemicals or environmental pollutants such as polycyclic aromatic hydrocarbons in fish, veterinary drugs in animal tissue and milk [26], and hormone esters in muscle tissues. QuEChERS and its variations have also been used for the determination of xenobiotics, mycotoxins, veterinary drugs, environmental or industrial contaminants, and nutraceutical products [27].

1.2.3 Swedish Extraction Technique (SweEt) and Other Fast Sample Preparation Methods

The SweEt method [28] was developed by the Swedish National Food Agency. It is a LLE technique that uses ethyl acetate to differentiate the polar impurities from less polar residues of pesticides or other chemicals. Based on the SweEt method, food samples are classified into four categories: fruit and vegetable, cereals, animal origin A, and animal origin B with high fat. For fruit, vegetable, cereals, or animal origin A matrices, the sample cleanup is filtration–centrifugation or centrifugation–filtration prior to injection for GC–MS/MS or liquid chromatography coupled with tandem mass spectrometry (LC–MS/MS) analysis. For animal origin B matrix, an additional gel permeation chromatography (GPC) cleanup step to remove the coextracted fat from the extracts and solvent exchange step is needed prior to GC–MS or LC–MS injection. The method can cover multiresidues or single group of residue(s). The method uses smaller volumes of solvent and provides extracts that are compatible with GC or LC injection methods. It eliminates complicated cleanup steps (except animal origin B samples with high fat) and introduces very low concentrations of matrix components such as proteins and sugar. The method has been used to determine pesticides in fruits, vegetables, cereals, and products of animal origin [28].

QuEChERS and SweEt are general methods for multipesticide residue screening. Based on the same principles of LLE and SPE, many other methods were recently developed for other analytes such as special groups of pesticide residues or veterinary drugs.

A set of methods was developed to analyze pesticide residues that could not be covered in large groups of multiresidue analysis [29]. An example is the analysis of polar pesticides such as paraquat and mepiquat. In the method, stable isotopically labeled internal standards were added to samples before extraction. For dry samples, water was added to the sample first and then methanol with 1% formic acid was used to extract the samples. After centrifugation and filtration, the extracted solutions were injected into LC–MS/MS for quantification. For the analysis of paraquat and diquat, H_2O:MeOH (1:1) with. 0.05 M HCl was used as the extraction solution.

An efficient acetonitrile extraction method followed by using a C-18 SPE cartridge for cleaning up the extracted solution was developed and fully validated to detect tetracycline and seven other groups of veterinary drug residues in eggs by LC–MS/MS [30]. The method can detect 1–2 ng/g of 40 drugs from eight different classes.

1.2.4 Turbulent Flow Chromatography

TFC was introduced in the late 1990s as a technique for the direct injection of biological fluids into a small-diameter column packed with 30 μm spherical porous particles [31]. A high flow rate mobile phase runs through the column to form a turbulent flow. Then, the eluents are directed to an analytical column or waste controlled by a switch valve. The first column (turbo flow column) runs SPE, which can be reversed phase, hydrophilic interaction liquid chromatography (HILIC), size exclusion, or some other modes. The second column runs regular HPLC separation. Today, TFC has been developed as an automated online high-throughput sample preparation technique that makes use of high flow rates in 0.5 or 1.0 mm internal diameter columns packed with particles of size 30–60 μm. These large particle columns allow much higher flow rate with lower backpressure. The smaller analytes diffuse more extensively than larger molecules (e.g., proteins, lipids, and sugars from the matrix) into the pores of the sorbent. The larger molecules do not diffuse into the particle pores because of high flow rate and are washed to waste. The trapped analytes are desorbed from the TFC column by back-flushing it with an organic solvent and the eluate can be transferred with a switching valve onto the analytical LC–MS/MS system for further separation and detection.

Compared with traditional SPE, TFC reduces the time required for off-line sample preparation from hours to minutes because it uses reusable extraction columns in a closed system. It also allows automatic removal of proteins and larger molecules in complex mixtures by combining turbulence, diffusion, and chemistry. TFC technology also allows a broad selection of stationary phases for different matrices. For example, melamine and eight veterinary drugs, belonging to seven different classes, were detected by TFC–LC–MS/MS in milk [31,32].

1.2.5 Pressurized Liquid Extraction

PLE is a rapid extraction of solid/semisolid matrices using organic solvents or water by applying high temperatures (up to 200 °C) and high pressures (up to 1500 psi) to keep solvents in a liquid state above their atmospheric boiling points to increase solvation power and change extraction kinetics. Raised temperature can also disrupt the strong solute–matrix interactions. The process reduces solvent consumption and operating time so as to increase the extraction efficiency. The automated PLE system can automatically load up to 24 samples in one batch. The sample cell is of different sizes, such as 1, 5, 10, 22, 34, 66, and 100 ml. Azamethiphos, avermectins, carbamates, and benzoylurea pesticides as well as chemotherapeutic agents in seaweeds were determined using PLE and separation of analytes by LC–MS/MS [33]. The applications of PLE in the analysis of food samples have been comprehensively summarized by Mustafa and Turner [34].

1.2.6 Automated 96- and 384-Well Formatted Sample Preparation as well as Automated SPE Workstations

Although automated 96- and 384-well extractions (e.g., LLE and SPE) have been widely used for bioanalysis [35], they have not yet been widely applied for food safety analysis. The possible reasons are mainly attributed to the high cost of automated extraction equipment and more varieties and relatively larger sample size of food samples. Some new automated SPE workstations can handle a much wider range of sample sizes (1–6 ml/40 ml). Therefore, they can overcome some of the limitations. For example, an autosampler-compatible cartridge (Strata-X, 3 ml/200 mg, SPE cartridge) was applied in an automated SPE workstation to detect acrylamide in brewed coffee by LC–MS/MS [36]. We predict that the application of automated extraction systems will draw more and more attention for food safety analysis in the near future.

1.2.7 Solid-Phase Microextraction

SPME was introduced in the early 1990s as a simple and effective adsorption/absorption (based on the solid/liquid coating) and desorption technique. Instead of using a syringe to pick up and inject sample into a chromatography instrument, SPME uses a piece of bonded-phase capillary tube or metal/polymer fiber to load (adsorption) and introduce (desorption) sample into instrument. The device with a bonded-phase capillary tube is called in-tube SPME and the device with a bonded-phase fiber is called fiber SPME.

The capillary tube for in-tube SPME is like a short GC column. When the sample solution goes through the tube, the bonded phase is enriched in analytes through absorption/adsorption. After the solvent is dried by a gas flow, the sample becomes a film adsorbed on the surface of the tube, and is then desorbed with heat and introduced into the instrument. The fiber SPME uses the same steps of absorption/adsorption–desorption as does in-tube SPME. The difference is that the fiber can be immersed into a solution, which is called liquid immersion SPME, or be held above solutions or solid particles/powders to adsorb the vapor from such samples, which is called headspace SPME [37].

Because different surface coatings (bonded phases) have different selectivities to different compounds, choosing an appropriate SPME fiber or tube can differentiate these compounds from a sample matrix. Therefore, SPME can combine sampling, isolation, and enrichment in one step. SPME can be connected easily to a GC and LC system using available interfaces. Thus, SPME can reduce the time required for sample preparation and eliminate the use of large volumes of extraction solvents.

Besides the properties of surface coating, analytes, and sample matrix, the concentration of analytes is also an important factor for optimization with both tube and fiber SPME. The headspace sampling SPME is a little more complicated because of the heterogeneous phases in sample vials at the adsorption step: one factor is the distribution coefficient of the analyte in two phases (gas–solid or gas–liquid); another factor is the volume ratio of the two phases. These factors are affected by temperature, sample volume, and sample matrix. Since the introduction of SPME, it has become a practical, low-cost alternative for sample preparation for GC–MS. New surface coating materials extended SPME from small molecule to large molecule analysis, from food sample to blood or tissue samples, and from in-lab sample preparation to on-site sample preparation. Besides the application of SPME to GC–MS, SPME has been applied to analyze mycotoxins (ochratoxins A and B) in nuts and grain samples and insecticides in honey by LC–MS [38,39]. It is believed that SPME will become a practical alternative for sample preparation for LC–MS in the future.

1.2.8 Microextraction by Packed Sorbent

Microextraction by packed sorbent (MEPS) is a new development in the field of sample preparation and sample handling. It entails the miniaturization of conventional SPE packed-bed devices from milliliter bed volumes to microliter volumes. MEPS can be connected online to GC or LC without any modifications. In MEPS, ~1 mg of the solid packing material is packed inside a syringe (100–250 μl) as a plug or between the barrel and the needle as a cartridge. Sample preparation occurs on the packed bed. The bed can be coated to provide selective and suitable sampling conditions. The combination of MEPS and LC–MS is a good tool for screening and determining drugs and metabolites in blood, plasma, and urine samples [40].

MEPS has also been applied to food and beverage analysis, including the analysis of bioflavonoids from red wine, diterpene glycosides from tea extract, pesticides and PCB in fats, aflatoxin B_2 and M_2 metabolite trace analysis in milk, mycotoxin trace analysis in cereals, fatty acid methyl esters (long chain) in fermentation medium, omega-6 fatty acid in malt lipid, pigment anthocyanidins in wine, atrazine in cereals, sulfonamide trace analysis in meat, penicillin in dairy products, and cork taints in wine [41].

1.2.9 Liquid Extraction Surface Analysis

LESA was developed at Oak Ridge National Laboratory [42] to bring the benefits of nano-ESI/MS to surface analysis and to automate surface sampling for faster and more effective analyses. This approach mainly involves three steps. In step 1, a robot aliquots a sample of extraction and sprays solvent into a pipette tip. In step 2, the solvent in the pipette tip is dispensed/aspirated onto the sample surface (e.g., an apple skin) to perform extraction of any chemicals on the surface of the apple. The pipette tip diameter is 800 μm, which produces a surface area wetted with extraction solvent. In step 3, the pipette tips and the sample extract are robotically positioned at the inlet of the ESI chip for nano-ESI-MS analysis [43]. It has been applied to analyze pesticides on apples.

1.2.10 Headspace GC

Headspace analysis has been used for more than 30 years [44] and is still one of the most important sample preparation techniques for gas chromatography [45]. It is based on the principles of gas extraction, that is, on the partition of an analyte in a heterogeneous liquid–vapor system. A good example is that headspace gas chromatography–mass spectrometry (HS-GC–MS) has been successfully applied to rapidly detect benzene, toluene, ethylbenzene, *o*-, *m*-, and *p*-xylenes, and styrene in olives and olive oil [46].

1.2.11 Summary

Using these advanced extraction techniques and their automated analogs, as already discussed, coupled with LC–MS and GC–MS techniques, more analytes per unit time can be analyzed from an increasing range of matrices, thereby increasing throughput in food analyses.

1.3 FUTURE PERSPECTIVES

In addition to GC–MS and LC–MS techniques, other techniques such as near-infrared (NIR), nuclear magnetic resonance (NMR), and capillary electrophoresis have also been developed for high-throughput food safety analysis. A handheld unit based on NIR spectroscopy and chemometrics has been developed for the rapid (<5 min) detection and quantification of economic adulterants in foods, specifically melamine in skimmed milk powder, for potential field use [47]. A new NMR procedure has been developed for routine nontargeted and targeted analyses of foods [48]. Capillary electrophoresis combined with inductively coupled plasma mass spectrometry (CE-ICP-MS) has been developed as an analytical tool for the characterization of nanomaterials in dietary supplements. These nanoparticles are difficult to separate with other techniques such as asymmetric field flow fractionation and size exclusion chromatography, due to their smaller particle sizes (typically less than 20 nm) [49].

Compared with bioanalysis, high-throughput analysis for food safety using mass spectrometry-based techniques (LC–MS and GC–MS) is not popular and gets less attention. However, we predict that throughput for food safety analysis will be significantly improved with the use and development of automated sample preparation technologies, ultrahigh-performance liquid chromatography (UHPLC) and high-resolution MS.

ACKNOWLEDGMENT

The authors wish to thank Alexander J. Krynitsky for helpful discussions.

REFERENCES

1. Malik, A.K.; Blasco, C.; Picó, Y. Liquid chromatography–mass spectrometry in food safety. *J. Chromatogr. A* **2010**, 1217, 4018–4040.
2. Wu, M.T.; Wu, C.F.; Wu, J.R.; Chen, B.H.; Chen, E. K.; Chao, M.C.; Liu, C.K.; Ho, C.K. The public health threat of phthalate-tainted foodstuffs in Taiwan: the policies the government implemented and the lessons we learned. *Environ. Int.* **2012**, 44, 75–79.
3. Commission of the European Communities. **2000**. Available at http://ec.europa.eu/dgs/health_consumer/library/pub/pub06_en.pdf (accessed March 2, 2014).
4. European Commission Decisions. **2002**. Available at http://eur-lex.europa.eu. (accessed March 2, 2014).
5. The Food Safety Commission in the Cabinet Office. **2006**. Available at http://www.mhlw.go.jp/english/topics/foodsafety/positivelist060228/index.html (accessed March 2, 2014).
6. CODEX Alimentarius. Available at http://www.codexalimentarius.org/standards/. (accessed March 2, 2014).
7. Kellmann, M.; Muenster, H.; Zomer, P.; Mol, H. Full scan MS in comprehensive qualitative and quantitative residue analysis in food and feed matrices: how much resolving power is required? *J. Am. Soc. Mass Spectrom.* **2009**, 20, 1464–1476.
8. Garcia-Canas, V.; Simo, C.; Herrero, M.; Ibanez, E.; Cifuentes, A. Present and future challenges in food analysis: foodomics. *Anal. Chem.* **2012**, 84, 10150–10159.
9. Pico, Y.; Font, G.; Ruiz, M.J.; Fernandez, M. Control of pesticide residues by liquid chromatography–mass spectrometry to ensure food safety. *Mass Spectrom. Rev.* **2006**, 25, 917–950.
10. Fillatre, Y.; Rondeau, D.; Jadas-Hecart, A.; Communal, P.Y. Advantages of the scheduled selected reaction monitoring algorithm in liquid chromatography/electrospray ionization tandem mass spectrometry multi-residue analysis of 242 pesticides: a comparative approach with classical selected reaction monitoring mode. *Rapid Commun. Mass Spectrom.* **2010**, 16, 2453–2461.
11. Hoff, R.; Ribarcki, F.; Zancanaro, I.; Castellano, L.; Spier, C.; Barreto, F.; Fonseca, S.H. Bioactivity-based screening methods for antibiotics residues: a comparative study of commercial and in-house developed kits. *Food Addit. Contam.* **2012**, 29(4), 577–586.
12. European Food Safety Authority. *Method validation and quality control procedures for pesticide residues analysis in food and feed*. **2011**. Available at http://ec.europa.eu/food/plant/protection/pesticides/docs/qualcontrol_en.pdf (accessed March 2, 2014).
13. US Food and Drug Administration Office of Foods. *Guidelines for the validation of chemical methods for the FDA foods program*. **2012**. Available at http://www.fda.gov/downloads/ScienceResearch/FieldScience/UCM298730.pdf (accessed March 2, 2014).
14. Wang, P.G. *High-Throughput Analysis in the Pharmaceutical Industry*, 1st edition. New York: CRC Press; **2008**.
15. Zhang, K.; Wong, J.W.; Wang, P.G. A perspective on high throughput analysis of pesticide residues in foods. *Chin. J. Chromatogr.* **2011**, 29, 587–593.

16. Faeste, C.K.; Ronning, H.T.; Christians, U.; Granum, P.E. Liquid chromatography and mass spectrometry in food allergen detection. *J. Food Prot.* **2011**, 74, 316–345.
17. Di Stefano, V.; Avellone, G.; Bongiorno, D.; Cunsolo, V.; Muccilli, V.; Sforza, S.; Dossena, A.; Drahos, L.; Vekey, K. Applications of liquid chromatography–mass spectrometry for food analysis. *J. Chromatogr. A* **2012**, 1259, 74–85.
18. Smith R.M. Before the injection: modern methods of sample preparation for separation techniques. *J. Chromatogr. A* **2003**, 1000, 3–27.
19. Gilpin, R.K.; Zhou, W. Designing high throughput HPLC assays for small and biological molecules. In: Wang, P.G., editor. *High Throughput Analysis in the Pharmaceutical Industry*, 1st edition. New York: CRC Press; **2008**, pp. 339–353.
20. Zhang, L.J.; Liu, S.W.; Cui, X.Y.; Pan, C.P.; Zhang, A.L.; Chen, F. A review of sample preparation methods for the pesticide residue analysis in foods. *Cent. Eur. J. Chem.* **2012**, 10, 900–925.
21. Rostagno, M.A.; D'Arrigo, M.; Martinez, J.A. Combinatory and hyphenated sample preparation for the determination of bioactive compounds in foods. *Trends Anal. Chem.* **2010**, 29(6), 553–561.
22. Moreno-Bondi, M.C.; Marazuela, M.D.; Herranz, S.; Rodriguez, E. An overview of sample preparation procedures for LC–MS food samples. *Anal. Bioanal. Chem.* **2009**, 395, 921–946.
23. Lambropoulou, D.A.; Albanis, T.A. Methods of sample preparation for determination of pesticide residues in food matrices by chromatography–mass spectrometry-based techniques: a review. *Anal. Bioanal. Chem.* **2007**, 389, 1663–1683.
24. Kanable, S.; Mitchell, B.; Leuenberger, C.; Meinholz, E.; Dallman, M.; Vacha, E.; Richard, J.; Volkmann, C.; Ellefson, W. An automated method for accurate determination of EPA and DHA in marine oils found in today's supplement market. *125th AOAC Annual Meeting & Exposition*, New Orleans, LA, September 18–21, **2011**.
25. Anastassiades, M.; Lehotay, S.J.; Stajnbaher, D. Fast and easy multiresidue method employing acetonitrile extraction/partitioning and "dispersive solid-phase extraction" for the determination of pesticide residues in produce. *J. AOAC Int.* **2003**, 86(2), 412–431.
26. Keegan, J.; Whelan, M.; Danaher, M.; Crooks, S.; Sayers, R.; Anastasio, A.; Elliott, C.; Brandon, D.; Furey, A.; O'Kennedy, R. Benzimidazole carbamate residues in milk: detection by surface plasmon resonance–biosensor, using a modified QuEChERS (quick, easy, cheap, effective, rugged and safe) method for extraction. *Anal. Chim. Acta* **2009**, 654 (2),111–119.
27. Wilkowska, A.; Biziuk M. Determination of pesticide residues in food matrices using the QuEChERS methodology. *Food Chem.* **2011**, 125, 803–812.
28. Ekroth, S. Simplified analysis of pesticide residues in food using the Swedish ethyl acetate method (SweEt). *3rd Latin American Pesticide Residue Workshop (LAPRW 2011)*, Montevideo, Uruguay, May 8–11, **2011**.
29. Anastassiades, M.; Kolberg, D.S.; Mack, D.; Sigalova, I.; Roux, D. Multiclass, multiresidue analysis of pesticides typically analyzed by single-analyte methods. *47th Florida Pesticide Residues Workshop*, St. Pete Beach, FL, July 18–21, **2010**.
30. Chang, E.; An, H.; Wong, G.; Wang, K.; Cain, T.; Paek, H.C.; Sram, J. Sensitive and accurate multi-class multi-residue veterinary drug analytical method validation for shell egg using liquid chromatography tandem mass spectrometry. *2nd Annual FDA*

Foods Program Science and Research Conference, Silver Spring, MD, August 1–2, **2012**.

31. Stolker, A.A.M.; Peters, R.J.B.; Zuiderent, R.; DiBussolo, J.M.; Martins, C.P.B. Fully automated screening of veterinary drugs in milk by turbulent flow chromatography and tandem mass spectrometry. *Anal. Bioanal. Chem.* **2010**, **397**, 2841–2849.
32. Roacha, J.A.G.; DiBussolob, J.M.; Krynitskya, A., Noonana, G.O. Evaluation and single laboratory validation of an on-line turbulent flow extraction tandem mass spectrometry method for melamine in infant formula. *J. Chromatogr. A* **2011**, 1218, 4284–4290.
33. Lorenzo, R.A.; Pais, S.; Racamonde, I.; Garcia-Rodriguez, D.; Carro, A.M. Pesticides in seaweed: optimization of pressurized liquid extraction and in-cell clean-up and analysis by liquid chromatography–mass spectrometry. *Anal. Bioanal. Chem.* **2012**, 404, 173–181.
34. Mustafa, A.; Turner, C. Pressurized liquid extraction as a green approach in food and herbal plants extraction: a review. *Anal. Chim. Acta* **2011**, 703, 8–18.
35. Chang, M.S.; Kim, E.J.; El-Shourbagy, T.A. Evaluation of 384-well formatted sample preparation technologies for regulated bioanalysis. *Rapid Commun. Mass Spectrom.* **2007**, 21, 64–72.
36. Foster, F.D.; Stuff, J.R.; Pfannkoch, E.A. *Automated solid phase extraction (SPE)–LC–MS/MS method for the determination of acrylamide in brewed coffee samples.* **2012**. Available at http://www.gerstel.com/pdf/p-lc-an-2012-13.pdf.
37. Chen, Y.; Guo, Z.P.; Wang, X.Y.; Qiu, C.G. Sample preparation, *J. Chromatogr. A* **2008**, 1184, 191–219.
38. Saito, K.; Ikeuchi, R.; Kataoka, H. Determination of ochratoxins in nuts and grain samples by in-tube solid-phase microextraction coupled with liquid chromatography–mass spectrometry. *J. Chromatogr. A* **2012**, 1220, 1–6.
39. Blasco, C.; Vazquez-Roig, P.; Onghena, M.; Masia, A.; Pico, Y. Analysis of insecticides in honey by liquid chromatography–ion trap-mass spectrometry: comparison of different extraction procedures. *J. Chromatogr. A* **2011**, 1218, 4892–4901.
40. Abdel-Rehim, M. Microextraction by packed sorbent (MEPS): a tutorial. *Anal. Chim. Acta* **2011**, 701, 119–128.
41. Lahoutifard, N.; Dawes, P.; Wynne, P. *Micro extraction packed sorbent (MEPS): analysis of food and beverages.* Available at http://www.sge.com/uploads/ff/75/ff750c33411fa6370baa2937172cab7b/TP-0189-M.pdf. (accessed March 2, 2014).
42. Kertesz, V.; Van Berkel, G.J. Fully automated liquid extraction-based surface sampling and ionization using a chip-based robotic nanoelectrospray platform. *J. Mass Spectrom.* **2010**, 45(3), 252–260.
43. Eikel, D.; Henion, J. Liquid extraction surface analysis (LESA) of food surfaces employing chip-based nano-electrospray mass spectrometry. *Rapid Commun. Mass Spectrom.* **2011**, 25, 2345–2354.
44. Kolb, B.; Ettre, L. S. *Static Headspace-Gas Chromatography: Theory and Practice*, 2nd edition. New York: John Wiley & Sons, Inc.; **2006**.
45. Snow, N. H.; Bullock, G.P. Novel techniques for enhancing sensitivity in static headspace extraction-gas chromatography. *J. Chromatogr. A* **2010**, 1217, 2726–2735.
46. Gilbert-Lopez, B.; Robles-Molina, J.; Garcia-Reyes, J.F.; Molina-Diaz, A. Rapid determination of BTEXS in olives and olive oil by headspace-gas chromatography/mass spectrometry (HS-GC–MS). *Talanta* **2010**, 83(2), 391–399.

47. Ashour, A.; Mossoba, M.; Fahmy, R.; Hoag, S.W. Hand-held near infrared detector coupled with chemometrics for field use: rapid quantification of economic chemical adulterants in foods. *2nd Annual FDA Foods Program Science and Research Conference*, Silver Spring, MD, August 1–2, **2012**.
48. Lachenmeier, D.W.; Humpfer, E.; Fang, F.; Schutz, B.; Dvortsak, P.; Sproll, C.; Spraul, M. NMR-spectroscopy for nontargeted screening and simultaneous quantification of health-relevant compounds in foods: the example of melamine. *J. Agric. Food Chem.* **2009**, 57, 7194–7199.
49. Mudalige T.K.; Linder, S.W. Capillary electrophoresis/ICP-MS as an analytical tool for the characterization of nanomaterials in dietary supplements. *2nd Annual FDA Foods Program Science and Research Conference*, Silver Spring, MD, August 1–2, **2012**.

CHAPTER

2

SURVEY OF MASS SPECTROMETRY-BASED HIGH-THROUGHPUT METHODS IN FOOD ANALYSIS

LUKAS VACLAVIK, TOMAS CAJKA, WANLONG ZHOU, and PERRY G. WANG

2.1 INTRODUCTION

Mass spectrometry (MS) and hyphenated chromatographic techniques have been subjects of dramatic developments, resulting in the introduction of newer tools for the analysis of diverse food components previously separated using either gas chromatography (GC) or liquid chromatography (LC). In most cases, the analysis time is reduced by faster chromatographic separations combined with more selective and sensitive mass spectrometers. In addition, many laboratories place great emphasis on streamlining sample preparation by simplifying or omitting impractical, laborious, and time-consuming steps.

In addition to these chromatography-based approaches, a large number of direct MS techniques have become available. Their main advantages compared with conventional techniques (GC–MS and LC–MS) include the possibility of direct sample examination, minimal or no sample preparation requirements, and remarkably high sample throughput.

In this chapter, recent advances in the rapid analysis of food components employing MS as a primary detection tool (both with and without chromatographic separation) are discussed with the emphasis on high sample throughput.

2.2 TECHNIQUES EMPLOYING CHROMATOGRAPHIC SEPARATION

2.2.1 Gas Chromatography–Mass Spectrometry

In food analysis, GC is one of the key separation techniques for many volatile and semivolatile compounds. The separation power combined with a wide range of detectors, including MS, makes GC an important tool in the determination of various components in food crops and products.

High-Throughput Analysis for Food Safety, First Edition.
Edited by Perry G. Wang, Mark F. Vitha, and Jack F. Kay.

In practice, a GC-based method consists of the following steps: (i) isolating analytes from a representative sample (extraction); (ii) separating coextracted matrix components (cleanup); (iii) identifying and quantifying target analytes (determinative step), and if sufficiently important (iv) confirming results by an additional analysis.

In some cases, cleanup steps can be omitted, which is the case with the use of headspace techniques or the application of injection techniques such as direct sample introduction/difficult matrix introduction (DSI/DMI). In the latter case, the sample extract is placed in a microvial that is placed in an adapted GC liner. The solvent is evaporated and vented at a relatively low temperature. The injector is rapidly heated to volatilize the GC-amenable compounds, which are then focused at the front of a relatively cold GC column. The column then undergoes normal temperature programming to separate the analytes, followed by cooling to initial conditions. The microvial is removed and discarded along with the nonvolatile matrix components that it contains. Thus, only those compounds with the volatility range of the analytes enter the column [1,2].

The "dilute-and-shoot" approach frequently used in LC–MS is not fully applicable in GC–MS because the extracts usually contain many nonvolatile matrix coextracts that can negatively affect method performance [3,4]. In addition, repeated injections of nonvolatiles lead to their gradual deposition in the GC inlet and/or front part of the GC column. As a consequence, new active sites can give rise, which may be responsible for matrix induced signal diminishment. The observed phenomena include (i) gradual decrease in analyte responses, (ii) distorted peak shapes (broadening and tailing), and (iii) shifting retention times toward higher values [5]. However, despite these limitations, GC–MS remains an essential technique for fast and comprehensive screening of various food contaminants and naturally occurring organic compounds. In fact, for compounds with low ionization efficiency observed in LC–ESI-MS (e.g., organochlorine pesticides), GC–MS is a valuable and necessary alternative. Given the possibility of matrix interference, it is clear that sample preparation plays a crucial role in the reproducibility, sensitivity, and robustness of GC–MS methods [6].

2.2.1.1 Fast Gas Chromatography–Mass Spectrometry

Fast GC separation is generally desirable because the decreased time of analysis can increase sample throughput, and consequently, thus decreasing the laboratory operating costs per sample. Changing either the column geometry or the operational parameters is the strategy that may enable fast runs. In practice, a combination of both tactics is commonly employed [7–9].

- Reduction of column length is a simple, and the most frequently used approach, in fast GC–MS. In practice, a conventional GC column (usually 30 m) is replaced by a short column (usually 10 m), which in combination with other approaches significantly decreases GC analysis times [7].
- Use of a column with a small internal diameter (e.g., 0.10–0.18 mm) is another way to achieve faster GC analyses. Unfortunately, difficulties with introducing

larger sample volumes plus lower sample capacity limit their application in real-world analyses [7].

- Decreases in analyte retention factors, and thus faster GC analyses, can be also achieved by using a column with a thin film of stationary phase (e.g., 0.1 μm). However, reduced ruggedness and sample capacity are the trade-offs for increased speed [7].
- The most popular approach to fast GC in food analysis represents fast temperature programming. When using convection heating facilitated by a conventional GC oven at faster programming rates, heat losses from the oven to the surrounding environment may cause a poor oven temperature profile, and hence, lower retention time reproducibility. Reducing the effective size of the oven helps to improve reproducibility. Use of resistive heating is preferred because of very good retention time repeatability as well as very rapid cool-down rate, which results in higher sample throughput [7].
- Operating the column outlet at low pressures (low-pressure gas chromatography (LP–GC)) is another fast GC–MS alternative. The analyses are conducted on a megabore column (typically 10 m length × 0.53 mm internal diameter × 0.25–1 μm phase) connected through a connector to a short, narrow restriction column (2–5 m × 0.1–0.18 mm internal diameter) at the inlet. Using this GC configuration, the entire analytical column is kept under vacuum conditions while the inlet remains at usual column head pressures in GC. Because optimum carrier gas linear velocity is attained at a higher value because of increased diffusivity of the solute in the gas phase, faster GC separations can be achieved with a disproportionately smaller loss of separation power. The advantages of LP–GC involve: (i) reduced peak tailing and width, (ii) increased sample capacity of megabore columns, and (iii) reduced thermal degradation of thermally labile analytes [7].
- Replacing helium by hydrogen carrier gas results in increasing the speed of analysis as well as lower inlet pressure requirements. This results from the higher diffusivity of analytes in hydrogen, which allows higher operating linear velocities without increasing peak broadening. In practice, however, helium is usually used due to concerns about safety and inertness [8].

In most cases, an increase in separation speed leads to lower chromatographic resolution and/or sample capacity. The lower chromatographic resolution is not necessarily the limiting factor in speed because MS has the ability to distinguish between analytes that have differences in their MS spectra [10]. In addition, MS systems acquiring full mass spectra can benefit from automated spectral deconvolution of partially overlapped peaks on the basis of increasing/decreasing ion intensities in collected spectra [11]. With the exception of certain applications such as the separation of isomeric compounds, MS can resolve coeluting peaks spectrometrically. However, the detector must be able to record the narrower peaks with an acceptable precision, thereby providing reproducible quantitation and identification.

There are various types of mass analyzers available for the detection of ions in GC. For fast GC, however, elution of narrow chromatographic peaks dictates the requirements for mass spectrometers. In particular, it is necessary to collect mass spectra at high acquisition rates. Also, for the detection of analytes in (ultra)trace analysis, high sensitivity is required. To achieve these requirements, a quadrupole MS is typically operated in selected ion monitoring (SIM) mode, and triple quadrupole instruments are operated in multiple reaction monitoring (MRM) mode to achieve high sensitivity (SIM and MRM) as well as selectivity (MRM). The increase in selectivity in MRM mode is achieved through monitoring one or more characteristic product ions formed within collision-induced dissociation of the parent ion. While there is some probability that several analytes/matrix components will form ions with the same nominal mass, it is much less probable that two compounds will produce identical fragmentation pattern. In general, the use of these mass spectrometers limits not only the number of ions/transitions that can be monitored for each analyte but also the total number of analytes that can be analyzed to obtain acceptable detectability in (ultra)trace analysis. In contrast to these scanning instruments, a time-of-flight MS (TOFMS) allows acquisition of full mass spectra without the loss of sensitivity [11]. Currently, three types of TOFMS instruments differing in their basic characteristics are available: (i) high-resolution/accurate mass analyzers (7000 full width at half maximum (FWHM) providing only moderate acquisition speed (up to 20 spectra/s), (ii) unit-resolution instruments that feature high acquisition speeds (up to 500 spectra/s), and (iii) high-speed high-resolution/accurate mass analyzers permitting high acquisition speeds (up to 200 spectra/s) as well as high mass resolving power (50,000 FWHM).

Regarding ionization, in GC–MS, electron ionization (EI) and chemical ionization (CI) represent the fundamental ionization techniques. On the basis of the scientific literature abstracted in SciFinder Scholar, EI was used in ~95% of all food GC–MS applications, while the remaining applications (5%) employed CI.

EI is preferred not only for confirmation of target component identity through consistent ion abundance ratios but also for identification of unknowns and determination of molecular structure [12]. Unfortunately, EI fragmentation can be too extensive, leaving little or no trace of a molecular ion, which makes the determination of the molecular weight difficult or impossible. Use of low energy or "soft" ionization techniques such as positive chemical ionization (PCI) can enhance the detection of molecular ion-based species. Because little or no fragmentation occurs during PCI, this ionization technique is less suitable for confirmation. However, this is useful in some analyses because the ion corresponding to the molecular species (e.g., protonated molecule, adduct with reagent gas) is more intense and specific than lower mass fragment ions [13]. For a limited number of compounds, such as analytes containing a halogen atom, a nitro group, or an extended aromatic ring system, negative chemical ionization (NCI) can provide significant improvement in sensitivity (2 orders of magnitude or even greater) and selectivity compared to EI and PCI because only a limited number of analytes are prone to efficient electron capture during NCI [11].

2.2.1.2 Applications of Fast Gas Chromatography–Mass Spectrometry

Applications of fast GC approaches, such as the use of a column with a small internal diameter, the use of a column with a thin film of stationary phase, and low-pressure GC combined typically with fast temperature programming, were evaluated by several authors in the analysis of food and environmental contaminants (pesticide residues [4,9,14–18], brominated flame retardants [19,20], polycyclic aromatic hydrocarbons [21], and polychlorinated biphenyls [22] as well as naturally occurring food compounds (flavor compounds [23,24] and lipids underwent a transesterification in order to obtain the fatty acid methyl esters [25]).

In general, fast GC–MS has been demonstrated to increase the speed of analysis for GC-amenable analytes in various foods and provide more advantages over the traditional GC–MS approach, including high sample throughput with, in most cases, <10 min instrumental analysis time per sample (see an example in Figure 2.1).

The benefit of fast GC was further enhanced by rapid sample preparation. In particular, the fast and inexpensive QuEChERS (quick, easy, cheap, effective, rugged, and safe) extraction method [26] was employed for sample preparation in pesticide residue analysis, which reduced the total time needed for the processing of samples by a factor of ~5 and the analysis time by a factor of ~6 compared with "conventional" sample preparation approaches [9]. For the analysis of flavor compounds, the use of headspace solid-phase microextraction (HS-SPME) was shown to provide appropriate sample preparation and preconcentration. This solvent-free, inexpensive sampling technique enabled isolation of a wide range of analytes present in food crops and products by their extraction from its headspace and concentration in the fiber coating [27].

Regarding MS detection, the use of a single quadrupole MS operated in SIM represented a limiting factor since only two to three ions could be monitored to obtain acceptable detectability in (ultra)trace analysis [4,14]. In the case of a triple quadrupole MS operated in MRM, the initial identification of pesticide residues was based on MS/MS screening that monitored a single MS/MS transition (1 precursor ion → 1 product ion) of each target compound followed by the repeated analysis of potentially positive samples again using MS/MS to monitor two to three MS/MS transitions (1 precursor ion → two to three product ions) for each compound [17]. The disadvantages of this approach were the need to optimize MS/MS conditions, reanalyze the positive samples, and create many time segments in the method. However, with the development of new instruments allowing accurate acquisition with very low dwell times (1–5 ms), even two or more MS/MS transitions for >150 analytes with analysis time <10 min were possible [16].

The number of analytes was not a limiting factor in studies using a time-of-flight (TOF) analyzer (either high-speed TOFMS (HSTOFMS) or high-resolution TOFMS (HRTOFMS)) [11,15]. While the HSTOFMS instruments employed spectral deconvolution of the acquired GC–MS records, the use of HRTOFMS allowed the unbiased identification and reliable quantification of pesticide residues through the application of a narrow mass window (0.02 Da) for extracting analyte ions and the availability

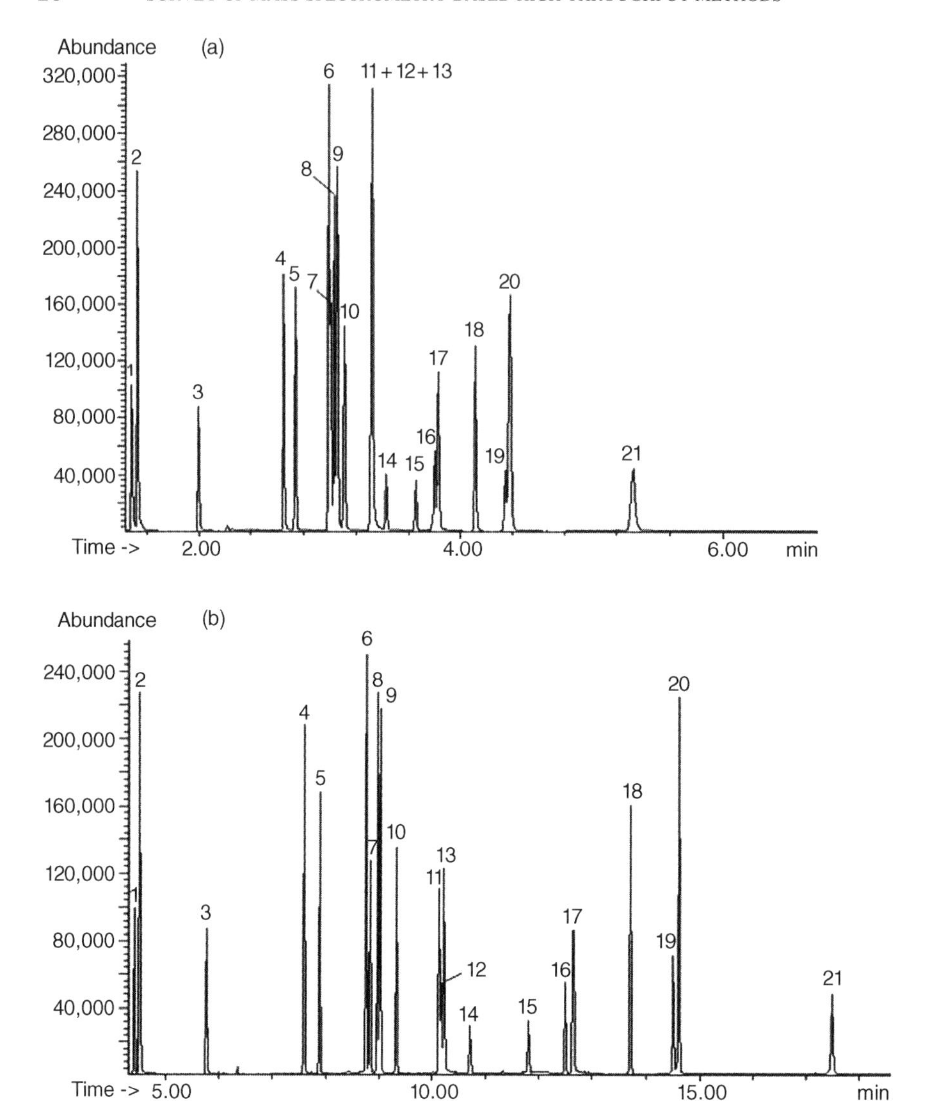

Figure 2.1. Chromatogram of standard (1 μl injection of 5 μg/ml pesticide mixture in toluene) at (a) the optimized LP-GC–MS conditions and (b) conventional GC–MS conditions. (1) Methamidophos, (2) dichlorvos, (3) acephate, (4) dimethoate, (5) lindane, (6) carbaryl, (7) heptachlor, (8) pirimiphos-methyl, (9) methiocarb, (10) chlorpyrifos, (11) captan, (12) thiabendazole, (13) procymidone, (14) endosulfan I, (15) endosulfan II, (16) endosulfan sulfate, (17) propargite, (18) phosalone, (19) *cis*-permethrin, (20) *trans*-permethrin, and (21) deltamethrin. Ref. [14], Figure 3, p. 299. Reproduced with permission of Elsevier Science Ltd.

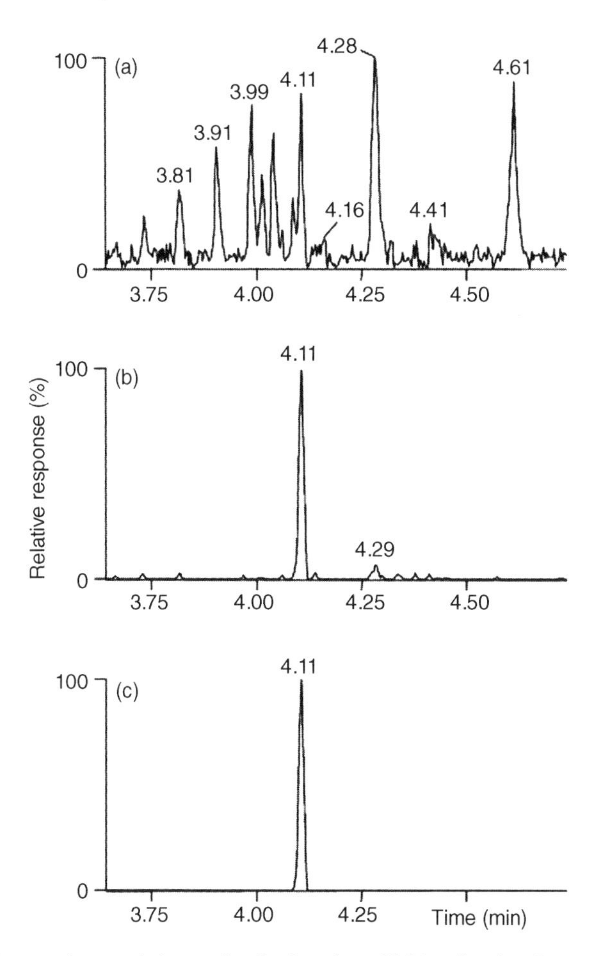

Figure 2.2. Influence of mass window setting for detection of 0.01 mg/kg phosalone (t_R = 4.11 min; *m/z* 182.001) in apple baby food extract prepared with the QuEChERS method. Using a 1 Da mass window gave peak-to-peak signal-to-noise (S/N) ratio of 6, but setting the mass window to 0.1 Da or even as low as 0.02 Da led to a S/N of 25 and 74, respectively. Ref. [9], Figure 3, p. 288. Reproduced with permission of Elsevier Science Ltd.

of full spectral information even at very low levels without spectral deconvolution (Figure 2.2) [11].

2.2.2 Liquid Chromatography–Mass Spectrometry

2.2.2.1 Advanced Column Techniques

Ultrahigh-Performance Liquid Chromatography (UHPLC) and Sub-2 μm Columns The basic concepts behind rapid analyses performed by high-performance liquid chromatography (HPLC) have not changed since HPLC was invented. What has changed is the development of better separation media (i.e., available columns) and hardware with higher pressure limits for using even smaller particles in combination with longer columns. Based on the theories of van Deemter et al., then Giddings, and finally Knox, the use of small particles is one of the best solutions in the quest to improve chromatographic performance [28]. Over the last three decades, the size of standard high-efficiency particles has decreased from the 5–10 μm range in the 1980s to the 3–5 μm range in the 1990s and to the 1–3 μm range more recently for use in ultrahigh-efficiency separations [29].

However, small particles induce a high pressure drop because, based on Darcy's law, the pressure drop is inversely proportional to the square of particle size at the optimum linear velocity. The traditional 400 bar pump systems have been the standard instrumentation since the 1970s [29]. In 2004, Waters introduced the ACQUITY UPLC™ System. A new ultrahigh pressure/performance era began with this launch. Most LC vendors have identified their modified HPLC systems as ultrahigh-pressure liquid chromatography (UHPLC) systems. UHPLC systems can reliably deliver solvents up to 1300 bar with routine operation in the 500–1000 bar range.

Compared with a traditional HPLC system, a UHPLC system has much lower extracolumn variance ranging from 4 to 9 μl^2. The HPLC system can contribute ~40–200 μl^2. In order to maximize sub-2 μm column efficiency, the UHPLC system should be well configured and operated in the optimized conditions such as using a smaller volume needle seat capillary, narrower and shorter connector capillary tubes, and a smaller volume detector cell. Otherwise, the column efficiency can lose as much as 60% [30].

The sub-2 μm particle columns and UHPLC systems offer much shorter analysis times with higher resolution and greatly reduced solvent consumption. Some separations can be performed within minutes or even seconds using UHPLC systems. UHPLC systems have been widely used for routine bioanalysis, food safety, and in other fields. Many applications and reviews have been published recently [31–34]. UHPLC has become one of the most advanced techniques for LC in the past decade.

Hydrophilic Interaction Liquid Chromatography Hydrophilic interaction liquid chromatography (HILIC) is a variation of normal-phase chromatography with the advantage of using organic solvents that are miscible with water. It uses polar materials such as amino, cyano, diol, and silanol as the stationary phase. Thus, HILIC is sometimes called "reverse reversed-phase" or "aqueous normal phase" chromatography. The HILIC concept was first introduced by Dr. Andrew Alpert in his 1990 paper [35]. A large number of papers on this subject have been published since then [36,37].

A hydrophilic stationary phase and an aqueous–organic solvent mobile phase with high organic solvent content are used in HILIC. Like normal phase LC, retention increases when the polarity/hydrophilicity of the analytes and/or the stationary phase increases. However, retention also increases as the polarity of the mobile phase decreases. The most common organic solvent for HILIC is acetonitrile due to its low viscosity [38]. Methanol, tetrahydrofuran, and other organic solvents can be used as well. It was reported that the use of methanol may result in poor chromatographic separation or lack of retention for some analytes [38,39].

Compared with reversed-phase liquid chromatography (RPLC), HILIC provides unique benefits for the separation of polar and/or hydrophilic compounds. Due to the high organic content in the mobile phases, HILIC results in a lower operating back pressure, which allows higher flow rates for high-throughput analysis. The high organic solvent concentration in the mobile phase also leads to a higher sensitivity for LC–MS analyses because the ionization efficiency can be significantly increased.

Although the mechanism of HILIC is still being debated [40], partitioning theory is well accepted, that is, HILIC involves the partitioning of an analyte between a predominantly polar organic mobile phase and a water-enriched layer of the mobile phase that is partially immobilized on the stationary phase. However, some scientists believe that the separation mechanism involves both partition and "adsorption" processes [39]. As with any other chromatographic separation mode, HILIC depends on the different interaction of solutes between the mobile phase and the stationary phase. These interaction forces include hydrogen bonding, which depends on the acidity or basicity of the solutes, electrostatic interactions, and dipole–dipole interactions, which rely on the dipole moments and polarity of molecules [40]. Based on the mechanism of HILIC, the formation and stability of water-enriched layers are very important. Usually, at least 3% water is needed in the mobile phase for sufficient hydration of the stationary-phase particles so that a stable water-enriched layer forms [38,41]. Because the mobile phase is an aqueous–organic solvent, it has good solubility for polar, hydrophilic analytes. Furthermore, these molecules are more retained and thus better separated in HILIC than they would be by conventional RPLC with water-rich mobile phases. Of specific note, HILIC columns found wide applicability for detecting melamine in different matrices during China's melamine milk crisis because melamine and its analog cyanuric acid are very polar and have very poor retention on regular reversed-phase columns [42,43]. The use of HILIC in a broad range of food analyses, including the detection of vitamins, marine toxins, and amino acids in a variety of matrices, has recently been detailed in two books [37,44].

Monolithic Columns Monolithic media have been used for various niche applications in GC and LC for a long time. Only recently did they acquire a major importance in HPLC. Unlike conventional particle-based construction, monolithic columns contain a continuous network (monolith) of porous silica or organic polymer.

Although the preparation and use of these types of columns was first reported in 1967 [45], they suffered from poor flow characteristics resulting in little interest in this idea as a feasible approach for producing separation media until the 1990s [29]. At this point, improvements in manufacturing approaches resulted in monoliths with better performance characteristics and usefulness for high-throughput assays. They are prepared using a sol–gel process, either *in situ* or in a manufacturing mold, which enables the formation of a highly porous material, containing both macropores and mesopores in its structure. Compared with particle-based columns, the most important advantage of monolithic columns is low back pressure due to macropores (2 μm) throughout the network that allows high flow rates with relatively flat van Deemter profiles. Because of this, monolithic columns can be operated either at much higher linear velocities or with a combination of gradient and flow programming, significantly improving throughput.

The advent of monolithic silica standard- and narrow-bore columns and of several families of polymer-based monolithic columns has considerably changed the HPLC field, particularly in the area of narrow-bore columns. By using monolithic columns, complex mixtures of biological molecules can be efficiently separated, and throughput in pharmaceutical and biotechnology laboratories has been significantly improved. Monolithic columns have been found to be especially useful for separating larger peptides, proteins, and polycyclic compounds [46]. Highly sensitive proteomics applications are also easily performed using these columns.

Core–Shell Columns It is interesting to note that some manufacturers and a number of researchers have rediscovered pellicular construction in the form of smaller 1.5–2.5 μm particle-based packings containing a very thin porous outer layer [29]. Nevertheless, the original pellicular work can be traced back to the late 1960s and was the packing of choice until the introduction of completely porous 5–10 μm materials in 1972 [46,47]. With the reintroduction of pellicular materials in combination with longer columns and higher solvent delivery pressures, it is possible to obtain highly efficient separations of complex proteomic mixtures. Modern porous shell packings are highly efficient (i.e., with improved solute mass transfer kinetics) because their outer microparticulate layer is only 0.25–1.0 μm thick with pores in the 300 Å range. Typically, these materials have surface areas in the 5–10 m^2/g range. Core–shell columns offer one option for users who want higher separation efficiency and higher sample throughput without a UHPLC system. The main disadvantage of core–shell columns, relative to regular columns, is their lower saturation capacity.

2.2.2.2 Mass Spectrometry

Targeted Analysis The application of MS for food safety analysis can be divided into targeted and nontargeted analyses. A targeted analysis is a conventional analysis based on developing a method with standards prior to the analysis and monitoring of real samples, and does not detect compounds not defined in the developed method [48]. The main techniques involve the application of LC–MS, such as using triple quadrupole mass spectrometry (QqQ-MS), quadrupole linear ion trap mass spectrometry (QLT-MS), TOFMS, and Orbitrap MS. Selected reaction monitoring (SRM), also called MRM mode, is still preferred for a quantitative analysis of known or targeted compounds using QqQ-MS due to its good selectivity, sensitivity, reproducibility, dynamic range, and quantifiability.

To overcome the limited number of compounds that can be simultaneously determined, which mainly depends on the scan speed/dwell time, scheduled MRM™ or similar techniques (e.g., dynamic MRM mode) have been widely used. Higher sensitivity and more robustness are achieved by applying these techniques compared with the more commonly used SRM mode. When these techniques are applied for an analysis, the whole data acquiring period is divided into different time segments. The partition of the time segments mainly depends on the retention times of

Table 2.1. Common Parameters of Mass Spectrometers Used in LC–MS

Mass Analyzer Type[a]	Resolving Power ($\times 10^3$)	Mass Accuracy (ppm)	m/z Range (Upper Limit) ($\times 10^3$)	Acquisition Speed (Spectra/s)	Linear Dynamic Range	Price
Quadrupole	3–5	Low	2–3	2–10	10^5–10^6	Lower
(Linear) Ion trap	4–20	Low	4–6	2–10	10^4–10^5	Moderate
Time-of-flight	10–60	1–5	10–20	10–50	10^4–10^5	Moderate
Orbitrap	100–240	1–3	4	1–5	5×10^3	Higher

Source: Ref. [49], Table 2, p. 4. Reproduced with permission of Elsevier Science Ltd.

[a]TOF and Orbitrap also include common hybrid configurations with quadrupole or linear ion trap as the first mass analyzer.

analytes. During the data acquisition process, the instrument acquires only the selected SRM data for the current time segment. Thus, the number of concurrent transitions is significantly reduced; a higher dwelling time can be applied for the analytes, which can result in higher sensitivity and more peak points for the analytes.

High resolution MS such as TOF and Orbitrap MS can be used to analyze virtually an unlimited number of compounds because they operate in full scan mode and can reconstruct any desired ion chromatogram using the same full scan data file [48]. Because accurate mass measurements are almost specific and universal for each target analyte regardless of the instrumentation used, a library search of high-resolution mass spectra may be performed using the libraries obtained by different mass spectrometers.

An overview of some common parameters of mass analyzers used in LC–MS is provided in Table 2.1. The mass resolving power (RP) characterizes the ability of the mass analyzer to separate two ions of similar m/z values. RP is defined as the m/z value of particular spectral peak divided by the peak FWHM. The older definition based on mass difference between two spectral peaks by a 10% valley is not used in current LC–MS practice. Mass resolution is the inverse of RP expressed in $\Delta m/z$ for a particular m/z (see Eq. 2.1) [11,49].

$$\mathrm{RP} = \frac{m/z}{\Delta m/z} \tag{2.1}$$

Mass accuracy (MA) is the deviation between measured and theoretical (exact) mass of an ion expressed in parts per million (ppm) (see Eq. 2.2) [11,49]:

$$\mathrm{MA\ (ppm)} = \frac{(m/z_{\mathrm{meas}}) - (m/z_{\mathrm{theor}})}{m/z_{\mathrm{theor}}} \times 10^6 \tag{2.2}$$

Although electrospray ionization (ESI) is a common ionization method, especially for relatively polar compounds, less polar or neutral analytes may have lower ionization efficiencies and lower sensitivities and thus require atmospheric pressure chemical ionization (APCI) or atmospheric pressure photoionization (APPI). Therefore, a multimode ionization source (e.g., ESI and APCI or ESI and APPI) is ideal for analyzing compounds with a wide range of chemical structures and properties. The flexible ionization capabilities of these multimode sources also minimizes the need for multiple LC–MS injections or repeated analyses of failed samples using an alternative ionization mode. For example, with ESI and APCI multimode ionization source, ESI (+), ESI(−), APCI(+), and APCI(−) ionization modes can be chosen and performed in a single LC–MS run to obtain maximum ionization and sample coverage. While perhaps not a high-throughput technique *per se*, eliminating the need for multiple analyses of the same sample clearly increases the number of different samples that can be analyzed.

Nontargeted Analysis Nontargeted analysis includes the possibility of detecting any compounds (i.e., compounds related or not related to contaminations such as pesticides, mycotoxins, drugs, and plasticizers) present in a sample. It offers the possibility of identifying unexpected contaminations, transformation products, and/or impurities [48]. Such analyses are more complicated because they require the identification of unknown compounds. Due to the complexity in identifying unknown compounds, high-resolution MS, such as TOF and Orbitrap MS, especially QTOF/Q-Orbitrap, are often employed for this purpose. Zhang et al. demonstrated the rapid screening and accurate mass confirmation of 510 pesticides at low ppb levels using UHPLC coupled to a high-resolution benchtop Orbitrap mass spectrometer [50]. QTOF/Q-Orbitrap provides not only accurate mass measurements and high full scan sensitivity but also additional features such as structure confirmation. The accurate product ion spectra can be obtained by performing MS/MS experiments using QTOF/Q-Orbitrap MS and used to search compound libraries to confirm the structures of compounds. This is crucial for analytes or their metabolites when the reference standards are not available. The number of applications of LC–QTOF/Q-Orbitrap for targeted and nontargeted analyses in food safety has recently increased [32]. For example, a UHPLC–QTOFMS system was used to perform targeted and nontargeted analyses of ~1000 organic contaminants, including residues and illicit substances, such as mycotoxins in food samples, cocaine and several metabolites in human urine, and pesticides, antibiotics, and drugs of abuse in urban wastewater [51].

Matrix Effects Although LC–MS and LC–MS/MS are sensitive and selective, they often suffer from matrix effects, especially with ESI ionization. Matrix effects are the alteration of ionization efficiency of target analytes in the presence of coeluting compounds in the same matrix. They can be observed as either a loss of the signal of the target (ion suppression) or a gain of the signal (ion enhancement). Matrix effects alter the detection capability, precision, and/or accuracy of measurements for the analytes of interest. Evaluating and minimizing ion suppression and enhancement are important considerations during method development and validation [31,38]. Two commonly used methods to evaluate matrix effects are (i) postcolumn infusion for qualitative evaluation [52] and (ii) postextraction spikes for quantitative evaluation [53]. To perform postcolumn infusion, an infusion pump delivers a constant amount of analyte into the LC stream entering the ion source of the mass spectrometer. A blank sample extract is injected under the same conditions as those for the assay. Any variation in ESI response of the infused analyte caused by an endogenous compound that elutes from the column is seen as a matrix effect. If no matrix effect exists, a steady ion response is obtained as a function of time because the analyte is infused into the MS at a constant flow (concentration). In the postextraction spike method, matrix effects are quantitatively accessed by comparing the response of an analyte in neat solution with the response of the analyte spiked into a blank matrix sample that has been carried through the sample preparation process [54].

Both the postcolumn infusion method and the postextraction spike method are not suitable for studies in which a representative blank sample matrix is not available. In these situations, recovery studies can be performed to evaluate the matrix effects [31]. To perform a recovery study, known amounts of analytes are added to a sample before the sample preparation, and the same procedure as that for a unspiked sample is carried out to prepare the spiked sample. If the recovery is lower than 70% or greater than 130%, the sample matrix is considered to cause serious ion suppression or enhancement, respectively. Because some samples contain analytes and some samples lack analytes, the recoveries for these two types of samples are calculated using different formulas. For samples lacking analytes, recoveries are calculated by the following formula (Eq. 2.3):

$$\text{Recovery (\%)} = \frac{\text{detected amount}}{\text{spiked amount}} \times 100 \tag{2.3}$$

For samples containing analytes, the recoveries are calculated as follows (Eq. 2.4):

$$\text{Recovery (\%)} = \left(\frac{\text{total amount} - \text{amount in sample}}{\text{spiked amount}}\right) \times 100 \tag{2.4}$$

There are various techniques used to minimize matrix effects for different matrices. The use of stable isotopically labeled analogs is a preferred technique to compensate for matrix effects because the analogs have very similar properties and almost the same retention times as the analytes due to their identical structures. However, the labeled analogues are not always commercially available and may be costly. Dilution is the simplest way to minimize matrix effects if the concentrations of the samples are high enough to withstand the dilution [31]. Other common sample preparation techniques such as QuEChERS, liquid–liquid extraction, solid-phase extraction (SPE), and solid-phase microextraction (SPME) have also been applied to minimize matrix effects. In addition to the use of the method of standard additions, other options include smaller injection volumes, using structurally similar unlabeled compounds that elute close to the compounds of interest as internal standards, and modification of chromatographic conditions. In many instances, a few approaches are combined to obtain suitable quantitative results [54,55].

The Role of Weighting Factors for Calibration Coefficients of determination (r^2) or correlation coefficients (r) have often been used as indicators of linearity for calibration curves, although they do not guarantee that the calibration curve fits the data well, that is, all data points across the curve have good accuracy [38]. In statistics, calibration curve data can be divided into homoscedastic data and heteroscedastic data. Homoscedastic data have similar standard deviations for the entire calibration range [56]. That is, the errors at the low end of the curve are close to the errors at the high end of the curve.

The calibration curve can be generated using calibration data and expressed using the following equation:

$$y = ax + b \tag{2.5}$$

where y and x are the response (signal intensity) and concentration of an analyte, respectively, a is the slope, and b is the intercept. The accuracy or error of each data point can be used to evaluate the quality, that is, linearity, of a calibration curve. The accuracy and error can be calculated as follows (Eqs. 2.6 and 2.7):

$$\text{Accuracy (\%)} = \frac{\text{calculated concentration}}{\text{theoretical concentration}} \times 100 \tag{2.6}$$

$$\text{Error (\%)} = \left(\frac{\text{calculated concentration} - \text{theoretical concentration}}{\text{theoretical concentration}}\right) \times 100 \tag{2.7}$$

where calculated concentration is back calculated concentration using the calibration curve equation. From Equations 2.6 and 2.7, the error and accuracy have the following relationship (Eq. 2.8):

$$\text{Error (\%)} = \text{accuracy} - 100 \tag{2.8}$$

In general, either error (%) or accuracy (%) for each data point can be automatically calculated by the vendors' software to show the quality of a calibration curve when a calibration curve is generated, that is, a, b, and r^2 or r are calculated.

For homoscedastic data, curve weighting is not necessary, that is, no weighting or equal weighting. Therefore, the r^2 or r can be safely used as an indicator of linearity for the calibration. Higher r^2 or r value is taken to indicate higher accuracy for the calibration curve.

For heteroscedastic data, however, the standard deviation increases with the concentration of an analyte. The absolute error is more or less proportional to the concentration. Because a calibration curve for LC–MS or GC–MS analyses often covers three to five orders of magnitude, most of the data are heteroscedastic. In this case, the value of r^2 or r cannot be used as a unique indicator of linearity. Higher r^2 or r

values do not mean higher accuracy for the calibration curve. This is because the large standard deviations of the points at the top of the curve dominate the calculation, meaning that, the value of calibration parameters including slope, intercept, and r^2 (or r) mainly depends on the points at the top of the curve (higher concentration points) [56].

In order to improve the accuracy, especially for the lower end points of a calibration curve, a suitable weighting factor, such as $1/x$ or $1/x^2$, is usually applied when generating the calibration curve, where x represents the concentration of an analyte. The F-test can be used to determine whether the data are homoscedastic or not, but this is unnecessary because the vendors' software provides the curve-fitting function (linear and quadratic) and selection of weighting factors (no weighting, $1/x$ and $1/x^2$, etc.) to calculate either error or accuracy for each calibration point. Users can easily select an appropriate weighting factor, which gives the least sum of the absolute values of the errors for all data points across the calibration curve.

For example, a set of calibration data for analyzing Ac-EEMQRR-amide acquired by LC–MS/MS were processed to generate a calibration curve using different weighting factors: equal weighting (i.e., no weighting), inverse of concentration ($1/x$), and inverse square of concentration ($1/x^2$), as given in Table 2.2, where x represents the concentration of Ac-EEMQRR-amide [38]. Even though correlation coefficients (r) were all greater than 0.999 for the three weighting factors, their errors were significantly different at the low end of the calibration curve. For example, when equal weighting (no weighting) was applied, the point of 2.0 ng/ml was not quantifiable. The errors were 37.4 and 16.4% for points of 10 and 20 ng/ml, respectively. When the weighting factors $1/x$ and $1/x^2$ were applied, all points were quantifiable and had lower errors. When the $1/x^2$ weighing factor was applied, the errors were all lower than 5%. Obviously, the $1/x^2$ weighing factor resulted in the minimum sum of the absolute values of the error. Therefore, the inverse square of concentration ($1/x^2$)

Table 2.2. The Effects of Weighting Factors for Calibration of Ac-EEMQRR-Amide Obtained by LC–MS/MS

Weighting Factor	Equal ($r = 0.9995$); $a = 0.00843$; $b = 0.0384$		$1/x$ ($r = 0.9995$); $a = 0.00860$; $b = 0.0325$		$1/x^2$ ($r = 0.9997$); $a = 0.00879$; $b = 0.0014$	
Theo. ng/ml	Cal. ng/ml	Error (%)	Cal. ng/ml	Error (%)	Cal. ng/ml	Error (%)
2.00	NQ	NA	1.79	−10.75	1.96	−2.08
10.00	6.26	−37.43	10.22	2.17	10.22	2.17
20.00	16.72	−16.42	20.47	2.33	20.25	1.25
50.00	48.68	−2.63	51.85	3.70	50.98	1.97
100.00	100.35	0.35	102.50	2.50	100.62	0.62
200.00	207.83	3.92	208.17	4.08	203.67	1.83
1000.00	992.00	−0.80	976.83	−2.32	956.83	−4.32

Source: Ref. [38], Table 2, p. 7960. Reproduced with permission of Elsevier Science Ltd.

Note: Theo. ng/ml: theoretical concentration in ng/ml; Cal. ng/ml: back-calculated concentration in ng/ml. NQ: not quantifiable; NA: not applicable; r: correlation coefficient; a: slope, b: intercept.

was applied for this study as it introduced the least errors. Evidently, it is not appropriate just to report r^2 or r without considering accuracy or error because a reported concentration at the lower end may be erroneous and misleading, especially for a wider calibration range. From the above example, it can be concluded that the selection of weighting factors for calibration curve is necessary. Selecting appropriate weighting factors for calibration curves has become a general practice for bioanalysis in the pharmaceutical industry to get acceptable analytical results.

2.3 DIRECT TECHNIQUES

2.3.1 Matrix-Assisted Laser Desorption/Ionization-Mass Spectrometry

Matrix-assisted laser desorption/ionization-mass spectrometry (MALDI-MS) enables detection of analytes varying by molecular weight and rapid examination of small quantities of complex samples, while exhibiting good tolerance toward contamination [57]. Although MALDI is largely used in biological studies to characterize large molecules such as proteins, peptides, or carbohydrate polymers [58], it has also been applied to the ionization of intermediate and low molecular weight compounds [59]. The unique features and versatility of MALDI have led to its use in diverse research fields, including food quality, authenticity, and safety control. MALDI-based mass spectrometry is undoubtedly an analytical technique that has significant potential for use in high-throughput food analysis.

2.3.1.1 Principles and Instrumentation

In MALDI, analytes are mixed with an excess of a UV-absorbing matrix compound (usually a small organic acid) and dried to allow formation of crystals on a sample plate. The sample plate is subsequently introduced into the MALDI ion source in which a vacuum is maintained. The sample–matrix crystals are exposed to a short pulse (few nanoseconds) from a laser and, as a consequence of heat energy absorption, transferred into the gaseous phase, where analyte ionization takes place. The ions are immediately extracted into the mass spectrometer. It should be noted that many laser pulses (typically 10–500) are summed to form the mass spectrum. The simplified schemes of a MALDI source and ionization process are shown in Figure 2.3. Due to the nature of the MALDI technique, it is widely used with reflectron-type TOF mass analyzers, which provide enhanced mass resolving power and enable analysis of ions in a wide range of m/z values. It should be noted that in 2000, atmospheric pressure (AP) MALDI was developed and enabled reduction in cost and enhanced ease of operation [60].

The MALDI ionization process is complex and is not yet completely understood. The most general mechanism described by Knochenmuss proposes a two-step process: (i) primary ionization after laser impact resulting in the formation of matrix-derived species, and (ii) secondary reactions taking part in the MALDI plume after evaporation of matrix–analyte crystals [61]. According to the photoionization/pooling model

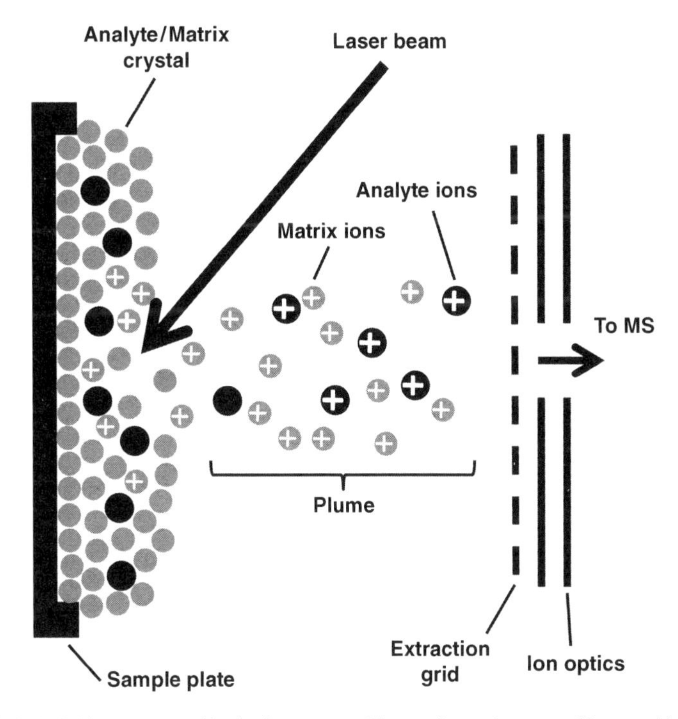

Figure 2.3. MALDI ion source and ionization process. The matrix–analyte crystal impacted by laser beam excites the matrix, which in turn ionizes the analyte. Analyte ions are collected, analyzed, and detected by the mass spectrometer.

proposed by Ehring et al., the absorbed energy migrates into the matrix molecules and focuses on the "pooling" event [62]. The matrix ions are formed as a result of the interactions between electrons excited to singlet and higher states. The secondary matrix–analyte reactions yielding analyte ions comprise proton transfer, electron transfer, and gas-phase cationization through interactions with alkali metals [63]. MALDI is a soft ionization technique, with mass spectra that are typically characterized by either $[M+H]^+$ and alkali metal adducts or $[M-H]^-$ ions. Depending on the properties of the matrix and analytes, as well as some other experimental conditions (e.g., matrix/analyte ratio, crystallization conditions, or sample deposition method), multiply charged species can also be formed [64].

2.3.1.2 Optimization of Key Parameters

In order to achieve the desired outcome of MALDI-MS analysis, a number of factors have to be considered. The most important parameters to be optimized are the following:

- Type of matrix.
- Isolation of target analytes and sample cleanup.

- Preparation and deposition of matrix–sample mixture.
- Laser parameters.

Matrix A MALDI matrix has to fulfill a number of requirements. The matrix compound has to absorb the laser wavelength, dissolve and/or cocrystallize with the sample, be stable under the vacuum conditions, induce codesorption of the sample components upon the laser impact while minimizing its thermal and chemical degradation, and, last but not least, promote the ionization of the target analyte(s). Given these requirements, it is apparent that the matrix choice is strongly linked to the nature of the analyte; hence, a matrix that works well for one analyte/sample type combination can be ineffective for another. A wide variety of matrices has been proposed for different applications [57]. An overview of some matrices used in food-related applications is provided in Table 2.3.

In addition, the choice of optimal matrix-to-analyte concentration ratio is crucial for effective MALDI ionization. Considering the principle of MALDI, the concentration of matrix has to be consistently maintained in excess with respect to the analyte. A 500–50,000-fold molecular ratio range is typically recommended. If the amount of analyte is too high or too low, no analyte-related mass spectra are observed [65]. On the positive side, using an appropriate concentration ratio can lead to diminution or complete elimination of matrix ions, which can interfere with analyte signals [63]. Serious matrix-related problems may arise when applying MALDI to low molecular weight compounds, because of abundant, low *m/z* matrix ion signals. To overcome this drawback, a matrix-free approach has been proposed. However, such procedures are limited by the stability of the target analytes as rapid thermal degradation occurs upon sample exposure to the laser beam [66]. As an alternative, ionic liquids (combinations of organic cations with a variety of anions) have been employed as matrices in order to obtain matrix-free mass spectra. The use of ionic liquids has other benefits, including the production of a much more homogeneous sample solution, greater vacuum stability, and, in many cases, higher signal intensities compared with conventional liquid and solid matrices [67]. Regardless of the type of matrix used, it is a common practice to dope the sample with solutions of acids, alkali metals, or silver salts to improve the ionization yield of analytes and enhance the formation of particular ion types [57].

Sample Preparation Because the time required for MALDI measurements is typically under 1 min, the time required to prepare the samples dictates the throughput of the whole analytical procedure. The process of sample preparation for MALDI consists of two steps: (i) pretreatment of sample, and (ii) deposition and drying of the matrix–sample mixture on the surface. While the first phase can comprise a simple procedure such as dilution with suitable solvent, in some cases, various extraction and/or cleanup strategies such as SPE or dialysis [68,69] have to be employed to isolate target analytes from the complex food material and to minimize undesired effects caused by interfering compounds (see Table 2.3).

Table 2.3. Overview of MALDI-MS Applications Relevant to Food Safety and Quality

Aim of the Study	Analytes	MALDI Matrix	Sample Pretreatment	Remark	Reference
Quantitation of glycoalkaloids in potatoes	α-Chaconine, α-solanine	2,4,6-Trihydroxyacetonephenone in methanol–water (1:1)	Extraction with methanol–water (1:1, v/v), centrifugation	Analytes detected as $[M+K]^+$ ions, tomatine used as internal standard, good correlation ($r^2 = 0.98$) with results obtained by HPLC method	[70]
Quantitation of gluten and gliadins in food	Gliadins	*trans*-3,5-Dimethoxy-4-hydroxycinnamic acid in 30% acetonitrile with 0.1% trifluoroacetic acid	Repeated ($n = 2$) extraction with 60% ethanol, centrifugation, dilution with extraction mixture	Good agreement with results of enzyme-linked immunoassay	[71]
Qualitative analysis of isoflavones in soy products	Daidzein, genistein, glycitein, and their glucosides	2,5-Dihydroxybenzoic acid in methanol	Extraction with a mixture of acetonitrile and HCl, filtration, C18S PE cleanup	Analytes detected as $[M+H]^+$ ions	[68]
Detection of oxidation products originated from triacylglycerols in vegetable oils	Oxidated triacylglycerols and their β-scission products	2,5-Dihydroxybenzoic acid in methanol with 0.1% trifluoroacetic acid	Dilution with chloroform	Analytes detected as $[M+Na]^+$ ions	[72]
Qualitative analysis of mycotoxins in peanuts	Aflatoxins B1, B2, G1, and G2	Triethylamine-α-cyano-4-hydroxycinnamic acid in methanol (ionic liquid) doped with 10 mM NaCl	Extraction with methanol, cleanup with combination of $CuSO_4$ and Celite, reextraction with chloroform	Analytes detected as $[M+Na]^+$ ions, LOD 50 fmol	[73]
Fish authentication based on protein profiles	Parvalbumins (fish muscle proteins with molecular	α-Cyano-4-hydroxycinnamic acid in acetonitrile–water mixture (1:1)	Extraction of muscle with 0.1% aqueous formic acid	Positive ionization mode, cytochrome C used as internal standard	[74]

(continued)

Table 2.3 (*Continued*)

Aim of the Study	Analytes	MALDI Matrix	Sample Pretreatment	Remark	Reference
	weight ~11 kDa)	with 0.1% trifluoroacetic acid			
Determination of geographical origin of honey based on protein fingerprint	Nonspecified proteins	α-Cyano-4-hydroxycinnamic acid in 30% acetonitrile with 0.1% trifluoroacetic acid	Extraction with water and removal of low molecular compounds by dialysis	Positive ionization mode, authentication based on comparison of sample mass spectra converted into barcodes	[69]
Detection and quantification of milk of cows, sheep, and goat	Milk proteins (α-lactalbumin, β-lactoglobulin, α_{S1}-casein, γ_1-casein, γ_2-casein)	*trans*-3,5-Dimethoxy-4-hydroxycinnamic acid in acetonitrile–water mixture (1:1)	Dilution of sample with 0.1% trifluoroacetic acid	Positive ionization mode, detection and quantification based on partial least squares (PLS) and kernel PLS regression models	[75]
Qualitative analysis of pesticides	Preselected pesticides ($n = 15$)	2,5-Dihydroxybenzoic acid or α-cyano-4-hydroxycinnamic acid in methanol	N/A, analysis of standards only	Analytes detected as $[M+H]^+$/$[M+Na]^+$/$[M+K]^+$ ions, LODs 0.001–0.5 ppm	[76]
Detection of hazelnut oil (HA) in extra virgin olive oil	Phospholipids	Tributylamine-α-cyano-4-hydroxycinnamic acid in methanol (ionic liquid)	Extraction with a mixture of matrix in methanol and chloroform, dilution	Analytes detected as $[M+H]^+$/$[M+Na]^+$/$[M+K]^+$ ions, detection of HA at 1% addition level	[77]
Detection of vegetable oils in bovine milk powder	Triacylglycerols	2,5-Dihydroxybenzoic acid in methanol	Extraction with *n*-hexane	Analytes detected as $[M+Na]^+$ ions	[78]

Several protocols have been developed for deposition of the matrix–sample mixture on the target surface. Each has its advantages and disadvantages. For MALDI-MS analysis, it is always beneficial to induce rapid formation of small crystals, because slow growth of large crystals can cause improper incorporation of analyte molecules. Slow growth can also compromise mass resolution because of differences in distance between the top and bottom of the crystal [65]. The oldest but still widely employed dried–droplet method is based on direct deposition of the sample mixed with a saturated matrix solution and drying under ambient conditions. The primary advantage of this method is its simplicity and tolerance to the presence of salts and buffers. On the other hand, relatively large, inhomogeneous, and irregularly distributed crystals are formed, which often results in the need for searching for the "sweet spots" on the sampling surface [79]. To reduce the size of the crystals, increase their homogeneity, and increase the speed of the procedure, the drying process can be accelerated by vacuum, a stream of nitrogen, or heating. Other methods that produce small and homogeneous crystals involve physical crushing [65], rapid evaporation of matrix solution applied to the surface in a volatile solvent [80], and electrospraying of the matrix–analyte mixture onto a grounded metal sample plate [81]. The throughput of sample preparation in MALDI can be significantly increased by the use of matrix-precoated layers, on which sample is directly deposited, either manually or using automated sample deposition systems [82].

In general, it is difficult to characterize the overall time requirements for sample preparation in MALDI-MS as this parameter is strongly application dependent and can span from few seconds (dilute-and-shoot approach) to several hours (isolation, cleanup, and enzymatic digestion).

Laser Parameters In order to achieve good sensitivity and mass resolving power, attention must also be paid to optimization of laser parameters. The most influential parameters are the laser wavelength and laser strength, which can be adjusted by changing the attenuation factor (the higher the attenuation, the lower the laser strength). Although the most widely employed nitrogen laser (radiation at 337 nm) usually provides good results in most applications, other alternatives such as tunable Nd:YAG or excimer lasers are also available. Because the best analytical results can be obtained only at laser wavelengths that correspond to a high absorption of the matrix, optimization of this parameter can significantly improve the sensitivity of a particular method [83]. For each type of sample, there is a minimum laser strength that is required for production of ions. At settings above this threshold, a pronounced increase of ionization yield can be observed until a plateau is reached. Contrarily, using a laser strength that is too high typically induces fragmentation of analyte ions. The strength of the laser is also linked to the mass resolving power. There is a relatively narrow range of laser strength values that provide superior mass resolving power [65]. The mass resolving power can be further improved by employing delayed extraction of ions, which can compensate for variations in velocities of ions with the same m/z values caused by uneven energy distribution during laser impact.

2.3.1.3 *MALDI-MS in High-Throughput Analysis of Food*

As can be seen in Table 2.3, MALDI-MS has been used to tackle various aspects of food analysis, such as food authenticity control or analysis of natural components and contaminants. Although most of the applications aim at qualitative analysis (screening, profiling, and fingerprinting), quantitation with MALDI is also possible with the use of suitable internal standard [70,71]. In the following paragraphs, several examples of the use of MALDI-MS for high-throughput food analysis are presented. It is obvious that with workflows employing MALDI-MS, a significant time reduction can be achieved. Although sample pretreatment and MALDI matrix preparation may be in some cases more time consuming compared with procedures involved with conventional techniques, the increase in sample throughput is enabled by a considerably lower analysis time (typically 1 min versus tens of minutes in case of HPLC–MS).

In a study by Catharino et al., MALDI-TOFMS was used to rapidly screen for multiple aflatoxins in peanuts [73]. To minimize matrix-related ions, an ionic liquid (triethylamine-α-cyano-4-hydroxycinnamic acid solution in methanol) was employed as the MALDI matrix. In order to eliminate the formation of multiple ions of target analytes, thus increasing the sensitivity of measurement, a 10 mM solution of NaCl was added to the sample prior to analysis. As a result, only $[M+Na]^+$ adducts of target aflatoxins were observed and LODs of 50 fmol were achieved. Only a single ion was detected at *m/z* 101, which was not interfering with signals of target aflatoxins. An example of mass spectra obtained for both a standard and real peanut samples are provided in Figure 2.4. The authors predicted that the proposed method would be easily applicable to other mycotoxins and matrices.

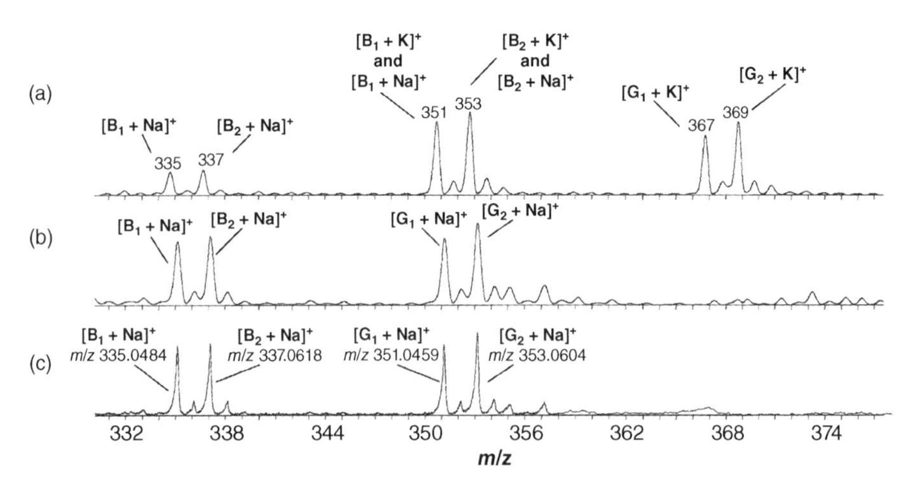

Figure 2.4. MALDI-TOF mass spectrum using triethylamine-α-cyano-4-hydroxycinnamic acid ionic liquid as the matrix of (a) an equimolar mixture of aflatoxins B_1, B_2, G_1, and G_2 (25 pg of each analyte); (b) an equimolar mixture of aflatoxins B_1, B_2, G_1, and G_2 (25 pg of each analyte spiked with 1 μl of a NaCl solution (10 mM); (c) aflatoxins detected as their Na^+ adducts from fungus-contaminated peanuts. Ref. [73], Figure 2, p. 8156. Reproduced with permission of American Chemical Society.

Madla et al. investigated the use of MALDI-MS for determination of pesticides [76]. The technique enabled the detection of 12 of 15 tested compounds below their maximum residue limits (MRLs) and good linearity was observed for responses of analytes in the concentration range of 0.001–50 mg/kg. However, different matrices (either CHCA or DHB) had to be used to achieve acceptable sensitivity and the method was not evaluated for real food matrices.

A promising approach to discrimination of fish species differing by commercial value was reported by Mazzeo et al. [74]. Following a simple and rapid sample extraction procedure with 0.1% aqueous formic acid (the overall time requirements for sample preparation were below 15 min per sample), highly specific MALDI mass spectrometric profiles containing biomarkers identified as parvalbumins (fish muscle proteins) were obtained from 25 fish species. Interestingly, the biomarkers exhibited a remarkable stability, as the heat-treated sample extracts provided mass spectra identical to those obtained for untreated samples. These results indicate that the method might be applicable to the authenticity control of cooked fish products as well as to raw fish-based commercial foods.

Calvano et al. used MALDI-TOFMS as a tool for the detection of adulteration of extra virgin olive oil (EVOO) with hazelnut oil based on the profile of phospholipids (PLs) [77]. Hazelnut oil is frequently used to adulterate olive oil due to its chemical similarity to olive oil in terms of triacylglycerols, sterols, and fatty acids composition. The authors developed a rapid and selective extraction procedure using an ionic liquid (arising from tributylamine and α-cyano-4-hydroxycinnamic acid) to isolate the PL fraction from subject oil. An identical ionic liquid was also used as the MALDI matrix. Such an experimental setup enabled a significant increase of PL signal and elimination of ions formed from other oil components. With regard to distinct differences between PL concentrations in olive oil (30–60 μg/kg) and hazelnut oil (10–20 g/kg), detection of adulteration levels as low as 1% was possible.

Garcia et al. developed a MALDI-based robust method for high-throughput forensic screening of milk powder adulteration by cheap vegetable oils and fats based on the profile of triacylglycerols [78]. With regard to simple sample preparation (sample shaken with *n*-hexane for 1 min followed by centrifugation for 2 min) and direct examination without prior chromatographic separation, the proposed method showed significant time-saving advantages compared with traditional methods using GC or LC (overall procedure time below 10 min per sample). Additionally, the method also has the potential for identification of the type of material used for adulteration.

2.3.2 Headspace (Solid-Phase Microextraction)-Mass Spectrometry E-Nose

The electronic nose based on the mass detector referred to as HS-MS and HS-SPME-MS e-nose (also referred to as a chemical sensor) is able to carry out analyses in very short times and with minimum sample preparation. In this instrumental setup, the volatile compounds are extracted from the headspace above the sample. Headspace methods include static headspace, dynamic headspace (purge and trap), and SPME [84].

SPME offers several important features including (i) fast analysis by reduction of sample preparation, (ii) minimization of the use of solvents, (iii) unattended operation via robotics (if a fully automated option is available), and (iv) the elimination of maintenance of the liner and column because contamination by nonvolatiles does not occur as much as it does with liquid injections. However, several factors influence the quality of generated data. Specifically, the profile of volatiles obtained is dependent on the type, thickness, and length of the SPME fiber used, as well as on the incubation and extraction time, temperature, and addition of salts [27]. In addition, the relative concentration of analytes in the headspace does not reflect the relative concentrations in the sample because of the differences in volatility of compounds. Taking into account all these factors, it is essential to analyze the samples under well-defined and constant conditions [84].

After desorption in a hot GC injection port and separation by a GC column, isolated volatiles are introduced in the ion source of a mass spectrometer, where they are fragmented, typically using EI at 70 eV. The fragments of all the volatile compounds are recorded as the abundance of each ion of different mass-to-charge ratios (m/z) (Figure 2.5).

Up to now, HS (SPME)-MS e-nose has been applied mainly in food authenticity studies. In most cases, the intensities of particular fragments (m/z) are subsequently submitted to multivariate data analysis for statistical evaluation. In general, equilibration times (HS-MS) and incubation and extraction times (HS-SPME-MS) ranged between 10 and 60 min, followed by desorption and acquisition (3–10 min) of MS fingerprints of isolated volatile compounds.

Vera et al. conducted a study to classify and characterize a series of beers according to their production site and chemical composition [85]. The analyzed beer samples were of the same brand but obtained from four different factories. The results obtained in this study enable consideration of the HS-MS (e-nose) as a potential aroma sensor because it is capable of discriminating and characterizing the samples according to their predominant aromas with the help of multivariate analysis.

Mildner-Szkudlarz and Jelen demonstrated the potential of HS-MS (e-nose) as a rapid tool for volatile compounds analysis with subsequent multivariate data analysis (PCA) for differentiation between EVOO samples adulterated with hazelnut oil [86]. This method allowed detection of olive oil adulteration with different contents of hazelnut oil ranging from 5 to 50% (v/v). Figure 2.6 shows average spectrum of hazelnut oil, pure EVOO, and EVOO with 5 and 50% (v/v) of adulteration obtained using HS-SPME-MS technique. Several groups of ions can be observed that grouped around ion m/z 43, 55, 70, and 83 for EVOO and m/z 43, 60, 74, and 96 for hazelnut oil. Changes in specific ion intensities in the HS-SPME-MS spectrum could be very cautiously correlated with the changes of particular components in pure and adulterated oils detected using HS-SPME-GC-MS.

HS (SPME)-MS e-nose has also been used to characterize and identify cheeses [87] and the country of origin of tempranillo wines [88], to study off-flavors in milk [89], and to detect unwanted fungal growth in bakery products [90].

2.3.3 Ambient Desorption/Ionization-Mass Spectrometry

The introduction of ambient desorption/ionization methods enabled a great simplification and an increase in speed of MS-based measurements. Unlike the conventional

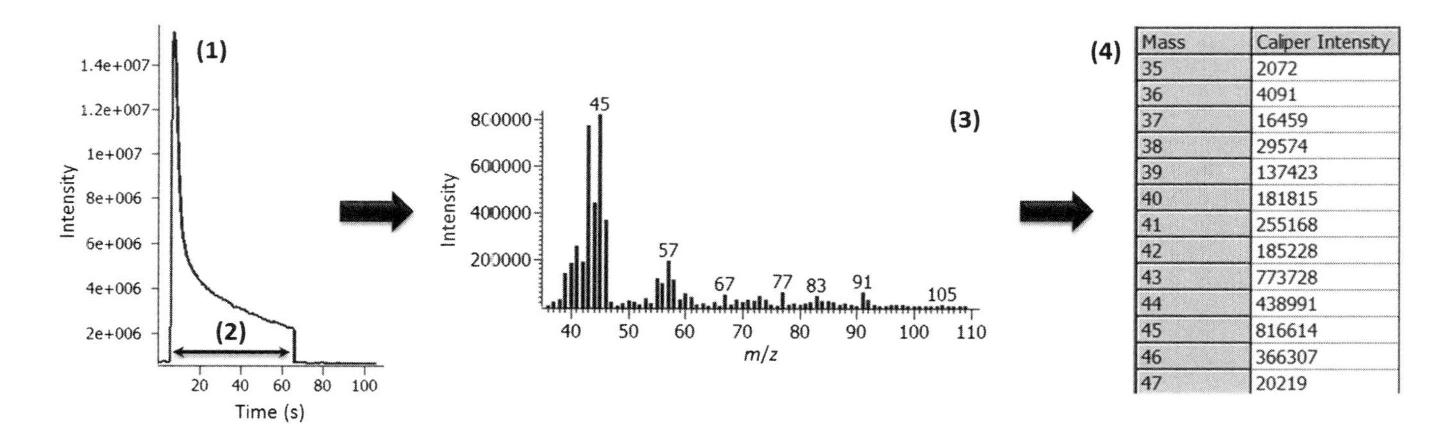

Mass	Caliper Intensity
35	2072
36	4091
37	16459
38	29574
39	137423
40	181815
41	255168
42	185228
43	773728
44	438991
45	816614
46	366307
47	20219

Figure 2.5. Analysis using MS e-nose. (1) Peak corresponding to total ion current (TIC); (2) averaging the peak; (3) fingerprint of EI mode; (4) table containing intensities of particular ions (fragments).

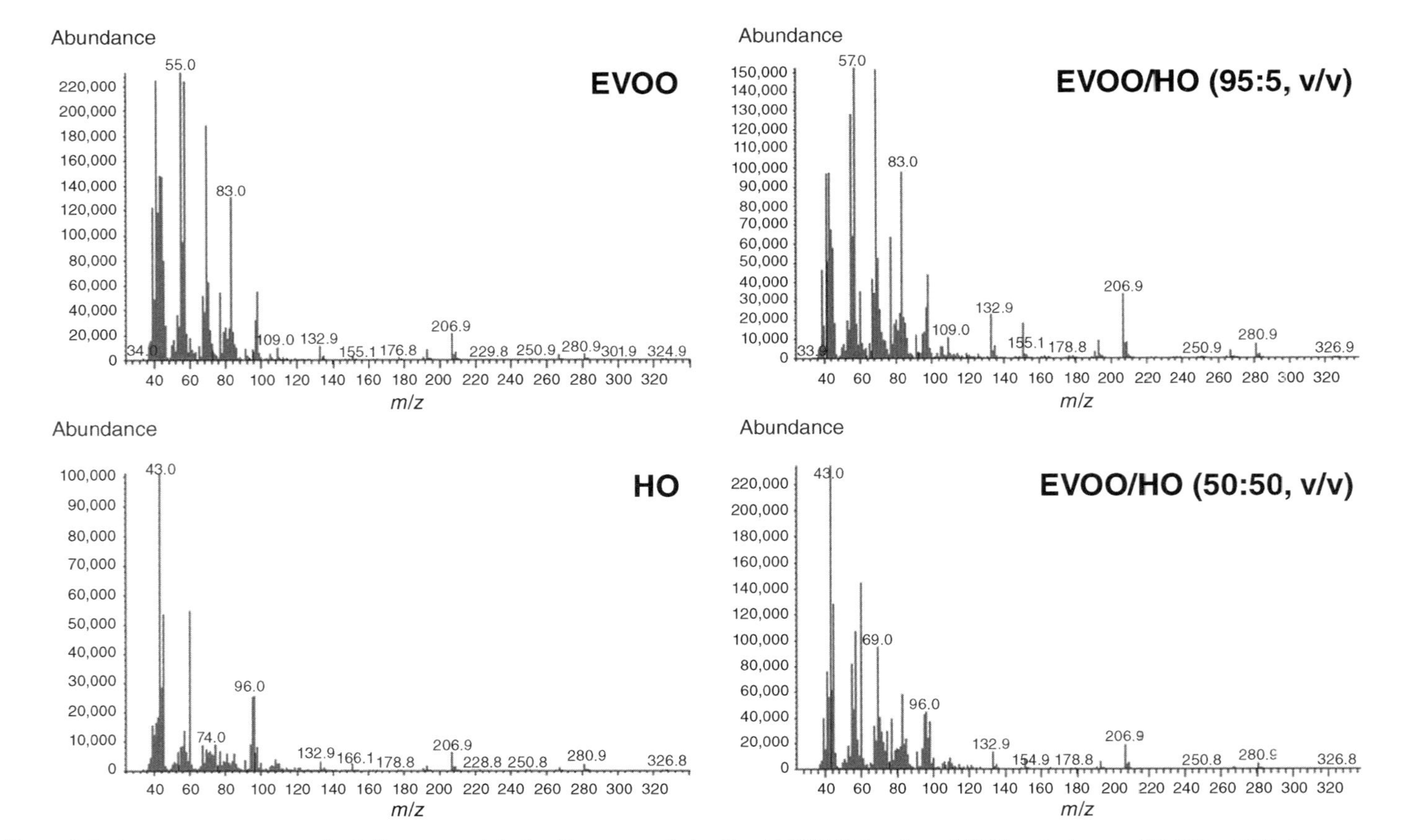

Figure 2.6. Average mass spectrum of volatile compounds isolated from extra virgin olive oil (EVOO), hazelnut oil (HO), and mixtures of EVOO and HO obtained using HS-SPME-MS Ref. [86], Figure 3, p. 758. Reproduced with permission of Elsevier Science Ltd.

ionization techniques, these novel approaches enable straightforward examination of various objects while requiring little or no sample pretreatment and significantly improving the overall throughput of these methodologies compared with methodologies that require multiple or complex sample preparation steps. Sample interrogation can be performed in an open environment (i.e., at atmospheric pressure) by introducing the sample into the ionization region and exposing it to a stream of desorbing and/or ionizing medium. Analyte ions arising from the ionization processes related to ESI, atmospheric pressure chemical ionization (APCI), or atmospheric pressure photoionization (APPI) are subsequently transferred through the open air to the inlet of the mass spectrometer [89]. More than 30 ambient ionization techniques (including some variants) have been developed and described to date. It should be noted, however, that not all of them have been widely used or commercialized. The pioneering techniques of desorption electrospray ionization (DESI) [92] and direct analysis in real time (DART) [93] remain the most established. Considering the ease of use and high throughput, ambient MS has been recognized as holding great potential for rapid characterization of food components, detection of various contaminants and fingerprinting/profiling [94–96], and in many other fields of analytical chemistry [97–99]. The ambient ionization techniques are anticipated to be applied in the field as powerful tools for early detection of various hazards related to food.

The following sections provide an overview of the most widely used ambient desorption/ionization techniques that have been applied to various aspects of food quality/safety and list example applications documenting both advantages and limitations of these techniques. Additionally, experimental parameters that significantly influence the outcome of DESI- and DART-based analyses are discussed.

2.3.3.1 Principles and Instrumentation

Desorption Electrospray Ionization DESI, which was introduced in 2004 by Takats et al., remains the most popular and widely used of all ambient desorption/ionization techniques [92]. Ionization in DESI experiments occurs when electrospray-generated charged droplets of solvent (typically a mixture of water and organic solvent) are directed toward the sample components deposited on a sampling surface. The ionization mechanism is not yet completely understood, but it is believed to be predominantly a multistage charged droplet pickup process. In the first phase, the sampling surface is prewetted by the initial solvent droplets and dissolution of sample components takes place. Subsequently, the surface solvent layer is impacted by later arriving charged droplets to form microdroplets. The analytes in the multicharged microdroplets are ionized through processes taking part in conventional ESI, that is, a continuous decrease in droplets size and formation of analyte ions via "Coulombic explosion." In addition to droplet pickup, condensed-phase and gas-phase charge transfer processes are also probably involved in ionization of some analytes [97]. The ions thus formed are transferred into the mass spectrometer through an extended ion transfer line that links the gap between the ionization region and the MS system's atmospheric pressure interface. Because of the similarity in ionization mechanisms, DESI and ESI yield similar mass spectra that contain both singly and multiply charged ions and are particularly effective in ionization of polar analytes.

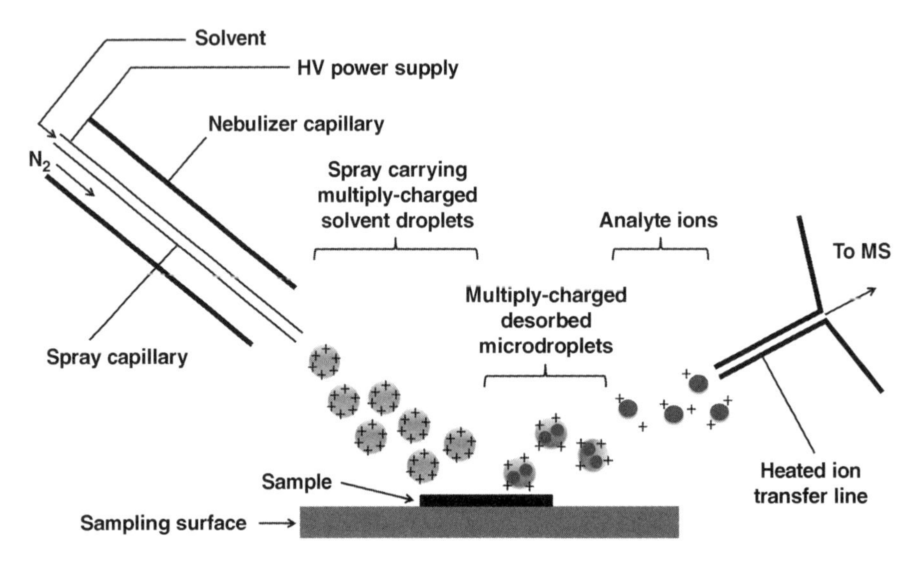

Figure 2.7. A scheme of DESI ion source and ionization process.

In standard DESI, a geometric setup similar to that depicted in Figure 2.7 is used. The sampling surface is mounted on a stage that can be freely moved in *x*-, *y*-, and *z*-directions and allows rapid examination of a number of samples. The sampling surface can be represented by either the sample itself (solid sample) or, in the case of liquid sample or sample extract, deposition onto a suitable nonconductive material followed by solvent evaporation. DESI can also be used for ionization of analytes after their chromatographic separation directly from silica TLC plates [100]. Compared with ESI, DESI has a greater number of critical experimental parameters that dramatically influence the outcome of analysis. The main DESI parameters relate to solvent (composition and pH), electrospray probe (e.g., solvent flow rate or spray voltage), source geometry (e.g., the incident and collection angle), and sampling surface [97]. Their optimization for food-related DESI applications is discussed in more detail in the following section.

In addition to the standard geometric configuration, several modifications have been developed to enable simplification and to improve sensitivity of DESI measurements. Chipuk and Brodbelt described the transmission mode (TM) DESI of liquid samples and solid residues from evaporated solvents [101]. Instead of deflecting the spray from a surface, in TM-DESI, the electrospray is directed through a mesh screen placed between the electrospray tip and the ion transfer line. Both incident and collection angles are 0°. Such geometry enables reduction in experimental variables while producing high-quality mass spectra. Another approach was introduced by Venter and Cooks who enclosed the DESI source in a pressure-tight enclosure with fixed spatial relationship between the sprayer, surface, and the ion transfer line. Of the geometries tested, the combination of 90° incident and collection angles was found to be most favorable in terms of ease of operation and signal stability [102].

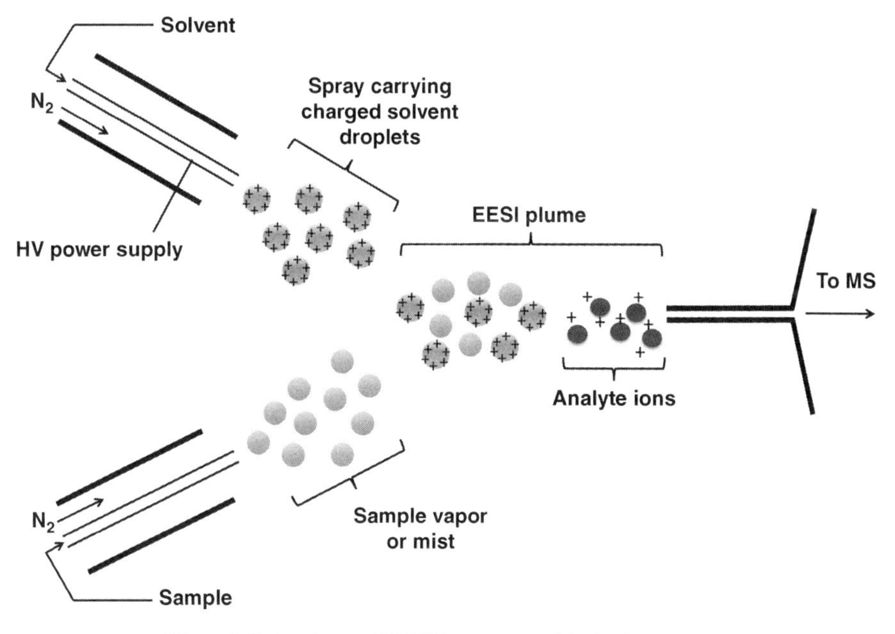

Figure 2.8. A scheme of EESI ion source and ionization process.

This so-called geometry-independent DESI enabled simpler and more efficient sampling compared with the conventional DESI setup.

Extractive Electrospray Ionization Extractive electrospray ionization (EESI) was primarily developed for ionization of (semi)volatile analytes in complex matrices and represents a variant of DESI [103]. The EESI source uses two separate sprayers: one is an auxiliary electrospray, which generates charged droplets of solvent, while the second acts as a nebulizer and is used to deliver the sample into the ionization region (Figure 2.8). The ionization process itself takes place in the EESI plume and involves liquid- and gas-phase interactions between the neutral analyte droplets/molecules and solvent-derived charged droplets/ions. The predominant ionization mechanism is probably based on condensed phase extraction of analyte molecules into the charged droplets and ESI processes leading to analyte ion formation [104]. Because the sample nebulization and ionization are separated in both space and time, EESI exhibits significantly higher tolerance to sample matrix and lowers the adverse impact of ion suppression effects on analyte intensities.

Liquids can be sampled with EESI by direct infusion and subsequent nebulization in the sampling sprayer. Alternatively, noninvasive neutral desorption (ND) of analytes can be performed by directing the neutral nitrogen gas stream onto the sample surface. As a consequence, sample droplets are formed through a micro-ejection mechanism: microdroplets are transported with carrier gas into the electrospray plume for ionization [105].

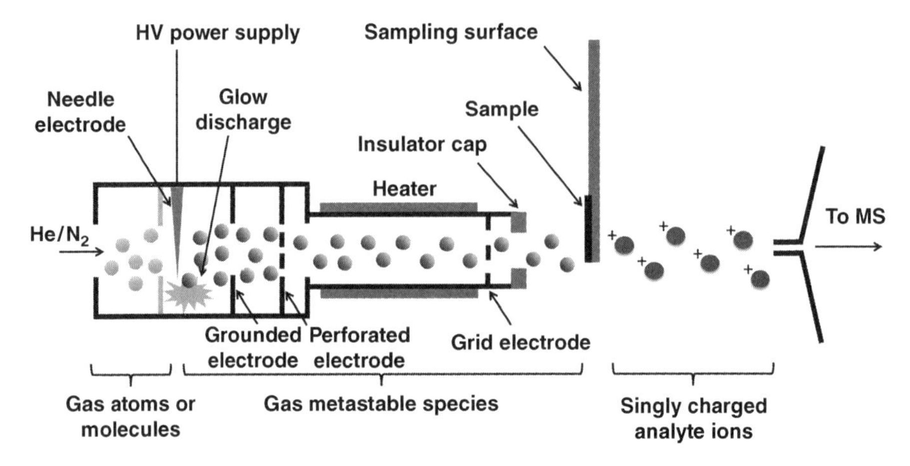

Figure 2.9. A scheme of DART ion source and ionization process.

Direct Analysis in Real-Time Ionization DART ionization, introduced in 2005 by Cody et al., is the second pioneering ambient ionization technique [93]. Like DESI, DART is commercially available and its popularity is rising. In DART, excited-state metastable atoms or molecules of gas (typically helium) are used as the medium for ionization. Gas metastables are formed via a glow discharge taking part in a compartment separated from the sample. In the next step, charged species are removed from the gas stream by passing through a perforated electrode, leaving only metastables. Gas can be optionally heated and directed to the sample (see Figure 2.9). The grid electrode at the exit of the DART gun serves as an ion repeller, which prevents ion–ion recombination resulting in signal loss. In the sampling region between the ion source exit and mass spectrometer inlet, the metastable species interact either directly with the (thermo)desorbed analyte molecules or with atmospheric components to form reactive species that further ionize the analytes. The major mechanism of DART ionization in positive ion mode involves Penning ionization of atmospheric water and nitrogen and subsequent proton transfer to analyte resulting in $[M+H]^+$ ion formation. In negative ion mode, negatively charged oxygen clusters formed by thermal electrons deprotonate molecules of analytes. In addition to proton abstraction, electron capture, dissociative electron capture, and anion attachment processes can take part under DART negative ion mode settings [96]. DART is suitable for ionization of analytes with medium/low polarity and molecular weights below 1000 Da. DART mass spectra show common features with those obtained by APCI and APPI techniques. Contrary to DESI, multiply charged ions or metal–cation adducts are not formed [93]. For some analytes, generation of adduct ions, such as $[M+NH_4]^+$ or $[M+Cl]^-$, can be induced by introducing vapors of suitable dopant solvents into the ionization region. In DART-based experiments, the source optimization is typically simpler compared with DESI and is mainly limited to tuning the ionization gas temperature and adjusting the setup geometry.

Several in-house and commercial autosamplers have been developed for repeatable and high-throughput sampling of liquid and solid samples. The most common method of liquid sample introduction is its transfer into the metastable gas stream on a surface of a glass melting point capillary either by robotic arm or by scanning autosampler. Direct and automated desorption of sample deposits from various surfaces (e.g., tablets or TLC plates) can also be performed provided the angle-adjustable ion source is available. An alternative approach is represented by TM sampling, within which a porous material (stainless steel wire mesh, fabric, or foam swabs) is positioned between the DART source exit and the mass spectrometer inlet to serve as the desorption/ionization surface [106–108]. In a study by Krechmer et al., TM sampling was combined with ohmic heating of the metal screen surface while operating the DART ionization gas at ambient temperature. The control of surface temperature through modulation of electrical current flow enabled a significant increase in sample vaporization rate, thus increasing the analysis throughput, compared with a conventional DART setup [109].

Desorption Atmospheric Pressure Chemical Ionization Desorption atmospheric pressure chemical ionization (DAPCI) is based on desorption of the sample surface by a heated gas stream containing reagent species (electrons, protons, hydronium ions, solvent ions, metastables, etc.) generated in an atmospheric pressure corona discharge [110]. The dominant mechanism of ion formation is similar to that taking part in a conventional APCI source in which gas-phase ion/molecule reactions seem to play the crucial role in the ionization process [111]. The primary ions generated in corona discharge collide with the solvent molecules to form secondary ions that transfer charge to the analytes emitted from the sample surface. The setup employing a supply of solvent (Figure 2.10) can be avoided if there is a sufficient concentration of atmospheric water present in the ionization region (H_3O^+ ions are then mainly involved in gas-phase ion reactions) [91]. DAPCI provides superior ionization yield for nonpolar compounds of rather lower molecular weight, thus offering an ionization method that is orthogonal to DESI.

Atmospheric Pressure Solids Analysis Probe Ionization Atmospheric pressure solids analysis probe (ASAP) is another APCI-like technique similar to DAPCI. The liquid or solid sample is loaded on a glass probe and inserted into a conventional APCI source to be exposed to a stream of hot nitrogen gas without solvents [112]. As a result of vaporization, the sample components are transferred into the gas phase and further carried to the discharge needle region where they are ionized through corona discharge-based APCI processes (Figure 2.11). The type of ions formed in ASAP ionization is strongly influenced by the environment in the ion source, especially by the humidity. Depending on the conditions, either $[M+H]^+/[M-H]^-$ or radicals ions ($M^{+\bullet}$ and $M^{-\bullet}$) are the predominant species observed [94]. Like DESI and DART, the ASAP technique has been commercialized and can be attached to most LC–MS systems. However, no option for automated sample introduction is currently available.

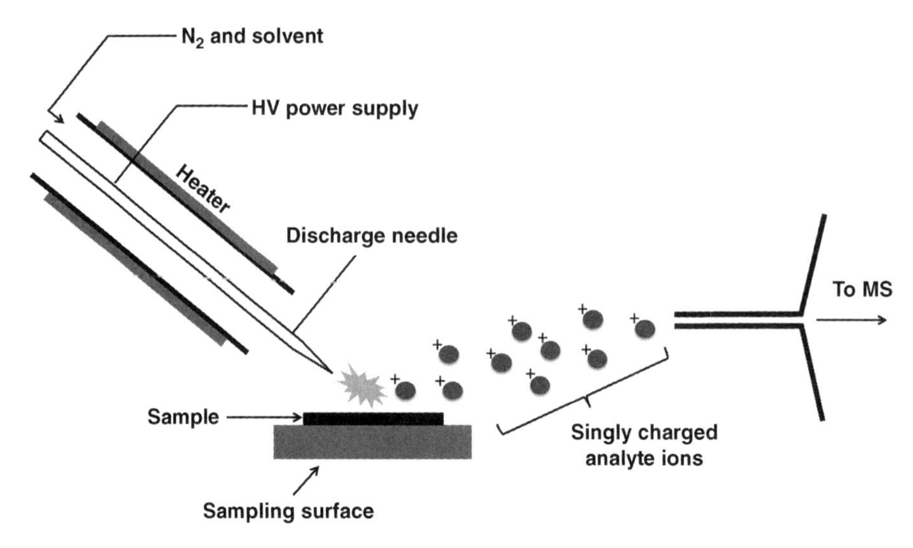

Figure 2.10. A scheme of DAPCI ion source and ionization process.

Desorption Atmospheric Pressure Photoionization In desorption atmospheric pressure photoionization (DAPPI), a heated nebulizer microchip is used to mix UV-absorbing solvent (typically toluene) with nitrogen gas and produce a hot vapor jet that is directed toward the sample deposited on a surface attached to a linear

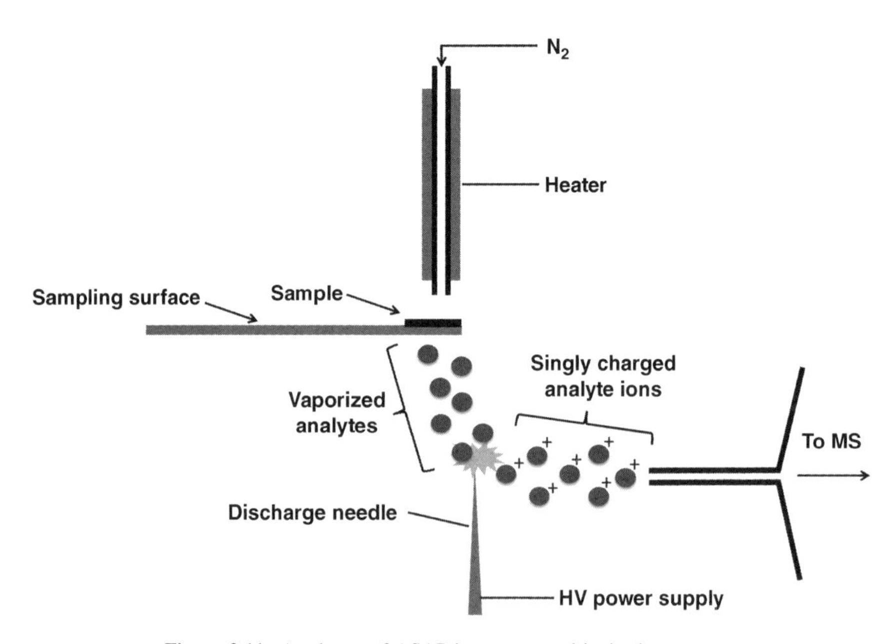

Figure 2.11. A scheme of ASAP ion source and ionization process.

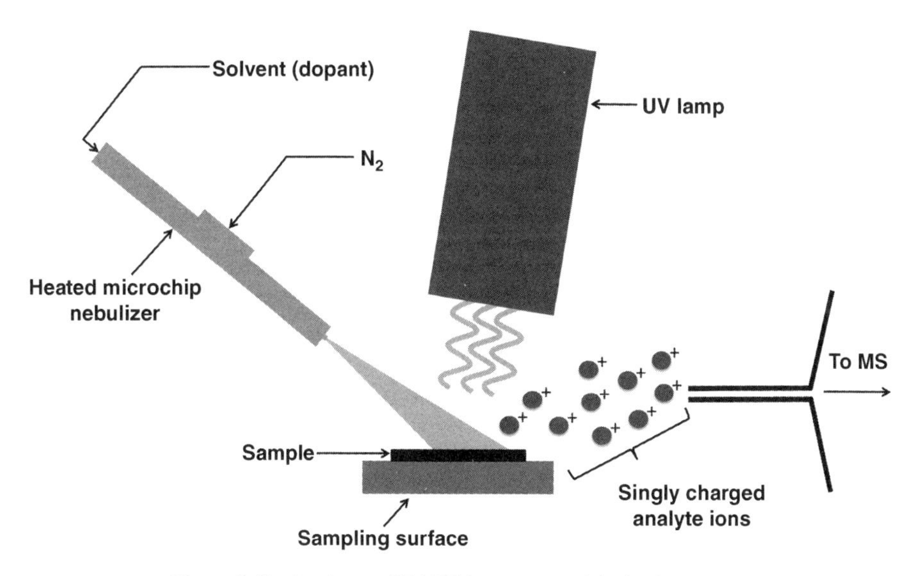

Figure 2.12. A scheme of DAPPI ion source and ionization process.

xyz-stage. The thermally desorbed vapors, containing both dopant and sample components, are subsequently irradiated by UV light and photoionization of analytes occurs (Figure 2.12). While the UV-absorbing analytes can be ionized directly, compounds lacking the chromophore group cannot undergo direct ionization and are ionized through molecule–ion interactions with dopant and atmospheric water ions [113]. The DAPPI technique was shown to be capable of ionizing both polar and nonpolar compounds. The nature of the particular analyte and the dopant solvent dictates the type and intensity of ions formed in DAPPI. In positive ion mode, the spraying solvents that yield radical cations upon photoionization (e.g., toluene) can be used for ionization of low-polarity, low proton affinity analytes ($M^{+\bullet}$ ions), while the solvents generating the proton-donating reactive species (e.g., acetone, methanol, or hexane) can protonate high proton affinity compounds to form $[M+H]^+$ ions. In negative ion DAPPI, solvents with ionization energies below the energy of UV lamp photons provide the best ionization efficiencies in the formation of $[M-H]^-$ and $M^{-\bullet}$ ion. Other important factors affecting the ionization yield are related to source geometry and thermal conductivity of the sampling surface. Materials with low thermal conductivity, such as poly(methyl methacrylate) (PMMA) or polytetrafluoroethylene (PFTE), can be locally heated to higher temperatures, which lead to improved efficiency of the thermodesorption process [114].

2.3.3.2 *Optimization of DESI-MS and DART-MS-Based Methods*

Although widely perceived as a simple and straightforward approach, successful application of ambient MS to an analytical problem typically requires careful optimization of many parameters. This need is even more pronounced in cases in which highly complex samples such as foods are to be analyzed. The aim of the following section is to address the key factors affecting the performance of ambient MS-based methods employing DESI and DART techniques. In addition to optimization of parameters directly related to ambient ionization, requirements on mass spectrometric detection and sample preparation for both qualitative and quantitative high-throughput analyses are briefly discussed.

DESI Source Parameters As mentioned above, a remarkably high number of parameters affect the performance of DESI-based analytical workflow. These parameters can be classified as follows [97]:

- Geometric parameters (incident angle α, collection angle β, tip-to-surface distance d_1, and MS inlet-to-surface distance d_2).
- Spray parameters (spray capillary voltage, nebulizer gas pressure, and solvent flow rate).
- Chemical parameters (composition of sprayed solvent).
- Surface parameters (potential, composition, and temperature).

Fortunately, some "gold standard" practices and settings, which represent a good start for DESI optimization in particular applications, are already available in the literature. Some of these useful settings are provided in Table 2.4.

The geometric setup in DESI significantly affects the sensitivity of measurements. The α and d_1 parameters directly affect the ionization process and their optimal values

Table 2.4. The Overview of Typical Parameter Settings in DESI and DART Experiments [91,95–97,115]

Parameter	Setting range
DESI	
Incident angle (α)	30–70°
Collection angle (β)	5–30°
Tip-to-surface distance (d_1)	1–10 mm
MS inlet-to-surface distance (d_2)	1–5 mm
Spray capillary voltage	2–6 kV[a]
Nebulizer gas pressure	8–12 bar
Solvent flow rate	2–5 μl/min
DART	
Ion source exit-to-MS inlet distance	5–25 mm
Ionization gas temperature	150–450 °C
Ionization gas flow	0.5–3.5 l/min
Discharge needle voltage	1–5 kV
Perforated electrode voltage	150–350 V[a]
Grid electrode voltage	50–150 V[a]

[a]Voltages represent either positive or negative potentials, depending on the ionization mode used.

might strongly differ for various analytes or analyte classes. The other two parameters (β and d_2) dictate the efficiency of the ion collection process. Another analyte-dependent parameter influencing sensitivity is the spray capillary voltage. As in the case of conventional ESI, the optimal polarity and voltage have to be tuned to obtain satisfactory results in the application of interest. The nebulizer gas and solvent flow setting are related to the size and velocity of droplets generated by the sprayer. High velocities and small droplets are favorable for the ionization yield (i.e., enhanced desolvation efficiency and more secondary microdroplets formed). However, settings that are too high result in signal drop for analytes ionized exclusively through the droplet pickup process, as charged solvent evaporation takes place prior to the impact with the surface [95,97].

In most applications, a mixture of water and either methanol or acetonitrile is used as the spray solvent. The type and content of organic solvent (typically ranging from 50 to 80%) can significantly influence the sensitivity of measurements [95]. For some hydrophobic compounds, the beneficial effect of the use of nonaqueous solvents has been reported [116]. To further improve the ionization yield, various pH-adjusting additives such as formic and acetic acid or buffers (ammonium formate/acetate) can be added to the spray mixture. In a study of the trace analysis of 16 multiclass representative agrochemicals in food, the composition of the solvent spray significantly affected the signal for some of the compounds [117]. The authors reported a remarkable increase of ionization yield for organophosphorus insecticides (malathion and isofenphos-methyl) when a spray solvent containing methanol was replaced with acetonitrile. Because the aim of the study was to analyze multiple compounds, a compromise spray solvent providing acceptable sensitivity for all analytes had to be used (acetonitrile/water, 80/20 (v/v) with 1% formic acid, in this case). The effect of the solvent pH was also observed by Hartmanova et al., who employed DESI for direct profiling of anthocyanins in dried wine droplet deposited on a coarsened glass plate. When the pH of the spray liquid solvent (methanol/water, 75/25 (v/v)) was lowered by addition of formic acid (optimal content 0.2%), the anthocyanins were detected as flavylium cation acidobasic form ($[M]^{+\bullet}$) and substantial improvement in the quality of the mass spectra was observed [118].

An undoubtedly attractive feature of DESI is the potential to adjust the measurement selectivity. In so-called reactive DESI, a suitable reagent is mixed into the spray mixture to induce catonization of problematic (i.e., low proton affinity) compounds, or even to perform *in situ* derivatization of the analytes. Reactive DESI using silver trifluoroacetate was demonstrated to enlarge the technique's application scope in the analysis of strobilurin fungicides [95]. While only 50% of analytes could be ionized in regular DESI, silver catonization enabled detection of all target strobilurins.

Surface properties also play a vital role in the ionization process. For effective ionization, nonconductive materials have to be employed for sampling in order to avoid neutralization of charged species. Insulators such as glass, PFTE, or PMMA are the most frequently used materials. The electrostatic properties of an insulator material are very important for signal stability. For example, PTFE, which is an electronegative polymer, provides superior signal stability in negative ion mode, while the PMMA polymer is more suitable for positive ion mode. The chemical nature and roughness of the DESI substrate also affect both ionization efficiency and homogeneity of the sample (when it is prepared by deposition from a solvent). Materials with high surface roughness, for which low affinity is provided by analytes, should be used [97].

In high-throughput DESI measurements, it is important to achieve sufficient spatial resolution, as sample-to-sample cross-contamination can occur during analysis of sample series deposited close to one another. The spatial resolution can be improved by tuning of α and d_1 parameters and/or by decreasing both the internal diameter of the spray capillary and the flow rate. Additional attention has to be paid to sample preparation prior to DESI of powder or dust samples due to potential MS system contamination problems. The use of double-sided adhesive tape or rinsing with methanol followed by analysis of dried droplets has been proposed to handle such sample types. Regardless of the type of sample, the mass spectra should always be background corrected with the use of records obtained from "blank" surfaces [95,97].

DART Source Parameters The following parameters should be considered when optimizing DART analyses:

- DART source geometry (sample position and ion source exit-to-MS inlet distance).
- Ionization gas parameters (type, temperature, and flow).
- DART source voltages (discharge needle voltage and perforated and grid electrode voltages).
- Dopants (type and method of introduction into the ionization region).

Settings likely to be used in a typical DART-MS application are provided in Table 2.4.

The position of the sample in the DART ionization region is a critical factor that influences sensitivity. Liquid samples spread on the surface of a glass capillary should always be placed slightly off the axis between the source exit and the MS inlet so that the gas stream is not blocked. Alternatively, the sample can be moved through the gas stream in the perpendicular direction. Harris et al. studied the impact of solid sample (tablet) position on the signal intensity and found that the highest ion transmission was observed when placing the tablet in an upright position close to the DART gun exit [119]. Because there are no comprehensive data from which some generally applicable settings for positioning of samples can be derived, a case-to-case optimization of this parameter should always be performed [96,115].

The most frequent ionization gas employed in DART is helium. However, the use of nitrogen, neon, and argon has also been considered. Because the metastables derived from various gases have different energies, their ability to directly ionize atoms and molecules present in the ionization region differs greatly. Only species with lower ionization energies can be directly ionized. The helium 2^3S excited-state metastable species has an energy of 19.8 eV, which is high enough to induce

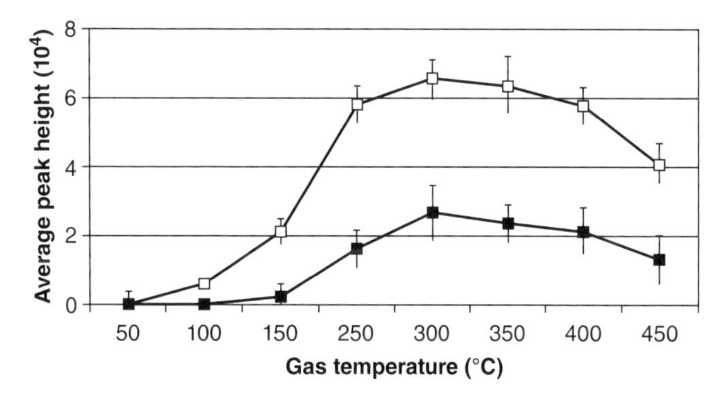

Figure 2.13. The impact of ionization gas temperature on signal intensity of melamine (□) and cyanuric acid (■) in powder extract. Error bars are standard deviations ($n = 5$) Ref. [96], Figure 2, p. 207. Reproduced with permission of Elsevier Science Ltd.

formation of charged water clusters, thus further ionizing the analytes. Nitrogen- and argon-derived metastables are of lower energy and can ionize only some compounds, thus limiting the application scope [115]. Nevertheless, this phenomenon can be used for selective ionization of target analytes while avoiding ion formation of other sample components [120]. Additional obstacles to wider application of nitrogen are the need for higher electric fields for metastable formation and possible oxidation of analytes during ionization [121].

The temperature of the ionization gas is often the key factor affecting the results in DART-based experiments. The optimal gas temperature for a particular analyte depends on its physicochemical properties, such as boiling point, polarity, and molecular weight. A gas temperature that is too low will not facilitate thermodesorption of nonvolatile analytes, while a gas temperature that is too high can lead to rapid volatilization and signal drop due to insufficient acquisition rate of the mass spectrometer. In addition, analyte thermal degradation, extensive ion fragmentation, or even sample pyrolysis can occur under high ionization gas temperatures [96,122]. Typically, a bell-shaped curve is obtained when plotting the ion intensity against the gas temperature used, as shown in Figure 2.13 for melamine and cyanuric acid spiked into milk powder extract. From the practical point of view, it is important to note that the actual temperature in the ionization region is different from that of the gas heater as a consequence of mixing with the cooler ambient atmosphere (see Figure 2.14). This temperature difference is even more pronounced when higher gas flow rate settings are applied [123].

Regarding the ionization gas flow, an increase in signal intensity due to promoted thermodesorption process was reported at higher helium gas flows. When operating DART at high flow rates, one should be aware of the risk of MS system contamination, as the sample can be easily blown off the sampling surface. The upper gas flow limit is also determined by the stability of the vacuum system of the mass spectrometer because high gas flow rates can increase the pressure in the atmospheric pressure interface and cause automatic shutdown of the instrument. To overcome this drawback, a gas ion separator flange enclosing the MS inlet and providing additional pumping has to be

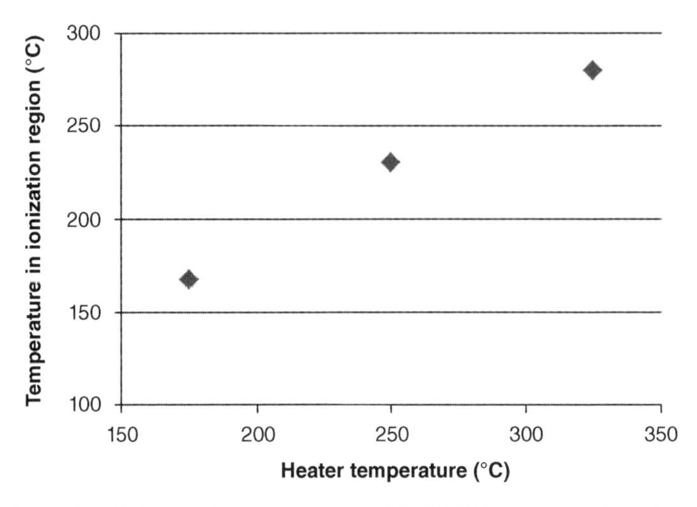

Figure 2.14. Relationship between heater temperature of the DART ion source and actual temperature in the ionization region at helium gas flow 2 L/min [123].

employed. Such a hardware setup can also improve the sensitivity of measurement by sweeping the analyte-laden carrier gas to the MS inlet region, thus reducing the potential for the gas to drift away into the surrounding atmosphere [124].

In most applications published to date, the impact of DART electrode voltages on the sensitivity has not been studied in detail and settings similar to those shown in Table 2.4 were used. The reports in which optimization of these parameters was carried out provide rather contradictory results [115]. For example, the signals of some organometallic compounds were intensified by increasing the discharge needle voltage (up to 4 kV) and were also strongly dependent on other electrode potential settings [125]. Improved sensitivity was obtained for melamine when a relatively low grid electrode potential of 50 V was used [120,126]. On the other hand, no significant impact of electrode voltages was observed in a study of the analysis of pharmaceuticals in biological fluids. In any case, the optimal voltage settings may vary for various compound classes and should be tuned when performing targeted analysis of a few analytes.

As already mentioned, formation of adduct ions in DART can be facilitated by allowing dopant vapors to access the ionization region. Depending on the particular analyte, the use of a dopant can yield adducts that are of higher intensity than the pseudomolecular ion or even facilitate ionization of compounds that would otherwise not provide any ions under standard DART setup. For this purpose, aqueous ammonia, dichloromethane, or trifluoroacetic acid solution is typically used to induce formation of $[M + NH_4]^+$, $[M + Cl]^-$, and $[M + CF_3COO]^-$ ions, respectively. The vapor introduction can be achieved either by placing an autosampler vial containing dopant in the proximity of the ionization region or by adding it to the sample [96]. The use of dopants has been demonstrated by Vaclavik et al., who employed ammonia to enhance formation of $[M + NH_4]^+$ from triacylglycerols in olive oil [127]. The adduct ion intensities increased approximately by one order of magnitude compared with $[M+H]^+$ and detection of minor triacylglycerols was enabled. In another study, formation of $[M+Cl]^-$ ions of some poorly ionizing *Fusarium* mycotoxins in cereal extracts was achieved in negative DART ionization mode. The characteristic isotope profile of the chlorine-containing ion could be used for confirmation of identity [128].

Mass Spectrometric Detection Both DESI and DART ionization sources can be relatively easily coupled to any of the currently available LC–MS systems. These novel techniques represent versatile and attractive alternatives to conventional ESI and APCI, which are typically used with separation techniques. Considering the fact that in ambient MS all components (both analytes and matrix) present in the sample are ionized almost simultaneously, high requirements are laid on the MS detection to provide desired selectivity and sensitivity that fit the purpose for a particular application. Different types of mass analyzers have specific features that make them more or less suitable to deal with diverse tasks faced in analysis of food by ambient MS. The triple quadrupole and (linear) ion trap analyzers operated in MRM mode provide superb sensitivity and selectivity in cases of targeted measurements of relatively low numbers of analytes. When full spectral information and rapid data acquisition are required (e.g., in food profiling and fingerprinting), the use of high-resolution instruments using TOF or Orbitrap mass analyzers is preferred. Additionally, the identity of analytes can be estimated based on accurate mass measurements and elemental formula estimation. In this respect, hybrid instruments capable of high-resolution tandem spectra (MS/MS) acquisition provide higher degrees of confidence in identification of unknowns. It is worth noting that in practice, either low- or high-resolution systems are applied to similar applications, largely due to limited access to the other instrumentation.

The selectivity of measurement often plays a critical role in ambient MS of complex samples. The presence of isobaric interferences in food samples or extracts can complicate both qualitative and quantitative analyses, potentially resulting in false positive results. While MS/MS or MS^n can overcome this drawback only in targeted analysis, the use of (ultra)high mass resolving power instruments that mitigate the loss in spectral peak capacity represents a more generally applicable option [95,96]. An example of the benefit of high resolving power was provided in a study by Cajka et al., who compared medium high-resolution TOF and Orbitrap mass analyzers in DART-based analysis of dithiocarbamate fungicides in fruit extracts [129]. The mass resolving power of the TOF analyzer (~5000 FWHM) was not sufficient to entirely separate the signal of the analyte (thiram) from matrix interferences. On the other hand, an Orbitrap mass analyzer operated at 25,000 FWHM allowed complete spectral separation even if the intensity of the analyte was lower than that of the interferences. One should be aware that the mass resolving power provided by the Orbitrap is linked to the acquisition speed, which might not be sufficient for good desorption peak characterization at ultrahigh mass resolving power settings.

A major drawback often encountered during MS-based analysis of complex samples is the suppression of analyte signals caused by the sample matrix. While in LC–MS this phenomenon can be diminished to some extent by chromatographic

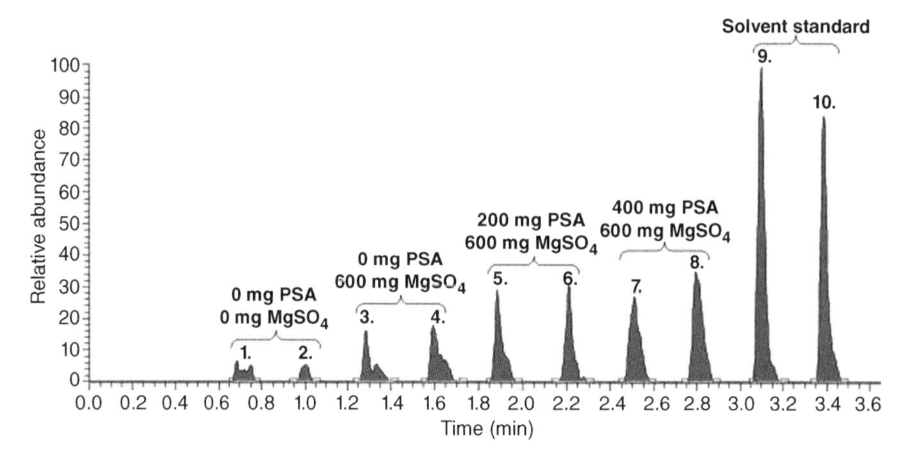

Figure 2.15. The impact of dispersive SPE cleanup employing PSA and $MgSO_4$ on deoxynivalenol (DON) (*m/z* 331.0943 ± 4 ppm) signal intensity in wheat extract (spike 500 μg/kg). Given sorbent amounts were used for 4 ml of acetonitrile extract containing equivalent 800 mg of matrix; solvent standard concentration was 100 ng/ml Ref. [128], Figure 2, p. 1956. Reproduced with permission of Elsevier Science Ltd.

separation, the impact of matrix effects in ambient MS is typically more severe. The need for characterizing and preferably minimizing matrix effects is of high concern in qualitative and quantitative food analyses of analytes occurring at low concentration levels. In such applications, the use of some sample preparation steps that enable discrimination of at least some sample matrix is practically unavoidable. Because these procedures represent a bottleneck of the whole analytical workflow, rapid and simple protocols are typically followed [95,96]. The effect of matrix on signal intensity was reported in studies concerned with both DART and DESI; in the latter case, mainly for some pharmaceuticals, Kauppila et al. documented severe signal suppression of dobutamine. Even if a diluted urine solution was analyzed, no signal of the target analyte could be detected. Apparently, DESI ionization was obstructed by the urine matrix [130]. To overcome signal suppression of pesticides in DESI-based direct surface analysis of fruits, a sample surface extraction with acetonitrile and analysis of dried extract were performed. The signal intensity was approximately half of that observed in pure solvent [129]. Another strategy employing a modified QuEChERS procedure was used in the analysis of mycotoxins in cereals [128]. The amount of sorbent (primary secondary amine, PSA and magnesium sulfate, $MgSO_4$) used in the dispersive SPE step of the crude acetonitrile extract was optimized for the most effective cleanup (see Figure 2.15). A signal increase of 12–39% was achieved for target analytes (100% intensity in pure solvent).

Quantification DESI and DART are currently perceived mainly as qualitative tools. However, several studies have documented that they can also be applied to semi-quantitative or even quantitative analysis [95,96]. In quantitative applications, relatively high signal fluctuation in repeated analyses (as high as 50%), as well as matrix effects, has to be overcome by using suitable internal standards (preferably isotope-labeled analogs of the analyte). It should be noted that this typically applies only to the analysis of liquid samples. Quantification in solid samples is much more complicated or even impossible due to uneven distribution of analytes and other problems related to preparation of homogeneous standard material. Thorough validation, in-batch analysis of quality control samples, and comparison of results with those of established methods have to be performed to obtain reliable quantitation and demonstrate that the use of ambient MS-based method is fit for purpose.

A precision of 15% and linearity of 0.99 over two orders of magnitude were reported by Garcia-Reyes et al. for DESI analysis of imazalil spiked into an orange extract together with labeled d_5-imazalil (internal standard). The results were in excellent agreement with those obtained using an LC–MS method. The authors noted that such performance characteristics cannot be considered a standard practice in food analysis and in most cases DESI is able to deliver only semiquantitative information on analyte concentration level [117].

The capabilities of DART in quantitative analysis of food have been recently demonstrated for melamine and cyanuric acid in milk powder [126], mycotoxins [128], some pesticides [131], and isoflavones (see Table 2.5) [132]. Without exception, an internal standard had to be employed to obtain acceptable precision.

2.3.3.3 Applications of Ambient Desorption/Ionization Techniques in High-Throughput Analysis of Food

An overview of selected applications of the above described ambient ionization techniques in high-throughput food analysis is provided in Table 2.5.

Edison et al. described a surface swabbing technique coupled to TM-DART-HRMS for the rapid screening of pesticide residues in fruits [138]. Rather than using a fixed ionization temperature, a gradient from 100 to 350 °C over 3 min was used to achieve a minimal separation of analytes based on volatility differences. Of the 132 pesticides involved in the study, 86% of target compounds could be consistently detected at levels of 2 ng/g (per apple and orange) and 10 ng/g (per grape). The identification of analytes was performed based on accurate mass measurements facilitated by the Orbitrap mass analyzer. The results of the procedure were found to be comparable in terms of identification of pesticides with those obtained by the LC–MS method and greatly increased the sample throughput by reducing sample preparation and analysis time.

ASAP-HRMS instrumentation was used in another study concerned with screening of strobilurin pesticides [94]. Direct detection of azoxystrobin in ground wheat samples was possible by stirring the glass ASAP probe among the ground solid sample and introducing it into the ASAP source. During analysis of blank and contaminated wheat ($n = 20$ each) containing 0.3 mg/kg of azoxystrobin, ASAP-MS enabled detection of all positive samples at 95% confidence interval.

An interesting application of noninvasive neutral desorption (ND) sampling and EESI-MS was reported by Chen et al., who applied this technique to fruit maturity and quality assessment [135]. The mass spectral fingerprints obtained by EESI of bananas,

Table 2.5. Overview of Selected Ambient Mass Spectrometry Applications Relevant to Food Safety and Quality

Technique	Aim of the Study	Analytes	Sample Pretreatment	Remark	Reference
DESI	Qualitative and quantitative analyses of pesticides on fruit/vegetable surface and in sample extracts	Ametryn, amitraz, atrazine, azoxystrobin, bitertanol, buprofezin, imazalil, isofenphos-methyl, malathion, nitenpyram, prochloraz, spinosad, terbuthylazine, thiabendazole, thiacloprid	None, QuEChERS extraction procedure with dispersive SPE cleanup	LODs 1–90 μg/kg, quantitation only for imazalil, comparison of results with LC–MS/MS data	[117]
DESI	Qualitative and semi-quantitative analysis of natural sweeteners in *Stevia* leave fragments and dietary supplement	Steviolbioside, rubusoside, stevioside, rebaudiosides (A–F), dulcoside A	None	Quantitation only for rebaudioside D, concentration of other analytes estimated based on relative abundance in the mass spectra	[133]
DESI	Profiling of anthocyanins in wine and wine grape slices	Multiple anthocyanins	Acidification with formic acid	Homemade nano-DESI source	[118]
DESI	Qualitative analysis of triacylglycerols and their oxidation products in vegetable oils and chocolate	Triacylglycerols and their oxidation products	None	Ammonium acetate added to spray solvent resulted in higher signal intensities of triacylglycerols and less complex mass spectra	[134]
EESI	Fingerprinting for differentiation of maturity and quality of bananas, grapes, and strawberries	Nonspecified analytes	None	Noninvasive sampling using neutral desorption, data visualization with PCA	[135]

EESI	Fingerprinting of cheese products and their differentiation according to type	Nonspecified volatile and nonvolatile components of cheese	None	Noninvasive sampling using neutral desorption, data visualization, and classification with PCA	[136]
EESI	Fingerprinting of beer for discrimination according to type	Esters, free fatty acids, volatile and nonvolatile organic acids, amino acids	None	Noninvasive sampling by bubble bursting, data visualization with PCA	[137]
DART/DESI	Quantitative analysis of strobilurin fungicides in wheat	Azoxystrobin, picoxystrobin, dimoxystrobin, kresoxim-methyl, pyraclostrobin, trifloxystrobin	Extraction with ethyl acetate, filtration and evaporation	Quantification with DART, confirmation with DESI. LOQs in the range of 5–30 μg/kg, comparison of results with LC–MS/MS data	[131]
DART	Discrimination between quality grades of olive oil, detection of adulteration with hazelnut oil	Triacylglycerols, phenolic compounds	Dilution with toluene or extraction with methanol–water mixture (80:20, v/v)	Sample discrimination with linear discriminant analysis	[127]
DART	Qualitative analysis of melamine in milk powder	Melamine	None	Argon ionization gas with acetylacetone and pyridine dopants were used for selective ionization of melamine	[120]
DART	Quantitative analysis of *Fusarium* mycotoxins in cereals	Deoxynivalenol, nivalenol, zearalenone, acetyldeoxynivalenol, deepoxy-deoxynivalenol, fusarenon-X, altenuene, alternariol, alternariol methyl ether,	Modified QuEChERS extraction procedure with dispersive SPE cleanup	LOQs 50–150 μg/kg, comparison of results with LC–MS/MS data	[128]

(continued)

Table 2.5 (*Continued*)

Technique	Aim of the Study	Analytes	Sample Pretreatment	Remark	Reference
		diacetoxyscirpenol, sterigmatocystin			
DART	Quantitative analysis of melamine and cyanuric acid in milk-based products	Melamine and cyanuric acid	Extraction with methanol–5% aqueous formic acid (1:1, v/v)	LOQs 450 μg/kg (melamine), 1200 μg/kg (cyanuric acid), comparison of results with LC–MS/MS data	[126]
DART	Screening for pesticides on the surface of fruits	Multiclass pesticides ($n = 132$)	Sample surface swabbed with foam disk	Transmission mode analysis of foam swabs, detection of 86% of analytes at levels 2–10 ng/g	[138]
DART	Fingerprinting-based discrimination among Trappist and non-Trappist beer brands	Nonspecified analytes	Degassing	Sample discrimination with the use of artificial neural networks, linear discriminant analysis, and partial least squares discriminant analysis	[139]
DART/DESI	Qualitative and quantitative analyses of dicarbamate fungicides in fruits	Thiram and ziram	Modified QuEChERS extraction procedure without dispersive SPE cleanup	LOQs (DART) 0.5–1.0 mg/kg, with DESI only thiram was detected	[129]
DART	Control of sample cleanup efficiency of fish and shrimp samples	Triacylglycerols	Modified QuEChERS procedure followed by silica minicolumn cleanup	Comparison between various cleanup procedures for GC–MS determination of polychlorinated biphenyls, polybrominated diphenyl ethers, and polycyclic aromatic hydrocarbons	[140]

DART	Quantitative analysis of isoflavones in soybeans	Daidzein, glycitein, and genistein	Extraction with acidified 96%, v/v ethanol, heating under reflux, neutralization	To determine total content of isoflavones, acidic hydrolysis was performed, comparison of results with LC–MS data	[132]
DAPCI	Demonstration of DAPCI capabilities in food analysis (detection of adulterants, pesticides, and biogenic amines)	Sudan dyes, atrazine, purescine, and cadaverine	None	–	[105]
DAPCI	Fingerprinting-based differentiation among tea products	Nonvolatile and semivolatile compounds	None	Data visualization and sample discrimination with PCA	[141]
DAPCI/DESI	Qualitative and quantitative analyses of melamine in milk powder and liquid milk	Melamine, cyanuric acid, and melamine–cyanurate complex	None	LODs (DAPCI) 1.6×10^{-11} g mm^2 (milk powder), 1.3×10^{-11} g mm^2 (liquid milk); to enable detection of analytes by DESI, sample has to be heated or agents that weaken the melamine–protein complex has to be added	[142]
DAPCI	Fingerprinting-based differentiation of dried sea cucumber products according to habitat	Low molecular compounds	None	Data visualization and sample discrimination with PCA and soft independent modeling of class analogy (SIMCA)	[143]

(continued)

Table 2.5 (*Continued*)

Technique	Aim of the Study	Analytes	Sample Pretreatment	Remark	Reference
ASAP	Qualitative analysis of carotenoids in a spinach leaf	Canthaxanthin, apocarotenal, and β-carotene	None	–	[112]
ASAP	Qualitative and semi-quantitative analyses of strobilurin pesticides in cereals and extracts of oranges	Azoxystrobin, dimoxystrobin, epoxiconazole, fluoxastrobin, picoxystrobin, pyraclostrobin, and trifloxystrobin	None, extraction with methanol, modified QuEChERS extraction procedure with and without dispersive SPE cleanup	LODs at maximum residue limits of respective strobilurin fungicides in wheat grains	[94]
DAPPI	Qualitative analysis of pesticides on an orange peel	Aldicarb, carbofuran, ditalimfos, imazazil, methiocarb, methomyl, oxamyl, pirimicarb, and thiabendazole	None	LODs of pesticides (standards) 30–300 pg, imazalil detected on orange peel surface	[144]
DAPPI/DESI	Analysis of nonpolar compounds in standards and food products	Palmitic acid, linoleic acid, α-tocopherol, phylloquinone, tributyrin, trilinolenin testosterone, cholesterol, phosphatidylcholine di-18:1, phosphatidylethanolamine di-14:0, phosphatidylserine di-14:0, phosphatidylinositol 16:0/18:2, sphingomyelin 18:1	None	DAPPI provided efficient ionization for neutral, less polar, and ionic lipids, DESI was more suitable for the analysis of the large and labile lipids	[145]

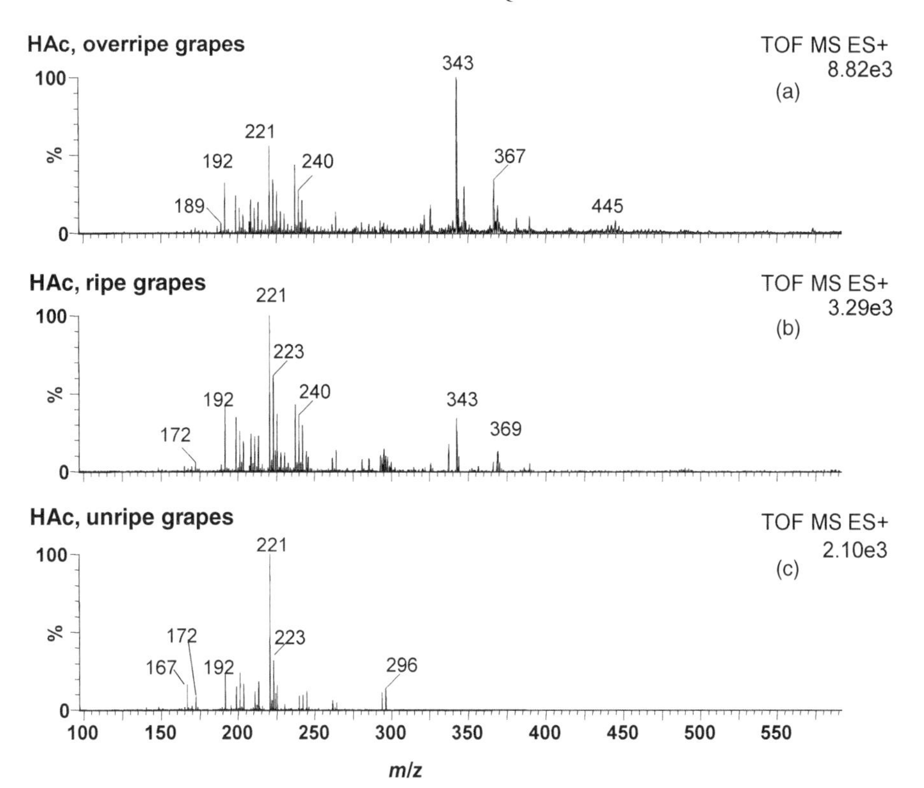

Figure 2.16. EESI-QTOF mass spectra of grapes at different maturity stages, showing differentiation patterns: (a) overripe grapes; (b) normally ripe grapes; and (c) unripe grapes Ref. [135], Figure 5, p. 1452. Reproduced with permission of American Chemical Society.

grapes, or strawberries contained signals of both volatile and nonvolatile compounds in the mass region *m/z* 100–1000. Clear differences between mass spectra obtained by analysis of samples at various ripening stages were observed, as demonstrated in Figure 2.16 for grapes. Due to the complexity of the records, multivariate statistical analysis employing principal component analysis was used for further visualization.

DESI-MS was used by Jackson et al. to directly characterize constituents of dietary supplements containing *Stevia* leaf extracts [133]. Among other constituents, such as oligosaccharides and polysaccharides, characteristic diterpene glycosides, which are responsible for sweet taste, were detected. The compliance of commercial products with declared composition could be confirmed this way.

Vaclavik et al. used DART-MS for authenticity assessment of extra virgin olive oil based on statistical analysis of data representing profiles of triacylglycerols and some phenolic compounds [127]. In addition to differentiation of various olive oil grades, detection of adulteration by hazelnut oil at levels as low as 6% (v/v) was demonstrated. High-throughput analysis was achieved due to simple sample preparation and automated sample introduction in front of a DART ion source. The extraction of phenolic compounds from a single sample was performed in less than 5 min. Under optimized conditions, the time required for analysis of one sample was below 1 min.

Cajka et al. demonstrated the applicability of DART-MS to chicken meat metabolomics for the retrospective control of feed fraud [146]. Samples representing meat of chickens fed by feed with and without the addition of banned chicken bone meal (5–8%, w/w) were extracted using a procedure enabling simultaneous isolation of polar and nonpolar metabolites. The multivariate analysis of the DART records facilitated differentiation of sample groups and highlighted marker metabolites that were more abundant in the group of chickens fed with the feed adulterated with chicken bone meal.

DART-MS instrumentation was used to directly monitor the transfer of matrix coextracts (mainly lipids) during the optimization of partition-based sample cleanup in a study reported by Kalachova et al. that focused on determining of polychlorinated biphenyls, polybrominated diphenyl ethers, and polycyclic aromatic hydrocarbons in fish and shrimp [140]. DART-MS was demonstrated to be a very efficient tool for the rapid determination of lipids and other ionizable impurities with analysis time of 30 s.

2.4 CONCLUDING REMARKS

Over the past few years, there has been substantial progress in technologies employing MS in rapid food analysis. In this context, a wide range of analytical methods involving GC–MS, LC–MS, and methods without the chromatographic separation have been reported to detect, identify, quantify, and confirm various naturally occurring as well as xenobiotic substances in food chain. These techniques have also been demonstrated as straightforward fingerprinting or profiling tools for food authenticity assessment. The development of advanced LC and MS technologies as well as automation of related sample preparation process has paved the way for high-throughput analysis for food safety, especially with the popularization of UHPLC, sub-2 μm columns, and high-resolution MS.

The introduction of direct MS techniques and specifically ambient desorption ionization techniques such as DESI and DART coupled with MS has brought the promise of simple, high-throughput qualitative and quantitative analyses of both major and minor (trace) components in various food matrices. However, thorough validation and carefully designed quality assurance and quality control procedures are still urgently needed when employing these techniques, because the lack of a chromatographic separation step makes direct MS techniques more prone to false (negative or positive) findings.

ACKNOWLEDGMENTS

L.V. and W.Z. acknowledge the support by an appointment to the Research Participation Program at the Center for Food Safety and Applied Nutrition administered by the Oak Ridge Institute for Science and Education through an interagency agreement between the U.S. Department of Energy and the U.S. Food and Drug

Administration. The authors wish to thank Jeanne I. Rader and Alexander J. Krynitsky for helpful discussions.

REFERENCES

1. Patel, K.; Fussell, R.J.; Goodall, D.M.; Keely, B.J. Evaluation of large volume-difficult matrix introduction-gas chromatography–time of flight-mass spectrometry (LV-DMI-GC–TOF-MS) for the determination of pesticides in fruit-based baby foods. *Food Addit. Contam.* **2004**, 21, 658–669.
2. Cajka, T.; Mastovska, K.; Lehotay, S.J.; Hajslova, J. Use of automated direct sample introduction with analyte protectants in the GC–MS analysis of pesticide residues. *J. Sep. Sci.* **2005**, 28, 1048–1060.
3. Fajgelj, A.; Ambrus, A. *Principles and Practices of Method Validation*, 1st edition. Cambridge: Royal Society of Chemistry; **2000**.
4. Mastovska, K.; Hajslova, J.; Lehotay, S.J. Ruggedness and other performance characteristics of low-pressure gas chromatography–mass spectrometry for the fast analysis of multiple pesticide residues in food crops. *J. Chromatogr. A* **2004**, 1054, 335–349.
5. Hajslova, J.; Zrostlikova, J. Matrix effects in (ultra)trace analysis of pesticide residues in food and biotic matrices. *J. Chromatogr. A* **2003**, 1000, 181–197.
6. Cajka, T.; Sandy, C.; Bachanova, V.; Drabova, L.; Kalachova, K.; Pulkrabova, J.; Hajslova, J. Streamlining sample preparation and gas chromatography-tandem mass spectrometry analysis of multiple pesticide residues in tea. *Anal. Chim. Acta* **2012**, 743, 51–60.
7. Mastovska, K.; Lehotay, S.J. Practical approaches to fast gas chromatography–mass spectrometry. *J. Chromatogr. A* **2003**, 1000, 153–180.
8. Matisova, E.; Domotorova, M. Fast gas chromatography and its use in trace analysis. *J. Chromatogr. A* **2003**, 1000, 199–221.
9. Cajka, T.; Hajslova, J.; Lacina, O.; Mastovska, K.; Lehotay, S.J. Rapid analysis of multiple pesticide residues in fruit-based baby food using programmed temperature vaporiser injection–low-pressure gas chromatography–high-resolution time-of-flight mass spectrometry. *J. Chromatogr. A* **2008**, 1186, 281–294.
10. Mastovska, K. Recent developments in chromatographic techniques. In: *Comprehensive Analytical Chemistry*, Vol. 51. Amsterdam: Elsevier; **2008**.
11. Cajka, T.; Hajslova, J.; Mastovska, K. Mass spectrometry and hyphenated instruments in food analysis. In: *Handbook of Food Analysis Instruments*. Boca Raton, FL: CRC Press; **2008**.
12. Watson, J.T.; Sparkman, O.D. Introduction to Mass Spectrometry: Instrumentation, Applications and Strategies for Data Interpretation, 4th ed.; Chichester: John Wiley and Sons; **2007**.
13. Mattern, G.C.; Singer, G.M.; Louis, J.; Robson, M.; Rosen, J.D. Determination of several pesticides with a chemical ionization ion trap detector. *J. Agric. Food Chem.* **1990**, 38, 402–407.
14. Mastovska, K.; Lehotay, S.J.; Hajslova, J. Optimization and evaluation of low-pressure gas chromatography–mass spectrometry for the fast analysis of multiple pesticide residues in a food commodity. *J. Chromatogr. A* **2001**, 926, 291–308.

15. Koesukwiwat, U.; Lehotay, S.J.; Leepipatpiboon, N. High throughput analysis of 150 pesticides in fruits and vegetables using QuEChERS and low-pressure gas chromatography–time-of-flight mass spectrometry. *J. Chromatogr. A* **2010**, 1217, 6692–6703.
16. Koesukwiwat, U.; Lehotay, S.J.; Leepipatpiboon, N. Fast, low-pressure gas chromatography triple quadrupole tandem mass spectrometry for analysis of 150 pesticide residues in fruits and vegetables. *J. Chromatogr. A.* **2011**, 1218, 7039–7050.
17. Martinez Vidal, J.L.; Arrebola Liebanas, F.J.; Gonzalez Rodriguez, M.J.; Garrido Frenich, A.; Fernandez Moreno, J.L. Validation of a gas chromatography/triple quadrupole mass spectrometry based method for the quantification of pesticides in food commodities. *Rapid. Commun. Mass Spectrom.* **2006**, 20, 365–375.
18. Domotorova, M.; Matisova, E. Fast gas chromatography for pesticide residues analysis. *J. Chromatogr. A* **2008**, 1207, 1–16.
19. Covaci, A.; Voorspoels, S.; de Boer J. Determination of brominated flame retardants, with emphasis on polybrominated diphenyl ethers (PBDEs) in environmental and human samples: a review. *Environ. Int.* **2003**, 29, 735–756.
20. Dirtu, A.C.; Ravindra, K.; Roosens, L.; van Grieken, R.; Neels, H.; Blust, R.; Covaci, A. Fast analysis of decabrominated diphenyl ether using low-pressure gas chromatography–electron-capture negative ionization mass spectrometry. *J. Chromatogr. A* **2008**, 1186, 295–301.
21. Ziegenhals, K.; Hubschmann, H.-J.; Speer, K.; Jira, W. Fast-GC/HRMS to quantify the EU priority PAH. *J. Sep. Sci.* **2008**, 31, 1779–1786.
22. Cochran, J.W. Fast gas chromatography–time-of-flight mass spectrometry of polychlorinated biphenyls and other environmental contaminants. *J. Chromatogr. Sci.* **2002**, 40, 254–268.
23. Setkova, L.; Risticevic, S.; Pawliszyn, J. Rapid headspace solid-phase microextraction–gas chromatographic–time-of-flight mass spectrometric method for qualitative profiling of ice wine volatile fraction II: classification of Canadian and Czech ice wines using statistical evaluation of the data. *J. Chromatogr. A* **2007**, 1147, 224–240.
24. Risticevic, S.; Carasek, E.; Pawliszyn, J. Headspace solid-phase microextraction–gas chromatographic–time-of-flight mass spectrometric methodology for geographical origin verification of coffee. *Anal. Chim. Acta* **2008**, 617, 72–84.
25. Mondello, L.; Casilli, A.; Tranchida, P.Q.; Costa, R.; Chiofalo, B.; Dugo, P.; Dugo, G. Evaluation of fast gas chromatography and gas chromatography–mass spectrometry in the analysis of lipids. *J. Chromatogr. A* **2004**, 1035, 237–247.
26. Anastassiades, M.; Lehotay, S.J.; Stajnbaher, D.; Schenck, F.J. Fast and easy multiresidue method employing acetonitrile extraction/partitioning and "dispersive solid-phase extraction" for the determination of pesticide residues in produce. *J. AOAC Int.* **2003**, 86, 412–431.
27. Kataoka, H.; Lord, H.L.; Pawliszyn, J. Applications of solid-phase microextraction in food analysis. *J. Chromatogr. A* **2000**, 880, 35–62.
28. Novakova, L.; Vlckova, H. A review of current trends and advances in modern bioanalytical methods: chromatography and sample preparation. *Anal. Chim. Acta* **2009**, 656, 8–35.
29. Gilpin, R.K.; Zhou, W. Designing high throughput HPLC assays for small and biological molecules. In: *High-Throughput Analysis in the Pharmaceutical Industry.* New York: CRC Press; **2008**.

30. Fekete, S.; Fekete, J. The impact of extra-column band broadening on the chromatographic efficiency of 5 cm long narrow-bore very efficient columns. *J. Chromatogr. A* **2011**, 1218, 5286–5291.
31. Wang, P.G.; Zhou W.; Krynitsky, A.J.; Rader, J.I. Simultaneous determination of aloin A and aloe emodin in products containing *Aloe vera* by ultra performance liquid chromatography with tandem mass spectrometry. *Anal. Methods* **2012**, 3612–3619.
32. Di Stefano, V.; Avellone, G.; Bongiorno, D.; Cunsolo, V.; Muccilli, V.; Sforza, S.; Dossena, A.; Drahos, L.; Vekey, K. Applications of liquid chromatography–mass spectrometry for food analysis. *J. Chromatogr. A* **2012**, 1259, 74–85.
33. Nunez, O.; Gallart-Ayala, H.; Martins, C.P.B.; Lucci, P. New trends in fast liquid chromatography for food and environmental analysis. *J. Chromatogr. A* **2012**, 1228, 298–323.
34. Guillarme, D.; Ruta, J.; Rudaz, S.; Veuthey, J.L. New trends in fast and high-resolution liquid chromatography: a critical comparison of existing approaches. *Anal. Bioanal. Chem.* **2010**, 397, 1069–1082.
35. Alpert, A.J. Hydrophilic interaction chromatography for the separation of peptides, nucleic acids and other polar compounds. *J. Chromatogr.* **1990**, 499, 177–196.
36. Spagou, K.; Tsoukali, H.; Raikos, N.; Gika, H.; Wilson, I.D.; Theodoridis, G. Hydrophilic interaction chromatography coupled to MS for metabonomic/metabolomic studies. *J. Sep. Sci.* **2010**, 33, 716–727.
37. Wang, P.G.; He, W. *Hydrophilic Interaction Liquid Chromatography (HILIC) and Advanced Applications*. New York: CRC Press; **2011**.
38. Zhou, W.; Wang, P.G.; Krynitsky, A.J., Rader, J.I. Rapid and simultaneous determination of hexapeptides (Ac-EEMQRR-amide and H_2N-EEMQRR-amide) in anti-wrinkle cosmetics by hydrophilic interaction liquid chromatography–solid phase extraction preparation and hydrophilic interaction liquid chromatography with tandem mass spectrometry. *J. Chromatogr. A* **2011**, 1218, 7956–7963.
39. Hemstrom, P.; Irgum, K. Hydrophilic interaction chromatography. *J. Sep. Sci.* **2006**, 29, 1784–1821.
40. Nguyen, H.P.; Schug, K.A. The advantages of ESI-MS detection in conjunction with HILIC mode separations: fundamentals and applications. *J. Sep. Sci.* **2008**, 31, 1465–1480.
41. Dejaegher, B.; Mangelings, D.; Heyden, Y.V. Method development for HILIC assays. *J. Sep. Sci.* **2008**, 31, 1438–1448.
42. Ihunegbo, F.N.; Tesfalidet, S.; Jiang, W. Determination of melamine in milk powder using zwitterionic HILIC stationary phase with UV detection. *J. Sep. Sci.* **2010**, 33, 988–995.
43. Deng, X.J.; Guo, D.H.; Zhao, S.Z.; Han, L.; Sheng, Y.G.; Yi, X.H.; Zhou, Y.; Peng, T. A novel mixed-mode solid phase extraction for simultaneous determination of melamine and cyanuric acid in food by hydrophilic interaction chromatography coupled to tandem mass chromatography. *J. Chromatogr. B* **2010**, 878, 2839–2844.
44. Olsen, B.A.; Pack, B.W. *Hydrophilic Interaction Chromatography: A Guide for Practitioners*. Hoboken, NJ: John Wiley & Sons, Inc.; **2013**.
45. Kubin, M.; Spacek, P.; Chromecek, R. Gel permeation chromatography on porous poly (ethylene glycol methacrylate). *Coll. Czech. Chem. Commun.* **1967**, 32, 3881–3887.
46. Wang, P.G. *Monolithic Chromatography and Its Modern Applications*. Glendale: ILM Publications; **2011**.

47. Majors, R.E. High-performance liquid chromatography on small particle silica-gel. *Anal. Chem.* **1972**, 44, 1722–1726.
48. Malik, A. K.; Blasco, C.; Picó, Y. Liquid chromatography–mass spectrometry in food safety. *J. Chromatogr. A* **2010**, 1217, 4018–4040.
49. Holcapek, M.; Jirasko, R.; Lisa, M. Recent developments in liquid chromatography–mass spectrometry and related techniques. *J. Chromatogr. A* **2012**, 1259, 3–15.
50. Zhang, A.; Chang, J.S.; Gu, C.; Sanders, M. Non-targeted screening and accurate mass confirmation of 510 pesticides on the high resolution Exactive benchtop LC/MS Orbitrap. Available at https://static.thermoscientific.com/images/D14441~.pdf (accessed February 25, 2013).
51. Diaz, R.; Ibanez, M.; Sancho, J.V.; Hernandez, F. Target and non-target screening strategies for organic contaminants, residues and illicit substances in food, environmental and human biological samples by UHPLC–QTOF-MS. *Anal. Methods* **2012**, 4, 196–209.
52. Bonfiglio, R.; King, R.C.; Olah, T.; Merkle, K. The effects of sample preparation methods on the variability of the electrospray ionization response for model drug compounds. *Rapid Commun. Mass Spectrom.* **1999**, 13, 1175–1185.
53. Matuszewski, B.; Constanzer, M.; Chavez-Eng, C. Strategies for the assessment of matrix effect in quantitative bioanalytical methods based on HPLC–MS/MS. *Anal. Chem.* **2003**, 75, 3019–3030.
54. Van Eeckhaut, A.; Lanckmans, K.; Sarre, S.; Smolders, I.; Michotte, Y. Validation of bioanalytical LC–MS/MS assays: evaluation of matrix effects. *J. Chromatogr. B* **2009**, 877, 2198–2207.
55. Zhou, W.; Wang, P.G.; Ogunsola, O.A.; Kraeling, M.E.K. Rapid determination of hexapeptides by hydrophilic interaction LC–MS/MS for *in vitro* skin-penetration studies. *Bioanalysis* **2013**, 5, 1353–1362.
56. Kiser, M.M.; Dolan, J.W. Selecting the best curve fit calibration curve. *LC–GC N. Am.* **2004**, 22, 112–117.
57. Sporns, P.; Abel, D.C. MALDI mass spectrometry for food analysis. *Trends Food Sci. Technol.* **1996**, 7, 187–190.
58. Pan, C.; Xu, S.; Zhou, H.; Fu, Y.; Ye, M.; Zou, H. Recent developments in methods and technology for analysis of biological samples by MALDI-TOF-MS. *Anal. Bioanal. Chem.* **2007**, 387, 193–204.
59. Van Kampen, J.J.A.; Burgers, P.C.; de Groot, R.; Gruters, R.A.; Luider, T.M. Biomedical application of MALDI mass spectrometry for small-molecule analysis. *Mass Spectrom. Rev.* **2011**, 30, 101–120.
60. El-Aneed, A.; Cohen, A.; Banoub, J. Mass spectrometry: review of the basics: electrospray, MALDI, and commonly used mass analyzers. *Appl. Spectrosc. Rev.* **2009**, 44, 210–230.
61. Knochenmuss, R. Ion formation mechanisms in UV-MALDI. *Analyst* **2006**, 131, 966–986.
62. Ehring, H.; Karas, M.; Hillenkamp, F. Role of photoionization and photochemistry in ionization processes of organic molecules and relevance for matrix-assisted laser desorption ionization mass spectrometry. *J. Mass Spectrom.* **1992**, 10, 472–480.
63. Batoy, S.M.A.B.; Akhmetova, E.; Miladinovic, S.; Smeal, J.; Wilkins, C.L. Developments in MALDI mass spectrometry: the quest for perfect matrix. *Appl. Spectrosc. Rev.* **2008**, 43, 485–550.

64. Liu, Z.; Schey, K.L. Fragmentation of multiply-charged intact protein ions using MALDI TOF-TOF mass spectrometry. *J. Am. Soc. Mass Spectrom.* **2008**, 19, 231–238.
65. Sporns, P.; Wang, J. Exploring new frontiers in food analysis using MALDI-MS. *Food Res. Int.* **1998**, 30, 181–189.
66. Zenobi, R. Laser-assisted mass spectrometry. *Chimia* **1997**, 51, 801–803.
67. Armstrong, D.W.; Zhang, L.-K.; He L.; Gross, M.L. Ionic liquids as matrixes for matrix-assisted laser desorption/ionization mass spectrometry. *Anal. Chem.* **2001**, 73, 3679–3686.
68. Wang, J.; Sporns, P. MALDI-TOF MS analysis of isoflavones in soy products. *J. Agric. Food Chem.* **2000**, 48, 5887–5892.
69. Wang, J.; Kliks, M.M.; Qu, W.; Jun, S.; Shi, G.; Li, Q.X. Rapid determination of the geographical origin of honey based on protein fingerprinting and barcoding. *J. Agric. Food Chem.* **2009**, 57, 10081–10088.
70. Abell, D.C.; Sporns, P. Rapid quantitation of potato glycoalkaloids by matrix-assisted laser desorption/ionization time-of-flight mass spectrometry. *J. Agric. Food Chem.* **1996**, 44, 2292–2296.
71. Camafeita, E.; Alfonso, P.; Mothes, T.; Mendez, E. Matrix-assisted laser desorption/ionization time-of-flight mass spectrometric micro-analysis: the first non-immunological alternative attempt to quantify gluten gliadins in food samples. *J. Mass Spectrom.* **1997**, 32, 940–947.
72. Schiller, J.; Süß, R.; Petkovic, M.; Arnold, K. Thermal stressing of unsaturated vegetable oils: effects analysed by MALDI-TOF mass spectrometry, ^{1}H and ^{31}P NMR spectroscopy. *Eur. Food Res. Technol.* **2002**, 215, 282–286.
73. Catharino, R.R.; de Azevedi Marques, L.; Silva Santos, L.; Baptista, A.S.; Gloria, E.M.; Calori-Dominguez, M.A.; Facco, E.M.P.; Eberlin, M.N. Aflatoxin screening by MALDI-TOF mass spectrometry. *Anal. Chem.* **2005**, 77, 8155–8157.
74. Mazzeo, M.F.; De Giulio, B.; Guerrero, G.; Ciarcia, G.; Malorni, A.; Russo, G.L.; Siciliano, R.A. Fish authentication by MALDI-TOF mass spectrometry. *J. Agric. Food Chem.* **2008**, 56, 11071–11076.
75. Nicolaou, N.; Xu, Y.; Goodacre, R. MALDI-MS and multivariate analysis for the detection and quantification of different milk species. *Anal. Bioanal. Chem.* **2011**, 399, 3491–3502.
76. Madla, S.; Miura, D.; Wariishi, H. Potential applicability of MALDI-MS for low-molecular-weight pesticide determination. *Anal. Sci.* **2012**, 28, 301–303.
77. Calvano, C.D.; De Ceglie, C.; D'Accolti, S.; Zambonin, C.G. MALDI-TOF mass spectrometry detection of extra-virgin olive oil adulteration with hazelnut oil by analysis of phospholipids using an ionic liquid as matrix and extraction solvent. *Food Chem.* **2012**, 134, 1192–1198.
78. Garcia, J.S.; Sanvido, G.B.; Saraiva, S.A.; Zacca, J.J.; Cosso, R.G.; Eberlin, M.N. Bovine milk powder adulteration with vegetable oils or fats revealed by MALDI-QTOF MS. *Food Chem.* **2012**, 131, 722–726.
79. Strupat, K.; Karas, M.; Hillenkamp, F. 2,5-Dihydroxybenzoic acid: a new matrix for laser desorption-ionization mass spectrometry. *Int. J. Mass Spectrom.* **1991**, 111, 89–102.
80. Nicola, A.; Gusev, A.I.; Proctor, A.; Jackson, E.K.; Hercules, D.M. Application of the fast evaporation sample preparation method for improving quantitation of angiotensin II in MALDI. *Rapid Commun. Mass Spectrom.* **1995**, 9, 1164–1171.

81. Erb, W.J.; Owens, K.G. Development of a dual-spray electrospray deposition system for matrix-assisted laser desorption/ionization time-of-flight mass spectrometry. *Rapid Commun. Mass Spectrom.* **2008**, 22, 1168–1174.
82. Onnerfjord, P.; Ekstrom, S.; Berquist, J.; Nilsson, J.; Laurell, T.; Marko-Varga, G. Homogenous sample preparation for automated high throughput analysis with matrix-assisted laser desorption/ionization time-of-flight mass spectrometry. *Rapid Commun. Mass Spectrom.* **1999**, 13, 315–322.
83. Soltwisch, J.; Jaskolla, T.W.; Hillenkamp, F.; Karas, M.; Dreisewerd, K. Ion yields in UV-MALDI mass spectrometry as a function of excitation laser wavelength and optical and physico-chemical properties of classical and halogen-substituted MALDI matrixes. *Anal. Chem.* **2012**, 84, 6567–6576.
84. Cajka, T.; Hajslova, J. Volatile compounds in food authenticity and traceability testing. In: *Food Flavors: Chemical, Sensory and Technological Properties*. Boca Raton, FL: CRC Press; **2011**.
85. Vera, L.; Acena, L.; Guasch, J.; Boque, R.; Mestres, M.; Busto, O. Characterization and classification of the aroma of beer samples by means of an MS e-nose and chemometric tools. *Anal. Bioanal. Chem.* **2011**, 399, 2073–2081.
86. Mildner-Szkudlarz, S.; Jelen, H.H. The potential of different techniques for volatile compounds analysis coupled with PCA for the detection of the adulteration of olive oil with hazelnut oil. *Food Chem.* **2008**, 110, 751–761.
87. Peres, C.; Viallon, C.; Berdague, J.-L. Solid-phase microextraction–mass spectrometry: a new approach to the rapid characterization of cheeses. *Anal. Chem.* **2001**, 73, 1030–1036.
88. Cynkar, W.; Dambergs, R.; Smith, P.; Cozzolino, D. Classification of Tempranillo wines according to geographic origin: combination of mass spectrometry based electronic nose and chemometrics. *Anal. Chim. Acta* **2010**, 660, 227–231.
89. Marsili, R.T. SPME–MS–MVA as an electronic nose for the study of off-flavors in milk. *J. Agric. Food Chem.* **1999**, 47, 648–654.
90. Vinaixa, M.; Marín, S.; Brezmes, J.; Llobet, E.; Vilanova, X.; Correig, X.; Ramos, A.; Sanchis, V. Early detection of fungal growth in bakery products by use of an electronic nose based on mass spectrometry. *J. Agric. Food Chem.* **2004**, 52, 6068–6074.
91. Gross, J.H. *Mass Spectrometry*, 2nd edition. Heidelberg: Springer; **2011**.
92. Takats, Z.; Wiseman, J.M.; Gologan, B.; Cooks, R.G. Mass spectrometry sampling under ambient conditions with desorption electrospray ionization. *Science* **2004**, 306, 471–473.
93. Cody, R.B.; Laramee, J.A.; Durst, H.D. Versatile new ion source for the analysis of materials in open air under ambient conditions. *Anal. Chem.* **2005**, 77, 2297–2302.
94. Fussell, R.J.; Chan, D.; Sharman, M. An assessment of atmospheric-pressure solids-analysis probes for the detection of chemicals in food. *Trends Anal. Chem.* **2010**, 29, 1326–1335.
95. Nielen, M.W.F.; Hooijerink, H.; Zomer, P.; Mol, J.G.J. Desorption electrospray ionization mass spectrometry in the analysis of chemical food contaminants in food. *Trends Anal. Chem.* **2011**, 30, 165–180.
96. Hajslova, J.; Cajka, T.; Vaclavik, L. Challenging applications offered by direct analysis in real time (DART) in food-quality and safety analysis. *Trends Anal. Chem.* **2011**, 30, 204–218.

97. Takats, Z.; Wiseman, J.M.; Cooks, R.G. Ambient mass spectrometry using desorption electrospray ionization (DESI): instrumentation, mechanisms and applications in forensics, chemistry, and biology. *J. Mass Spectrom.* **2005**, 40, 1261–1275.

98. Alberici, R.M.; Simas, R.C.; Sanvido, G.B.; Romao, W.; Lalli, P.M.; Benassi, M.; Cunha, I.B.S., Eberlin, M.N. Ambient mass spectrometry: bringing MS into the "real world". *Anal. Bioanal. Chem.* **2010**, 398, 265–294.

99. Harris, G.A.; Galhena, A.S.; Fernandez, F.M. Ambient sampling/ionization mass spectrometry: applications and current trends. *Anal. Chem.* **2011**, 83, 4508–4538.

100. Van Berkel, G.J.; Ford, M.J.; Deibel, M.A. Thin-layer chromatography and mass spectrometry coupled using desorption electrospray ionization. *Anal. Chem.* **2005**, 77, 1207–1215.

101. Chipuk, J.E.; Brodbelt, J.S. Transmission mode desorption electrospray ionization. *J. Am. Soc. Mass Spectrom.* **2008**, 19, 1612–1620.

102. Venter, A.; Cooks, R.G. Desorption electrospray ionization in a small pressure-tight enclosure. *Anal. Chem.* **2007**, 79, 6398–6403.

103. Chen, H.; Venter, A.; Cooks, R.G. Extractive electrospray ionization for direct analysis of undiluted urine, milk and other complex mixtures without sample preparation. *Chem. Commun.* **2006**, 2042–2044.

104. Law, W.S.; Wang, R.; Hu, B.; Berchtold, C.; Meier, L.; Chen, H.; Zenobi, R. On the mechanism of extractive electrospray ionization. *Anal. Chem.* **2010**, 82, 4494–4500.

105. Chen, H.; Zheng, J.; Zhang, X.; Luo, M.; Wang, Z.; Qiao, X. Surface desorption atmospheric pressure chemical ionization mass spectrometry for direct ambient sample analysis without toxic chemical contamination. *J. Mass Spectrom.* **2007**, 42, 1045–1056.

106. Vaclavik, L.; Belkova, B.; Reblova, Z.; Riddellova, K.; Hajslova, J. Rapid monitoring of heat-accelerated reactions in vegetable oils using direct analysis in real time ionization coupled with high resolution mass spectrometry. *Food Chem.* **2013**, 138, 2312–2320.

107. Jones, C.M.; Fernandez, F.M. Transmission mode direct analysis in real time mass spectrometry for fast untargeted metabolic fingerprinting. *Rapid Commun. Mass Spectrom.* **2013**, 27, 1311–1318.

108. Perez, J.J.; Harris, G.A.; Chipuk, J.E.; Brodbelt, J.S.; Green, M.D.; Hampton, C.Y.; Fernandez, F.M. Transmission-mode direct analysis in real time and desorption electrospray ionization mass spectrometry of insecticide-treated bed nets for malaria control. *Analyst* **2012**, 135, 712–719.

109. Krechmer, J.; Tice, J.; Crawford, E.; Musselman, B. Increasing the rate of sample vaporization in an open air desorption ionization source by using a heated metal screen as a sample holder. *Rapid Commun. Mass Spectrom.* **2011**, 25, 2384–2388.

110. Williams, J.P.; Patel, V.J.; Holland, R.; Scrivens, J.H. The use of recently described ionization techniques for the rapid analysis of some common drugs and samples of biological origin. *Rapid Commun. Mass Spectrom.* **2006**, 20, 1447–1456.

111. Cotte-Rodriguez, I.; Mulligan, C.C.; Cooks, R.G. Non-proximate detection of small molecules by desorption electrospray ionization and desorption atmospheric pressure chemical ionization mass spectrometry: instrumentation and applications in forensics, chemistry, and biology. *Anal. Chem.* **2007**, 79, 7069–7077.

112. McEwen, C.N.; McKay, R.G.; Larsen, B.S. Analysis of solids, liquids, and biological tissues using solids probe introduction at atmospheric pressure on commercial LC/MS instruments. *Anal. Chem.* **2005**, 77, 7826–7831.

113. Haapala, M.; Pol, J.; Saarela, V.; Arvola, V.; Kotiaho, T.; Ketola, R.A.; Franssila, S.; Kauppila, T.J.; Kostiainen, R. Desorption atmospheric pressure photoionization. *Anal. Chem.* **2007**, 79, 7867–7872.

114. Luosujarvi, L.; Arvola, V.; Haapala, M.; Pol, J.; Saarela, V.; Franssila, S.; Kotiaho, T.; Kostiainen, R.; Kauppila, T.J. Desorption and ionization mechanisms in desorption atmospheric pressure photoionization. *Anal. Chem.* **2008**, 80, 7460–7466.

115. Chernetsova, E.S.; Morlock, G.E.; Revelsky, I.A. DART mass spectrometry and its applications in chemical analysis. *Russ. Chem. Rev.* **2011**, 80, 235–255.

116. Badu-Tawiah, A.; Bland, C.; Campbell, D.I.; Cooks, R.G. Non-aqueous spray solvents and solubility effects in desorption electrospray ionization. *J. Am. Soc. Mass Spectrom.* **2010**, 21, 572–579.

117. Garcia-Reyes, J.F.; Jackson, A.U.; Molina-Diaz, A.; Cooks, R.G. Desorption electrospray ionization mass spectrometry for trace analysis of agrochemicals in foods. *Anal. Chem.* **2009**, 81, 820–829.

118. Hartmanova, L.; Ranc, V.; Papouskova, B.; Bednar, P.; Havlicek, V.; Lemr, K. Fast profiling of anthocyanins in wine by desorption nano-electrospray ionization mass spectrometry. *J. Chromatogr. A* **2010**, 1217, 4223–4228.

119. Harris, G.A.; Fernandez, F.M. Simulations and experimental investigation of atmospheric transport in an ambient metastable-induced chemical ionization source. *Anal. Chem.* **2009**, 81, 322–329.

120. Dane, A.J.; Cody, R.B. Selective ionization of melamine in powdered milk by using argon direct analysis in real time (DART) mass spectrometry. *Analyst* **2010**, 135, 696–699.

121. Cody, R.B. Observation of molecular ions and analysis of nonpolar compounds with the direct analysis in real time ion source. *Anal. Chem.* **2009**, 81, 1101–1107.

122. Zhou, M.; McDonald, J.F.; Fernandez, F.M. Optimization of a direct analysis in real time/time-of-flight mass spectrometry method for rapid serum metabolomic fingerprinting. *J. Am. Soc. Mass Spectrom.* **2010**, 21, 68–75.

123. Harris, G.A.; Hostetler, D.M.; Hampton, C.Y.; Fernandez, F.M. Comparison of the internal energy deposition of direct analysis in real time and electrospray ionization time-of-flight mass spectrometry. *J. Am. Soc. Mass Spectrom.* **2010**, 21, 855–863.

124. Yu, S.; Crawford, E.; Tice, J.; Musselman, B.; Wu, J.-T. Bioanalysis without sample cleanup or chromatography: the evaluation and initial implementation of direct analysis in real time ionization mass spectrometry for the quantification of drugs in biological matrixes. *Anal. Chem.* **2009**, 81, 193–202.

125. Borges, D.L.G.; Sturgeon, R.E.; Welz, B.; Curtius, A.J.; Mester, Z. Ambient mass spectrometric detection of organometallic compounds using direct analysis in real time. *Anal. Chem.* **2009**, 81, 9834–9839.

126. Vaclavik, L.; Rosmus, J.; Popping, B.; Hajslova, J. Rapid determination of melamine and cyanuric acid in milk powder using direct analysis in real time-time-of-flight mass spectrometry. *J. Chromatogr. A* **2010**, 1217, 4204–4211.

127. Vaclavik, L.; Cajka, T.; Hrbek, V.; Hajslova, J. Ambient mass spectrometry employing direct analysis in real time (DART) ion source for olive oil quality and authenticity assessment. *Anal. Chim. Acta* **2009**, 645, 56–63.

128. Vaclavik, L.; Zachariasova, M.; Hrbek, V.; Hajslova, J. Analysis of multiple mycotoxins in cereals under ambient conditions using direct analysis in real time (DART) ionization coupled to high resolution mass spectrometry. *Talanta* **2010**, 82, 1950–1957.

129. Cajka, T.; Riddelova, K.; Zolmer, P.; Mol, H.; Hajslova J. Direct analysis of dithiocarbamate fungicides in fruit by ambient mass spectrometry. *Food Addit. Contam.* **2011**, 1372–1382.

130. Kauppila, T.J.; Wiseman, J.M.; Ketola, R.A.; Kotiaho, T.; Cooks, R.G.; Kostianen, R. Desorption electrospray ionization mass spectrometry for the analysis of pharmaceuticals and metabolites. *Rapid Commun. Mass Spectrom.* **2006**, 20, 387–392.

131. Schurek, J.; Vaclavik, L.; Hooijerink, H.; Lacina, O.; Poustka, J.; Sharman, M.; Caldow, M.; Nielen, M.W.F.; Hajslova, J. Control of strobilurin fungicides in wheat using direct analysis in real time accurate time-of-flight and desorption electrospray ionization linear ion trap mass spectrometry. *Anal. Chem.* **2008**, 80, 9567–9575.

132. Lojza, J.; Cajka, T.; Schulzova, V.; Riddellova, K.; Hajslova, J. Analysis of isoflavones in soybeans employing direct analysis in real-time ionization–high-resolution mass spectrometry. *J. Sep. Sci.* **2012**, 35, 476–481.

133. Jackson, A.U.; Tata, A.; Wu, C.; Perry, R.H.; Haas, G.; West, L.; Cooks, R.G. Direct analysis of *Stevia* leaves for diterpene glycoside by desorption electrospray ionization mass spectrometry. *Analyst* **2009**, 134, 867–874.

134. Gerbig, S.; Takats, Z. Analysis of triglycerides in food items by desorption electrospray ionization. *Rapid Commun. Mass Spectrom.* **2010**, 24, 2186–2192.

135. Chen, H.; Sun, Y.; Wortmann, A.; Gu, H.; Zenobi, R. Differentiation of maturity and quality of fruit using noninvasive extractive electrospray ionization quadrupole time-of-flight mass spectrometry. *Anal. Chem.* **2007**, 79, 1447–1455.

136. Wu, Z.; Chingin, K.; Chen, H.; Zhu, L.; Jia, B.; Zenobi, R. Sampling analytes from cheese products for fast detection using neutral desorption extractive electrospray ionization mass spectrometry. *Anal. Bioanal. Chem.* **2010**, 397, 1549–1556.

137. Zhu, L.; Hu, Z.; Gamez, G.; Law, W.S.; Chen, H.; Yang, S.; Chingin, K.; Balabin, R.M.; Wang, R.; Zhang, T.; Zenobi, R. Simultaneous sampling of volatile and non-volatile analytes in beer for fast fingerprinting by extractive electrospray ionization mass spectrometry. *Anal. Bioanal. Chem.* **2010**, 398, 405–413.

138. Edison, S.E.; Lin, L.A.; Gamble, B.M.; Wong, J.; Zhang, K. Surface swabbing technique for the rapid screening for pesticides using ambient pressure desorption ionization with high-resolution mass spectrometry. *Rapid Commun. Mass Spectrom.* **2011**, 25, 127–139.

139. Cajka, T.; Riddelova, K.; Tomaniova, M.; Hajslova, J. Ambient mass spectrometry employing a DART ion source for metabolomic fingerprinting/profiling: a powerful tool for beer origin recognition. *Metabolomics* **2011**, 7, 500–508.

140. Kalachova, K.; Pulkrabova, J.; Drabova, L.; Cajka, T.; Kocourek, V.; Hajslova, J. Simplified and rapid determination of polychlorinated biphenyls, polybrominated diphenyl ethers, and polycyclic aromatic hydrocarbons in fish and shrimps integrated into a single method. *Anal. Chim. Acta* **2011**, 707, 84–91.

141. Chen, H.; Liang, H.; Ding, J.; Lai, J.; Huan, Y.; Qiao, X. Rapid differentiation of tea products by surface desorption atmospheric pressure chemical ionization mass spectrometry. *J. Agric. Food Chem.* **2007**, 55, 10093–10100.
142. Yang, S.; Ding, J.; Zheng, J.; Hu, B.; Li, J.; Chen, H.; Zhou, Z.; Qiao, X. Detection of melamine in milk products by surface desorption atmospheric pressure chemical ionization mass spectrometry. *Anal. Chem.* **2009**, 81, 2426–2436.
143. Wu, Z.; Chen, H.; Wang, W.; Jia, B.; Yang, T.; Zhao, Z.; Ding, J.; Xiao, X. Differentiation of dried sea cucumber products from different geographical areas by surface desorption atmospheric pressure chemical ionization mass spectrometry. *J. Agric. Food Chem.* **2009**, 57, 9356–9364.
144. Luosujarvi, L.; Kanerva, S.; Saarela, V.; Franssila, S.; Kostiainen, R.; Kotiaho, T.; Kauppila, J. Environmental and food analysis by desorption atmospheric pressure photoionization-mass spectrometry. *Rapid Commun. Mass Spectrom.* **2010**, 24, 1343–1350.
145. Suni, N.M.; Aalto, H.; Kauppila, T.J.; Kotiaho, T.; Kostianen, R. Analysis of lipids with desorption atmospheric pressure photoionization-mass spectrometry (DAPPI-MS) and desorption electrospray ionization-mass spectrometry (DESI-MS). *J. Mass Spectrom.* **2012**, 47, 611–619.
146. Cajka, T.; Danhelova, H.; Zachariasova, M.; Riddellova, K.; Hajslova, J. Application of direct analysis in real time–mass spectrometry (DART–MS) in chicken meat metabolomics aiming at retrospective control of feed fraud. *Metabolomics* **2013**, 9, 545–557.

CHAPTER

3

QUALITY SYSTEMS, QUALITY CONTROL GUIDELINES AND STANDARDS, METHOD VALIDATION, AND ONGOING ANALYTICAL QUALITY CONTROL

DAVID GALSWORTHY and STEWART REYNOLDS

3.1 INTRODUCTION

This first introductory section outlines the elements of a quality system that are needed by all analytical laboratories, and not just those that are using rapid methods.

Quality systems have, in the last 20 years, become a requirement for demonstrating the competence of an organization to carry out a specific task or activity. In the area of chemical testing of food, this has been focused on the implementation of the ISO 17025 quality standard covering general requirements for the competence of testing and calibration laboratories. This is the standard that is applied by the accreditation body accrediting a laboratory.

Advantages of implementing a quality system include

- efficiency improvements,
- risk management,
- market access,
- best practice transfer, and
- due diligence and legal protection.

Core elements of a quality system are presented below.

3.1.1 Quality System Design

An efficient quality system design is critical to the successful introduction of a quality system and will help enormously with the maintenance of the system once it has been applied. All quality systems have a common architecture and how this is organized is very much down to the individual laboratory. This architecture can be described in terms of three layers for the system. The highest level is that of policy that defines how

High-Throughput Analysis for Food Safety, First Edition.
Edited by Perry G. Wang, Mark F. Vitha, and Jack F. Kay.

the laboratory interprets the specific requirements of the quality standard they are working to. These policy statements are normally collected together as the organization quality manual. Below the policy level are the specific procedures that describe exactly how each of the policies is applied and the way the organization operates. The lowest level in the quality system structure is that of records. The records provide evidence that the quality system is being well maintained and the quality of the data from the laboratory can be assured. These three levels should be linked and this can be achieved very efficiently using electronic hyperlinking of the documentation.

3.1.2 Procedures

Procedures can be either quality system procedures such as how to carry out an audit or operational procedures defining how, for example, a specific analytical method is performed. The procedures performed by the laboratory need to reflect exactly how operations are carried out and need to be regularly updated to incorporate changes that evolve with time. Procedures should be succinct and the use of flow diagrams to pictorially document the flow of each operation is a very efficient way of documenting the procedures.

3.1.3 Roles and Responsibilities

All quality standards are very clear that staff roles and responsibilities in an organization need to be clearly defined. For ISO 17025, this includes the Technical Management and Quality Manager roles. The defined roles and responsibilities need to match the competencies described in the training records of the individual staff.

3.1.4 Quality Manual

The quality manual describes the specific policies that relate to all the elements of the quality standard. The quality manual should be concise, ideally with no more than 30 pages in length. Each of the policy statements should clearly reference the specific procedure used to implement the policy.

3.1.5 Document Control

Critical to document control is the availability of the most up-to-date documentation for the staff carrying out the various tasks of the laboratory. Document control can be facilitated through hard copy and electronic means, but the burden of hard copy control is such that for all but the smallest laboratories, electronic control should be established. Off-the-shelf document control products are available but at a considerable cost. However, freely available software, such as Google Docs, can be very effectively used where resources are not available to buy a document control software product.

3.1.6 Control of Records

The establishment of a comprehensive records system is critical in order to demonstrate that the quality system is well controlled. Again, hard copy and electronic records are acceptable. The implementation of an electronic Laboratory Information Management System (LIMS) that allows the collection of all the sample workflow information from sample receipt to the final report is critical in most laboratories.

3.1.7 Audits

The internal audit system ensures that the quality system is effectively monitored on an ongoing basis and that staff are adhering to the requirements of the quality system. The internal audits can also be used as a vehicle for process improvement, highlighting inefficiencies and waste. Staff carrying out the audits should be trained and external training courses are readily available through the certification and accreditation bodies. The annual audit plan will define what aspects of the quality system will be covered. This will include all aspects of the quality system as well as witnessing specific procedures and activities being carried out by staff. Nonconformances recorded at the audits are normally actioned through the nonconforming work procedure and evidence of the effectiveness of these actions produced.

3.1.8 Validation of Methodology

The validation of the methodology used in terms of fitness for purpose is a critical component of compliance with ISO 17025. The process of producing validation data for chemical tests has now been clearly established and includes the demonstration of analytical specificity, sensitivity, and repeatability/reproducibility. In order to smoothen the accreditation process, it is strongly suggested that data from the validation should be collated into a report that would include the purpose of the analysis, the validation planning, the validation data, interpretation of the data including an estimate of the uncertainty of measurement, and a final statement of method fitness for purpose. Method validation is discussed in much greater detail later in this chapter.

3.1.9 Staff Competency

Demonstration of staff competency is a critical component for compliance with ISO 17025. Wherever possible, this should be through objective measurements of staff performance and quality control (QC) measures and proficiency testing, spiked recoveries, and repeat analysis of samples are examples of these. A training procedure should be produced and applied for training new staff as well as established staff carrying out new duties. Training records need to be established for all staff and these should reference the specific procedures staff are competent to carry out.

3.1.10 Internal Quality Control

Internal quality control (IQC) measures are used for the ongoing monitoring of method performance. The method performance quality of each batch of samples analyzed needs to be checked. The normal methods for this include the use of spiked recoveries or in-house reference materials and these are analyzed in parallel with the samples in the batch.

3.1.11 Method Performance Criteria

Method performance criteria are important because they allow laboratories to use any method of their choice. This is particularly important as many methods require specific instruments and/or software packages that may not be available to the analyst. This is of no consequence provided that the method meets the minimum required performance criteria and can be demonstrated to be "fit for purpose" with regard to the various analyte/commodity combinations for which it is to be used. When undertaking any analysis, it is essential to adopt a robust analytical quality control (AQC) system in order to be able to demonstrate that the method is under control and that the results are valid. Such an approach is discussed in more detail in CAC/GL 71-2009 [1].

Initial validation data have to have been generated according to an internationally recognized standard before any method is put into routine use. If the data are acceptable to an accreditation body, then the laboratory will become accredited for that method. If the scope of the method is to be extended, further validation data will need to be generated for the additional target analytes and/or commodities that are to be analyzed.

In addition to undertaking routine recovery experiments, preferably with each batch of sample, it is recommended that laboratories include "blind" check samples (samples that have been analyzed previously and for which the analyte concentrations are known). In addition to these "within-laboratory" checks, "between-laboratory" performance checks should also be undertaken. Participation in appropriate proficiency tests (PTs) or sending a few samples to another competent laboratory for analysis is highly recommended.

3.2 QUALITATIVE SCREENING METHODS

Qualitative or semiquantitative screening methods are most likely to be used where high throughput of samples is needed. Such methods have been developed by analysts for a wide range of chemical contaminants in food but are most commonly used to test for veterinary drug residues and, in more recent years, for pesticide residues. Residue analysts have developed their own class of analyte-specific method validation and AQC guidelines. At the same time, they have also created different terminologies to describe the important criteria that describe the various aspects of method performance. Therefore, it is neither helpful nor practical to attempt to produce a generic protocol or guidance document that might be adopted by all residue analysts. Having said that, Macarthur and von Holst [2] have recently published a protocol for the validation of qualitative methods that has a broad scope of application and could therefore be adopted by a wide range of analysts. The protocol concentrates more on the assessment of validation data produced by several laboratories rather than by a single laboratory. Because of the costs and difficulties in acquiring funding, validation across several laboratories or by collaborative trial is now quite rare. With such limitations in mind, this chapter concentrates only on method validation and AQC as applied by a single laboratory to food samples for pesticide and veterinary drug residues, but the principles are relevant to other contaminants.

Screening methods are generally used where a high and rapid throughput of samples is required to detect an analyte, or analytes, in commodities where the frequency of detection is likely to be relatively low. This permits limited laboratory resources to be reserved for samples that require more detailed examination. Screening methods can be classified by detection principle and in relation to veterinary drugs. The Community Reference Laboratories Guidelines for the Validation of Screening Methods for Residues of Veterinary Medicines have divided methods into three classes [3]:

1. *Biological methods* detect cellular responses to analytes (e.g., inhibition of bacterial growth, cellular effect, and hormonal effect). They do not allow identification of individual analytes.
2. *Biochemical methods* detect molecular interactions (e.g., antigens and proteins) between analytes and antibodies or receptor proteins (ELISA, RIA, etc.). These methods are usually selective for a family of analytes, but can also be analyte specific.
3. *Physicochemical methods* distinguish the chemical structure and molecular characteristics of analytes by separation of molecules (e.g., GC, HPLC, and UHPLC) and the detection of signals related to molecular characteristics (e.g., nowadays invariably mass spectrometry). They are able to distinguish between molecular structures and allow simultaneous analysis of several analytes; such methods are referred to as multiresidue methods (MRMs).

In general, single-analyte screening methods are rapid, easy to use, low-tech (based on immunoassay-type techniques such as dipsticks or sensors), low cost, and provide high throughput, that is, analysis of thousands of samples. They are more cost effective and increase efficiency relative to other techniques, reducing the requirement for more laborious confirmatory techniques. For contaminants such as pesticide residues, veterinary drug residues, and mycotoxins, high-technology (based on chromatography/mass spectrometry) MRMs are more often used. In these cases, they may provide a cost-effective way of greatly increasing the scope of analytes that are sought. They can also be considered to be rapid, especially when the time per number of analytes screened is considered. Such methods may be used to detect the presence of analytes that are not regularly found in samples of foods and therefore not sought routinely. These analytes might be nonapproved chemicals that are unexpected but could be present due to misuse.

3.2.1 Selectivity of Mass Spectrometry-Based Methods

Typically, quantitative MRMs are based on nominal mass quadrupole technologies, either GC–MS operated in selective ion monitoring (SIM) mode or tandem quadrupoles (MS/MS) operated in selected reaction monitoring (SRM) mode. Physicochemical qualitative screening multiresidue methods (QSMRMs) often employ high-resolution/accurate mass time-of flight (ToF-MS), ion trap, or hybrid instruments operated in full-scan mode. Although quantitative MRMs provide identification of individual analytes, only those analytes that have ions preprogrammed into the method will be detected. In contrast, qualitative MRMs do not provide sufficient information to meet quantification AQC criteria, but can, at least in principle, detect any analyte that is included in a spectral library or database of compounds that is linked to the mass spectrometer. In practice, the efficiency of operation of the method is largely dependent on the performance of the data processing software package associated with the library.

Methods involving the use of GC–MS(/MS) and/or LC–MS/MS most often base analyte identification on the generation of precursor and product ions. The "identification power" of the method is then dependent on selectivity. The measure of selectivity is the probability of any compound in a sample extract showing the same precursor ion, product ions, and retention time as an analytical standard of the same compound in the same extract matrix that has been analyzed using the same operating conditions. It is therefore obvious that the selection of specific characteristic ions is essential for unambiguous identification. This aspect of the method has not been addressed in any detail in any method validation guideline document. Berendsen et al. [4] have recently published a paper that describes how the probability of co-occurrence of a compound showing the same characteristics in LC–MS/MS can be estimated from empirical models, derived from three databases that included data on precursor ion mass, product ion mass, and retention time. This approach describes how the "identification power" of an LC–MS/MS method operated in MRM acquisition mode can be determined. A further paper by Kmellár et al. [5] similarly details the effects of operational parameters on pesticide residue analysis by LC–MS/MS. The influence of different mobile phase modifiers on retention time and detector response was found to be important. The use of mixed analytical standards containing 150 different pesticides was also found to produce significant suppression or enhancement of coelution effects.

3.2.2 Confirmatory Methods

A screening method is normally qualitative or semiquantitative and it must be validated to a certain minimum target analyte concentration to allow results to be reported. The method must also be validated to ensure that possible false-positive and false-negative results are minimized. To this end, tentative identifications from screening methods are normally followed up using another method that provides unequivocal confirmation of identity and may also be used for quantification purposes. Such follow-up methods are referred to as "confirmatory methods." Validation of confirmatory methods must establish a degree of confidence in detection, at and above a satisfactory reporting limit (RL) for each analyte that is to be sought. This RL, or screening detection limit (SDL), is likely to correspond to a threshold or action limit. Such action limits may correspond to regulatory limits or levels, such as maximum residue levels (MRLs) for pesticides, mycotoxins, and authorized veterinary drugs. For unauthorized analytes, such as certain veterinary drugs, the action limit would be the minimum required performance limit (MRPL) as specified in the particular EU regulation. Any residues that are detected using a qualitative screening multiresidue method should trigger further analysis using a fully validated quantitative method.

In this context, we are using the term "confirmatory *method*" to describe a method that is used to provide "confirmation of identity" of the target analyte(s). In other publications, "confirmatory *analysis*" is correctly defined as the analysis of a second subsample in order to check that this second result agrees with the original result (in terms of both identity and concentration). Selectivity/specificity is a key component of confirmatory methods. The methods must be capable of providing unequivocal identification of a target analyte from an exclusive signal response. Thus, such methods should have been developed through a suitable combination of analytical procedures such as cleanup, chromatographic separation, and spectrometric detection. Confirmatory methods are generally not considered to be high throughput and therefore validation and ongoing AQC requirements of confirmatory methods are not covered in this chapter.

3.2.3 Validation of Qualitative Screening Multiresidue Methods for Pesticide Residues in Foods

Within the confines of a book chapter, it is not feasible to produce validation procedures for QSMRMs for all the chemical contaminants that are covered in this book.

In this section, the focus will be on the use of these methods for analyzing pesticide residues in foods and veterinary drug residues in products of animal/fish origins. However, the same approaches to method validation and analytical quality control could also be applied to many other food contaminants. Indeed, some workers have already published QSMRMs that analyze pesticides, veterinary drugs, plant toxins, and mycotoxins in the same sample extract [6]. MRMs can be considered to be "high throughput" because large numbers of analytes can be detected at the same time. They are therefore normally used to improve efficiency by increasing analytical scope rather than improving the speed of analysis. The main challenge with validating multiresidue methods is to include a large number of target analytes, probably with differing physicochemical properties, and cover a wide range of food matrices. In addition to this, action levels, as specified by regulations/legislation, may be set at very low concentrations. A number of published methods based on GC–MS and LC techniques have appeared in the last few years for pesticide residues and plant toxins [7–9]. In these papers, the validation of the methods has demonstrated the huge amount of work that is necessary, not only to initially validate the methods, but also to maintain comprehensive AQC that is also essential to check continuing method performance.

There are two options:

i. Determine the responses of each analyte in blank samples and in samples spiked at the anticipated screening reporting limit (SRL). Thus, there will be a numerical output and a cutoff value below which there is a defined certainty that the analyte is below the SRL. The advantage of this approach is that normal statistics can be applied and the numbers of samples used for method validation can be limited to, say, 20. The disadvantage is that using GC–MS or LC–MS, the SRL will vary with each batch of samples that is subsequently analyzed and calibration standards will need to be included so that the SRL is reestablished. This is time consuming especially with respect to large numbers of analytes.
ii. If a numerical response is not used, then the situation becomes "detected" or "not detected." Hence, with no numerical response, normal statistics no longer apply and according to the recently published protocol for validation of Macarthur and von Holst [2], many more blank and spiked samples are necessary to give confidence for the number of pesticides now being analyzed during each multiresidue determination.

Qualitative screening multiresidue methods may be used in parallel with established validated quantitative MRMs in order to demonstrate the absence/presence of unexpected analytes. For efficiency, the same "generic" multiresidue extraction and cleanup procedure that is used for the quantitative MRMs may be used for the qualitative MRMs.

It has been recognized for a number of years that full-scan, high-resolution mass spectrometry (HRMS) offers a means of multianalyte detection of a wide range of contaminants. Recent publications provide examples of the use of HRMS technologies (ToF-MS, Orbitrap, etc.) to detect and identify pesticides [10], veterinary drugs [11], mycotoxins [12], and plant toxins [7].

3.3 ELEMENTS OF THE ANALYTICAL WORKFLOW

3.3.1 Sample Preparation

Laboratory samples should be prepared before processing according to regulations pertaining to "Parts of the product to which the action level applies." For example, for pesticide residues, Commission Regulation (EU) No. 212/2013 [13] lists the foods in groups and defines the parts of the products to which the MRLs apply. So this may involve removal of, for example, soil, crowns, stems, roots, decayed outer leaves, and so on for certain fruits and vegetables and removal of trimmable fat for meats.

3.3.2 Effects of Sample Processing

Method validation is usually performed using laboratory spiked samples, and these samples are normally spiked following sample preparation and processing. Sample processing can affect the quality of the final analytical results for two possible reasons. First, if the laboratory sample is poorly homogenized, then representative analytical portions cannot be abstracted. This is particularly significant when using modern MRMs where small analytical portions (often 10 g or less) are used. It is therefore recommended that the homogeneity of the laboratory sample be checked by undertaking replicate analyses on two or more analytical portions. Second, there could be stability issues with regard to certain analytes and/or analyte/commodity combinations. In particular for fruits and vegetables, homogenization at room temperature will disrupt plant cells and release enzymes that can react with and/or degrade certain analytes. In such cases, cryogenic milling [14] can not only significantly reduce analyte losses but also improve the homogeneity of the laboratory sample.

3.3.3 Extraction Efficiency

Spiked samples may not properly represent "real" samples containing incurred residues of the same analyte(s) because spiked analyte molecules are unlikely to have been in contact with the sample for long enough to allow any possible "binding" to the matrix to occur. For example, Matthews [15] demonstrated that only 52% of radiolabeled chlorpyrifos-methyl could be recovered as the parent compound from wheat grain after 5 months of storage. Obtaining extraction efficiency data to demonstrate the validation of MRMs is, however, not straightforward. This is due to the limited availability of samples containing incurred residues for many analyte/commodity combinations. Simple spiked extraction experiments are generally the only option available. It should also be noted that the effects of sample processing and extraction efficiency are not assessed by most proficiency tests.

3.4 INITIAL METHOD VALIDATION

In order to obtain results that are reportable, a SDL has to be determined and checked regularly. Initial method validation, as for quantitative MRMs, has to be performed for each analyte that is to be sought as well as for a typical example taken from each commodity group. Commodity groups have been defined in a number of AQC documents based on similarity in chemical composition (e.g., water, sugar, protein, fat/oil) and characteristics such as pH (Table 3.1). Some foods or crops, such as tea, coffee, cocoa, spices, and hops, are considered unique in terms of their composition and therefore need to be validated individually [16,17].

Table 3.1. Commodity Groups and Representative Commodities

Commodity Groups	Typical Commodity Categories	Typical Representative Commodities
1. High water content	Pome fruit	Apples, pears
	Stone fruit	Apricots, cherries, peaches
	Other fruit	Bananas
	Alliums	Onions, leeks
	Fruiting vegetables/ cucurbits	Tomatoes, peppers, cucumbers, melons
	Brassicas	Cauliflowers, Brussels sprouts, cabbage, broccoli
	Leafy vegetables and herbs	Lettuce, spinach, basil
	Stem and stalk vegetables	Asparagus, celery
	Forage and fodder crops	Alfalfa, fodder vetch, sugar beets
	Leaves of root and tuber vegetables	Fodder and sugar beet leaves
	Legumes	Peas, mange tout, broad beans, runner beans
	Fungi (mushrooms)	Champignons, chanterelles
	Roots and tubers	Carrots, potatoes, sweet potatoes
2. High acid and high water contents	Citrus fruit	Lemons, limes, mandarins, oranges
	Small fruits and berries	Blueberries, raspberries, strawberries, black, red, and white currants, grapes
	Other	Kiwifruit, pineapples, rhubarb
3. High sugar content and low water content	Honey, dried fruits	Honey, dried apricots, prunes, raisins, fruit jams
4a. High oil content and very low water content	Tree nuts	Brazil nuts, hazelnuts, walnuts
	Oilseeds	Oilseed rape, sesame, and sunflower seeds, peanuts, soybeans
	Pastes of nuts and oilseeds	Peanut butter, tahini, hazelnut spreads
	Vegetable oils	Olive oils, rapeseed oils, sunflower oils
4b. High oil content and intermediate water content	Oily fruits and their products	Avocados, olives, and pastes thereof
5. High starch and/or protein content and low fat and water contents	Dry legumes/pulses	Dried broad beans, haricot beans, lentils
	Cereal grains and products thereof	Barley, oats, rye and wheat grains, maize, rice, bread, crackers, breakfast cereals, pastas
6. "Difficult or unique" commodities		Cocoa beans and products thereof, coffee, hops, spices, teas

Table 3.1 (*Continued*)

Commodity Groups	Typical Commodity Categories	Typical Representative Commodities
7. Meat (muscle) and seafood	Red meat	Beef, game birds, horse, lamb, pork
	White meat	Chicken, duck, turkey
	Offals	Liver, kidney
	Fish	Cod, haddock, salmon, trout
	Crustaceans	Prawns, shrimps, scallops, crabs
8. Milk and milk products	Milk	Cow, ewe, and goat milk
	Cheese	Cow and goat cheese, feta
	Dairy products	Yogurt, cream
9. Eggs	Bird eggs	Chicken, duck, goose, quail
10. Fat from foods of animal origin	Fat from animals	Kidney fat, dripping, lard
	Milk fat	Butter
	Fish oils	Cod liver oil

This table has been taken directly from the SANCO document 12571/2013 [16] and the commodity groups reflect those given in the OECD publication [17] that provides guidance to registrants on how to validate a residue method for their new pesticide.

i. Commodity group 2 may be merged with commodity group 1, if a buffer is used to stabilize pH changes during the extraction step.
ii. If commodities from group 3 are mixed with water prior to extraction to achieve a water content of >70%, this group may be merged with group 1. The RL should be adjusted to account for smaller sample portions (e.g., if 10 g portions are used for commodities from group 1 and 5 g for commodities from group 3, the RL of group 3 should be twice the RL of group 1 unless the group 3 commodity has been successfully validated at a lower level).
iii. "Difficult commodities" need to be fully validated only if they are to be frequently analyzed. If they are analyzed only occasionally, validation may be reduced to just checking the RLs using spiked blank extracts.
iv. If methods to determine nonpolar pesticides in commodities from group 7 are based on extracted fat, these commodities can be merged with group 10.

From each commodity group, at least 20 different samples should be analyzed following spiking at the anticipated SDL. The samples should be selected to cover multiple commodities within the commodity group, with a minimum of two samples per commodity. If validation criteria fail, then it must be repeated at a higher SDL. Once in routine use, the method must also be subjected to ongoing method validation.

When particular analytes are not detected, then the results can be reported as below the SDL (in appropriate units), as validated and underpinned by ongoing AQC. If detected, an analyte can only be reported after a second confirmatory analysis has been undertaken using a method that provides identification and quantification of the residue. Table 3.2 gives the different parameters and criteria that must be considered for validation of qualitative and quantitative methods.

Certain minimum performance criteria need to be defined before a qualitative screening method can be validated. As opposed to a quantitative method, there are no

Table 3.2. Method Validation Parameters and Criteria

Parameter	How to Address	Criterion	Applicability to MRMs
Accuracy	Determine mean recovery from spikes	70–120%	Quantitative only
Linearity	Construct calibration curve	Residuals <±20%	Quantitative only
LoD	The lowest concentration where 95% confidence of detection of analyte(s) is achieved	Less than or equal to default MRL (0.01 mg/kg)	Qualitative
LoQ	The lowest concentration at which criteria for accuracy and precision are met		Quantitative only
Matrix effect	Comparison between detector response for standards made up in solvent and that for standards made up in sample matrix	No criteria. Matrix effects may vary from analyte to analyte as well as between samples	Qualitative and quantitative
Precision (RSD_r)[a]	Determine repeatability from replicate spikes analyzed in same batch of samples	≤20%	Quantitative only
Precision (RSD_R)	Determine reproducibility from replicate spikes analyzed on different days	≤20%	Quantitative only
Selectivity	Response should be attributable to the analyte	<30% LoD	Qualitative[b] and quantitative
	Check for any response in reagent and sample matrix blanks	<30% LoQ	
Robustness[a]	How often the method fails to meet the criteria that are applicable above		Qualitative and quantitative

Source: Ref. [16].

[a]It is not essential to address these parameters during initial method validation as they can be derived from ongoing QC data generated as the method starts to be used for routine analyses.

[b]No requirement has been set since any detect is supposed to be followed up by an additional confirmatory analysis. However, selectivity should be such that the number of false detects is low enough for efficient use in routine practice.

AQC requirements relating to recovery or linearity, only selectivity. As it is unlikely that any method can detect all possible analyte/commodity combinations, it has become widely accepted [16] that a 95% confidence level (i.e., analyte detected in 19 out of 20 samples) is sufficient. This means that the SDL of a method is the lowest concentration for which it has been demonstrated that a particular analyte can be detected (without necessarily meeting unambiguous identification criteria) in at least

95% of samples. Thus, a false-negative detection rate of 5% is accepted. With regard to false-positive detects, the method must be verified using an unspiked (blank) sample of the same commodity. There is no need to specify a criterion for the numbers of false-positive detects as long as a second sample analysis is to be undertaken using a second appropriate method for confirmation of identity.

Cost-effective implementation requires automatic data processing that requires little or no intervention from the analyst in processing raw data files. This is usually achieved by searching the files against a library containing a database of compounds and associated information such as retention time, chemical formula, adducts, isotopic patterns, and so on. Optimization of processing parameters and thresholds is critical. Mol et al. [8] demonstrated the performance and limitations of a QSMRM based on GC–MS (single quadrupole) detection in routine use on a variety of fruit and vegetable samples over a 12-month period. Their results clearly demonstrated the need for regular maintenance of the GC–MS system and ongoing AQC to check performance. In the same paper, the authors also tested a QSMRM based on UHPLC–ToF-MS on samples that had previously been analyzed using a quantitative MRM based on SRM by LC–MS/MS. The latter experiment also demonstrated the importance of optimizing the thresholds and tolerances of the software in order to match mass spectrometric and chromatographic information from the sample extracts with the information in the library. A further publication [10] describes the analytical capabilities of liquid chromatography with single-stage high-resolution mass spectrometry (Orbitrap) with respect to selective detection and identification of pesticides in 21 different fruit and vegetable samples. This paper clearly demonstrates that the performance of the method is highly dependent on the instrumentation and software that are employed. In this paper, high-resolution mass spectrometry allowed analyte detection based on the exact mass (±5 ppm) of the major adduct ion and of a second diagnostic ion. Using this two-ion approach, there were only 36 (0.3%) false-positive results from 11,676 pesticide/commodity combinations. The percentages of false negatives, assessed from 2730 pesticide/commodity combinations, were 13, 3, and 1% at the 0.01, 0.05, and 0.2 mg/kg concentrations, respectively.

The authors used the protocol for method validation as described in the SANCO document [16] to determine the SDLs for 130 pesticides. These were found to be 0.01 mg/kg for 86 pesticides, 0.05 mg/kg for 30 pesticides, and ≥0.2 mg/kg for 14 pesticides. This paper demonstrates that even when using a high-resolution mass spectrometer only 66% of the pesticides could be detected at the default MRL value of 0.01 mg/kg [6]. It was suggested that a relative tolerance on the ion ratio (intensity of ion relative to higher second ion) of ±50% would be more applicable than the ±30% as stipulated in the SANCO document. Also, the relative retention time could be reduced from 2 min as stipulated in the pesticide document [16] and ±2.5% as stipulated in the veterinary medicine document [18] to ±1%. Adopting these values would improve the identification ability of the method at lower concentrations and reduce the number of false negatives. However, the authors do warn that the criteria should be checked using data generated from other types of instruments such as ToF.

3.5 ONGOING ANALYTICAL QUALITY CONTROL

3.5.1 Internal Quality Control

IQC procedures are used to check the performance of the methodology in use. Wherever possible, each batch of samples analyzed should include some form of IQC monitoring to give feedback on both the method and analyst performance. Obviously, there needs to be a balance between the risk associated with things going wrong with the analysis and the level of the IQC included with each set of samples. The key to an effective IQC regime is the availability of QC samples that can be taken through the whole analytical process. Where available, the quality control material should be included with each batch of samples and the results assessed against the determined acceptable range of the material. This range should be established by the analysis of at least 20 samples. Data collected from the QC samples should, where possible, be graphed to identify any shifts or trends in the results. Where QC samples have failed, the decision-making process of acceptance or rejection of the samples associated with the QC sample should be clearly documented, particularly if the decision is made to accept their results and not reanalyze the samples. Table 3.3 lists the minimum frequencies of recovery checks that are required for screening method performance verification.

3.5.2 Proficiency Testing

Participation in independently organized PTs provides laboratories with an assessment of their own analytical performance and how this compares with others. For many laboratories, participation in relevant PTs is a mandatory requirement of their accreditation service and their customer(s).

Proficiency test providers have to organize each test in accordance with the international proficiency tests standard, ISO/IEC 17043. This standard ensures that the homogeneity and stability of the test materials are satisfactory so that participating laboratories know that they will be receiving representative samples

Table 3.3. Minimum Requirements for Method Performance Verification in Routine Use

	Method Performance Indicators	Other Analytes
Number of analytes	At least 10 analytes per detection system covering all critical points in the method	All analytes that are included in the validated scope of the method
Frequency	Every batch of samples	At least every 12 months, but preferably every 6 months
Level	Screening detection limit	Screening detection limit
Criterion	All method performance indicator analytes should be detected	All validated analytes should be detected

Source: Ref. [16].

of the prepared bulk test material. It also provides essential information on how participant's results should be statistically treated in order to provide an unbiased assessment of performance.

Laboratory performance in a proficiency test is dependent on (i) the skill and experience of the analyst, (ii) the analytical method used, and (iii) the equipment/instrumentation used. It should also be remembered that performance in any particular PT round pertains solely to the specific test material provided and the analyte(s) it contains. Participating laboratories are expected to use the same method that they would normally use to analyze the samples on a daily routine basis.

Once the PT provider has received all the participant's results and assessed them statistically, they will provide a report that allows the laboratory to see how they performed in comparison with the other laboratories. Within the report, the results from each laboratory are coded so that anonymity is retained. If in the report the performance is not deemed to be satisfactory, then the cause of the poor performance must be investigated and established. Remedial action must then be taken in order to negate the particular problems associated with the poor performance.

Most proficiency tests demand the use of quantitative methods and numerical results are required to be reported. However, for qualitative screening methods the PT provider will only request information on the identities of each analyte that has been detected in the test material. In this case, performance will be dependent on the robustness of identification technique used and the scope of the method. Any analyte that was present in the test material but was not detected and therefore not reported would be deemed to be a false-negative result. Obviously, the more the false-negative results, the poorer the performance. Reporting a false-positive result (reporting the presence of an analyte that was not present in the test material) is also an indicator of poor performance, but this may be less important as if an analyte is tentatively detected during screening analysis it will normally be followed up by a second analysis using a quantitative method.

3.6 VALIDATION OF QUALITATIVE SCREENING MULTIRESIDUE METHODS FOR VETERINARY DRUG RESIDUES IN FOODS

3.6.1 EU Legislation Covering Method Validation for Veterinary Drug Screening

Chapter 3.1.3 of the Annex to Commission Decision 2002/657/EC [18] describes a comprehensive approach for the initial validation of methods for different matrices (e. g., muscle, liver, and kidney) and different species (e.g., bovine, porcine, ovine, and poultry). Such a comprehensive approach makes the task of initial method validation time consuming and hence expensive. However, a more recent publication from the Community Reference Laboratories [3] provides a more pragmatic approach to method validation. This document provides practical guidance on how to validate screening methods based on biological, biochemical, and physicochemical detection principles. In this section, only the validation of multiclass methods using physicochemical screening techniques will be covered. For initial method validation, at least one analyte from each known chemical class or subclass should be selected. For example, in the case of quinolones, one acidic compound and one amphoteric compound could be chosen for the validation study. Even when analytes have similar physicochemical properties, they are likely to have different retention times and therefore coelute with different concentrations of matrix coextractives. Thus, they can become subject to different ion suppression or ion enhancement effects. For this reason, it is preferable to test all analytes and not just a selection. It should be noted that there are a number of different terminologies pertaining to veterinary drug residue analyses compared with pesticide residue analyses.

There are two commonly used approaches to determining specificity/selectivity and detection capability (CCβ): the "classical approach" and the "alternative matrix-comprehensive approach."

3.6.2 Determination of Specificity/Selectivity and Detection Capability (CCβ) Using the Classical Approach

The degree of confidence required and the ratio between the screening target concentration (STC) and the action/regulatory limit (AL) should be used to determine the number of "screen-positive" control samples (samples spiked at the STC) that should be tested. Specific examples of the use of these values are given below:

- If the STC is set at 50% of the AL or lower, the occurrence of one, or no, false-compliant result from the analysis of 20 replicate samples spiked at the STC is considered to be sufficient to demonstrate that CCβ is less than the AL.
- If the STC is set between 50 and 90% of the AL, the occurrence of no more than two false-compliant results from the analysis of 40 replicate samples spiked at the STC is considered to be sufficient to demonstrate that CCβ is less than the AL.
- If the STC is set between 90 and 100% of the AL, the occurrence of no more than three false-compliant results from the analysis of 60 replicate samples spiked at the STC is considered to be sufficient to demonstrate that CCβ is fit for purpose.

If the method is to be applied to a single matrix, for example, muscle, then the replicate sample spikes can be split into separate animal species, such as bovine, porcine, and poultry.

3.6.3 Establishment of a Cutoff Level and Calculation of CCβ

Validation of screening methods requires identification of a cutoff level at, or above, which the sample would be deemed to be "screen positive" and liable to further conformational analysis using the additional specificity of a physicochemical method.

The STC at which the blank matrix samples are to be spiked should ideally be at 50% of the AL. If this is not possible, a concentration between 50 and 100% of the AL can be chosen. Ideally, 60 blank samples and 60 spiked samples of one particular matrix should be taken in order to determine CCβ. The sample numbers may be reduced in accordance with the STC:AL ratio as stipulated in the bullet points above. Each target analyte should be tested separately if the method does not produce the necessary specificity to distinguish between analytes.

The blank and spiked samples should be analyzed on different days and preferably using different analysts using the predefined method. The detection capability CCβ of the method can be judged using the numbers of false-compliant results that are within the criteria specified above.

3.6.4 Determination of the Applicability

3.6.4.1 Same Matrix and Different Species

It cannot be assumed that CCβ will be the same, even for the same matrix between species, for example, bovine muscle and porcine muscle. Therefore, CCβ must also be established for all the analyte(s) for the additional species. However, provided that the AL is the same for all species and the matrix stays the same, then the numbers of samples tested can be reduced from 20 blank and 20 spiked to 5 blank and 5 spiked for each additional species.

3.6.4.2 Different Matrix and/or Different Species

Again for the same species it cannot be assumed that CCβ will be the same for different matrices, for example, bovine muscle and bovine liver. Therefore, CCβ must also be established for all the analyte(s) in each new matrix. CCβ can be determined for each new species/matrix combination by analyzing 20 blank samples, for example, porcine livers, and the same 20 blank porcine livers overspiked at the STC.

Ruggedness studies (effects of variations in methodology that might affect the results) are described in Commission Decision 2002/657/EC [18].

3.6.4.3 Continuous Verification Using Quality Control Samples

As with all methods, initial method validation must be supplemented with ongoing quality control to ensure that method performance remains acceptable. Each batch of analyses should include a "blank matrix" (screen-negative control sample) and a "spiked blank matrix" (screen-positive control sample). The spiking concentration should be at the STC. If the "spiked blank matrix" produces a negative result (i.e., it is below the cutoff level) or the "blank matrix" gives a positive result (i.e., it is greater than the cutoff level), then the batch of analyses should be discarded. Results from these QC samples should be monitored to verify that the screening method is working reliably and has a false-compliant rate of no more than 5% for all target analytes. A minimum of 20 QC results should be produced annually and reviewed to check that the method is continuously working reliably and that no more than 5% of the "spiked blank matrix" samples have fallen below the cutoff level.

3.7 CONCLUSIONS

Laboratory quality systems must be implemented in the residue analytical laboratory in order to ensure that the quality of the results produced meets the requirements of the client. Increasingly in today's global market, quality systems that are formally recognized through accreditation and/or certification are required to facilitate international trade by providing the data that establish equivalence of food safety standards with trading partners. Such systems also provide confidence in domestic food systems when applied in laboratories involved in monitoring and surveillance programs for antibiotic residues in food.

In implementing a quality system, it is essential to define the needs of the laboratory and the customer in order to balance the costs and benefits of the system. Putting in place and maintaining a quality system requires the full commitment of management and staff and the necessary resources in terms of infrastructure, equipment, and appropriately trained and experienced staff. A key issue is developing the right mindset, in which the laboratory staff accept the system and the procedures involved as necessary and beneficial to both the organization and its clients, and perform the necessary tasks routinely. The system should be implemented based on what is done in the laboratory, rather than what should be done, and should effectively control the laboratory procedures while remaining as simple as possible. It should also retain sufficient flexibility to change in response to changing client demands and to allow continuous improvement.

The generation of both initial validation and ongoing quality control data is essential in order to be able to demonstrate that a method is fit for the purpose for which it is being used. Not only is it a requirement of accreditation bodies, but it is also essential for customer confidence. This is true for all methods, be they single- or multianalyte, high throughput, qualitative, and semiquantitative or quantitative. It ensures that an acceptable standard of quality and comparability of analytical results can exist between laboratories. Methods have been validated and utilized by analysts that have developed expertise and experience in specific fields of chemical contamination in foods. Over the years, different groups of analysts have become more specialized, so different terminologies have evolved. This makes it impossible to define a set of generic guidelines that would be acceptable to all analysts. It is therefore prudent to adopt and follow the AQC guidelines for the particular chemical contaminant(s) that is applicable to one's own specific needs.

REFERENCES

1. CAC/GL 71-2009. Guidelines for the Design and Implementation of National Regulatory Food Safety Assurance Programme Associated with the Use of Veterinary Drugs in Food Producing Animals.
2. Macarthur, R.; von Holst, C. A protocol for the validation of qualitative methods of detection. *Anal. Methods* **2012**, 4, 2744–2754.

3. Community Reference Laboratories (CRLs). Guidelines for the Validation of Screening Methods for Residues of Veterinary Medicines (Initial Validation and Transfer). **2010**. Available at http://ec.europa.eu/food/food/chemicalsafety/residues/Guideline_Validation_Screening_en.pdf (accessed June 5, 2014).
4. Berendsen, B.J.A.; Stolker, L.A.M.; Nielen, M.W.F. The (un)certainty of selectivity in liquid chromatography–tandem mass spectrometry. *J. Am. Soc. Mass Spectrom.* **2013**, 24, 154–163.
5. Kmellár, B.; Pareja, L.; Ferrer, C.; Fodor, P.; Fernández-Alba, A.R. Study of the effects of operational parameters on pesticide residue analysis by LC–MS/MS. *Talanta* **2011**, 84, 262–273.
6. Mol, H.J.G.; Plaza-Bolaños, P.; Zomer, P.; de Rijk, T.C.; Stolker, A.A.M.; Mulder, P.P.J. Toward a generic extraction method for simultaneous determination of pesticides, mycotoxins, plant toxins and veterinary drugs in feed and food matrices. *Anal. Chem.* **2008**, 80(24), 9450–9459.
7. Mol, H.J.G.; van Dam, R.C.J.; Zomer, P.; Mulder, P.P.J. Screening of plant toxins in food, feed and botanicals using full-scan high resolution (Orbitrap) mass spectrometry. *Food Addit. Contam. Part A* **2011**, 28(10), 1405–1423.
8. Mol, H.J.G.; Reynolds, S.L.; Fussell, R.J.; Štajnbaher, D. Guidelines for the validation of qualitative screening multi-residue methods used to detect pesticides in food. *Drug Test. Anal.* **2012**, 4(Suppl. 1), 10–16.
9. Mol, H.J.G.; van der Kamp, H.; van der Weg, G.; van der Lee, M.; Punt, A.; de Rijk, T.C. Validation of automated library-based qualitative screening of pesticides by comprehensive two-dimensional gas chromatography/time of flight mass spectrometry. *J. AOAC Int.* **2011**, 94, 6.
10. Mol, H.J.G.; Zomer, P.; de Koning, M. Qualitative aspects and validation of a screening method for pesticides in vegetables and fruits based on liquid chromatography coupled to full scan high resolution (Orbitrap) mass spectrometry. *Anal. Bioanal. Chem.* **2012**, 403, 2891–2908.
11. Vanhaecke, L.; Bussche, J.V.; Wille, K.; Bekaert, K.; De Brabander, H.F. Ultra-high performance liquid chromatography–tandem mass spectrometry in high-throughput confirmation and quantification of 34 anabolic steroids in bovine muscle. *Anal. Chim. Acta* **2011**, 700, 70–77.
12. Rahmani, A.; Jinap, S.; Soleimany, F. Qualitative and quantitative analysis of mycotoxins. *Compr. Rev. Food Sci. Food Saf.* **2009**, 8, 202–251.
13. Commission Regulation (EU) No. 212/2013. Available at http://eurlex.europa.eu/LexUriServ/LexUriServ.do?uri=OJ:L:2013:068:0030:0052:EN:PDF (accessed June 5, 2014).
14. Fussell, R.J.; Hetmanski, M.T.; Macarthur, R.; Findlay, D.; Smith, F.; Amrus, A.; Brodessor, P.J. Measurement uncertainty associated with sample processing of oranges and tomatoes for pesticide residue analysis. *J. Agric. Food Chem.* **2007**, 55, 1062–1070.
15. Matthews, W.A. An investigation of non-solvent extractable residues of [^{14}C] chlorpyrifos-methyl in stored wheat. *Pestic. Sci.* **1991**, 31(2), 141–149.
16. SANCO/12571/2013. Guidance Document on Analytical Quality Control and Validation Procedures for Pesticide Residues Analysis in Food and Feed. Available at http://www.eurl-pesticides.eu/library/docs/allcrl/AqcGuidance_Sanco_2013_12571.pdf (accessed June 5, 2014).

17. ENV/JM/MONO(2007)17. Guidance Document on Pesticide Residue Analytical Methods. OECD Environment, Health and Safety Publications, Series on Testing and Assessment No. 72 and Series on Pesticides No. 39. Available at http://search.oecd.org/officialdocuments/displaydocumentpdf/?cote=env/jm/mono%282007%2917&doclanguage=en (accessed June 5, 2014).
18. Commission Decision of 12 August 2002 implementing Council Directive 96/23/EC concerning the performance of analytical methods and interpretation of results (2002/657/EC). Available at http://eurlex.europa.eu/LexUriServ/LexUriServ.do?uri=OJ:L:2002:221:0008:0036:EN:PDF (accessed June 5, 2014).

CHAPTER

4

DELIBERATE CHEMICAL CONTAMINATION AND PROCESSING CONTAMINATION

STEPHEN LOCK

4.1 INTRODUCTION

Over the last few years, there have been a number of scares due to food contamination. This contamination can be divided into two types: that of malicious or deliberate chemical contamination to gain a financial benefit and, alternatively, the inadvertent contamination of food during processing and packaging. In each case, the contamination has generally been a result of small organic chemicals.

In recent years, there have been two major cases of malicious contamination that have made the news headlines. One of the earliest cases of this type of contamination involved the use of Sudan dyes to enhance the visual characteristics of chili spice in order to achieve a premium market price. In 2003, the Food Standard Agency (FSA) in the UK confirmed that Sudan dyes had been found in a number of relishes, chutneys, and seasonings containing chili powder [1] and this continued to be a food safety concern [2]. The Sudan dyes, which are normally used in shoe polish and waxes, are known carcinogens and are therefore banned from food in the UK and EU states. The contamination was traced to a sample of chili powder that had been imported into Europe from India and used in the processing of various products. The dyes were used to enhance the color of the spice and therefore increase the revenue for the product by artificially changing its appearance. Following a surveillance exercise by the FSA, food contamination was discovered causing large food withdraws [3] and costing the UK food manufacturing industry millions of pounds sterling (GBP). As a result, the Indian authorities (Spices Board) [4] cancelled the licenses of the five Indian exporters involved, and ordered the mandatory preship inspection of all future chili consignments leaving the country.

In the instance of the Sudan dye contamination in Europe, the health of consumers was put at risk but there were no adverse effects reported and no fatalities resulted. Sadly, in the case of melamine contamination of food in 2007, this was not the case. Melamine is not a naturally occurring chemical but is a substance used in a variety of industries, including the production of resins and foams, cleaning products, fertilizers, and pesticides. Melamine contains a high level of nitrogen and is used to artificially

High-Throughput Analysis for Food Safety, First Edition.
Edited by Perry G. Wang, Mark F. Vitha, and Jack F. Kay.

increase the nitrogen content of products. The nitrogen content of processed food products is often used as a surrogate measure of the protein levels in the food. Therefore, a product with higher detectable levels of nitrogen would command a higher market value. Traditionally, the Kjeldahl method [5], which converts all the nitrogen in a sample into ammonia, is used to determine nitrogen in food. The ammonia is then measured to determine the nitrogen content. Although the method is simple, the drawback is that it does not differentiate the source of the nitrogen and therefore is open to the possibility of fraudulent adulteration with a cheap chemical source of high nitrogen content, such as melamine or cyanuric acid. In several cases in China, melamine was added to pet food and even baby food to increase its value. In the U.S. following the death and illness of several cats and dogs that had eaten the contaminated food, the melamine scandal led to the withdrawal of pet food products in 2007 [6]. As a result of the investigation by the Food and Drug Administration (FDA) and the United States Department of Agriculture (USDA) in 2008, a U.S. company, its president and chief executive officer, and two Chinese nationals and the businesses they operated were indicted by a federal grand jury for their roles in importing melamine-tainted products into the U.S. [7]. However, this did not just affect pets. Melamine was also used as an adulterant in baby food and in 2008, it was reported that over 1200 babies had been affected in China, and some of the babies had died. As a result of the ensuing Chinese investigation, a number of criminal prosecutions were brought, resulting in the execution of two people, three received life imprisonments, and several senior government officials lost their jobs [8]. In cases of malicious contamination, fast analysis is essential to identify the contaminated food, remove the source from the market to reduce health risk to the consumer, and in the case of melamine, to save lives.

The melamine case is probably the worst of its kind to hit food manufacturing in recent years. However, most food contamination is not a result of malicious contamination, but rather of accidental contamination during food processing. The financial effect to any food manufacturer can still be significant, resulting in large product recalls and losses in revenue. Of all the food we consume, one of the most heavily regulated is infant formula. As such, this staple is heavily tested and food contamination can be readily uncovered. One example of nonintentionally added substances (NIAS) in food was the presence of isopropyl thioxanthone, or ITX, in Nestle infant formula in 2005 [9], which caused the withdrawal of millions of liters of infant formula in Italy. ITX came from an ink used in the packaging of infant formula, and although it was determined not to be harmful at the levels found, it was still an undesirable contaminant [10]. ITX represents one in a series of undesirable packaging contaminants that include plasticizers, monomers, and phthalates as well as other inks and photoinitiators used in packaging manufacture. When this type of contamination occurs, tracing the source is paramount to reduce financial impact on the company. This requires fast analysis with quick sample turn around.

Undesirable compounds can also be formed as a result of the food manufacturing process itself. The most famous case of this type of contamination occurred in 2002 when acrylamide, a known carcinogen, was found by Swedish scientists to be present in a variety of processed foods [11]. These findings were soon confirmed by other research groups [12–14] and, together with major stakeholders, efforts are still ongoing to build greater understanding of acrylamide concerning the mechanism of its formation in foods, the risks associated for consumers, and possible strategies to lower acrylamide levels in foodstuffs.

Chemical contamination of any sort is typically covered by government legislation. Globally, the World Health Organization has put in place a food monitoring program commonly known as the GEMS/Food Programme [15]. GEMS/Food informs governments, the Codex Alimentarius Commission, and other relevant institutions on levels and trends of chemical contaminants in food and their contribution to dietary exposure. However, it is up to regional and national bodies and governments to establish food legislation, which together with continual food surveillance programs helps reduce food contamination.

In the U.S., food safety is under the control of the U.S. FDA. The Federal Food, Drug, and Cosmetic Act (FFDCA) [16] provides the FDA with broad regulatory authority over food that is introduced or delivered for interstate commerce. In particular, Section 402(a)(1) of the FFDCA states that a food is deemed to be adulterated if it contains any poisonous or deleterious substances, such as chemical contaminants, which may ordinarily render it harmful to health. Under the provision of the FFDCA, the FDA oversees the safety of the U.S. food supply (domestic and imports), in part, through its monitoring programs for contaminants in food and the assessment of potential exposure and risk. In January 2011, the Food Safety Modernization Act (FSMA) [17] was signed with the aim to ensure the U.S. food supply is safe by shifting the focus of federal regulators from responding to contamination to preventing it.

In Europe, the basic principles of EU legislation on contaminants were laid out in the European Law in 1993 [18]. This legislation stipulates that food containing a level of contaminant that is unacceptable from a public health viewpoint, for example, toxic, cannot be placed on the market. Maximum limits are set for the contaminants of greatest concern, either due to their toxicity or due to their potential prevalence in the food chain. These limits are set on the basis of scientific advice provided by the European Food Safety Authority (EFSA). It is then up to the authorities of an EU Member State to ensure that they comply with the legislation. For imported foodstuffs, the country of origin is responsible for compliance, and this is controlled at the EU borders and by market sampling. In this way, legislation, together with random sampling and surveillance testing, helps to reduce the frequency of food scares. Member States that identify a risk as a result of a surveillance exercise can temporarily suspend or restrict production of products. They then have to notify the EU Commission and other Member States as to their reasons in order to prevent the further distribution of tainted food. This is helped by the RASFF (Rapid Alert System for Food and Feed) that distributes this information rapidly to the EFSA and other relevant institutions and notifies these institutions of potential contaminants. Here again, the need for high-throughput, rapid screening of samples is evident.

Where contaminant identity can be known, for example, NIASs such as packaging materials, legislation has been easier to put in place. Regulation (EC) No. 1935/2004 [19] sets out the law on chemical migration from all materials and articles in contact with food. It includes provisions for materials and articles expected to come into contact with foods or to transfer their constituents to food (e.g., printing inks and adhesive labels). However, the regulation does not include covering or coating substances that are part of the food and that may be eaten with it, such as sausage skin. These general laws are supplemented by specific laws governing particular materials, such as food contact plastics (Regulation 10/2011) [20] and "active and intelligent" food contact materials (Regulation No. 450/2009) [21]. In the 2009 regulations, "a maximum level of 0.01 mg/kg in food for the migration of a non-authorized substance through a functional barrier" was set in place for infant formula together with the list of authorized substances. The 2011 regulations go further, listing actual active substances that may be present in packaging and their permitted migration levels in different types of food. Where a substance is listed as "not detected" from the scientific studies, it is taken that the limit of detection of this substance would be 0.01 mg/kg of food. Acrylamide is actually not in this category as specific regulatory limits are still being investigated and an ongoing surveillance exercise in Europe is taking place. Maximum limits for other contaminants in food have been set by Commission Regulation (EC) No. 1881/2006 [22], which came into force on March 1, 2007. Maximum limits in certain foods were set for the following contaminants: nitrate, mycotoxins, and metals, but it also included contaminants such as 3-monochloropropane-1,2-diol (3-MCPD), dioxins and dioxin-like polychlorinated biphenyls (PCBs), and polycyclic aromatic hydrocarbons (e.g., benzo (α)pyrene).

When a serious instance of food contamination occurs, specific regulations may be put in place to prevent future contamination. In the case of the contamination of chili products with Sudan I, a new legislation was introduced within the European Community (2003/460/EC) [23]. This states that all hot chili and hot chili products imported into the Community in whatever form, intended for human consumption, should be accompanied by an analytical report provided by the concerned importer or food business operator demonstrating that the consignment does not contain Sudan I. In the absence of such an analytical report, the importer established in the Community shall have the product tested to demonstrate that it does not contain Sudan I. In case of pending availability of the analytical report, the product shall be detained under official supervision. The legislation also states that Member States shall conduct random sampling and analysis of hot chili and hot chili products at import or already on the market.

In the U.S., the thresholds for chemical contamination in food are regulated by Code of Federal Regulations (CFR) titles and several U.S. laws. For example, action levels for poisonous or deleterious substances in food are established and revised according to criteria specified in 21 CFR 109 and 21 CFR 509 and are revoked when a regulation establishing a tolerance for the same substance and use becomes effective.

21 CFR 109 actually includes limits for polychlorinated biphenyls with a lowest limit of 0.2 ppm in infant and junior foods [24].

21 CFR 170.39 covers substances used in food-contact articles (e.g., food packaging or food processing equipment) and the FDA states that a substance used in a food-contact article (e.g., food packaging) that migrates into food, or

that may be expected to migrate, will be exempted from regulation as a food additive because it becomes a component of food at levels that are below the threshold of regulation if the substance satisfies certain criteria (e.g., low-potential carcinogenicity and estimated dietary exposure among others) [25].

21 CFR 175 covers indirect food additives from adhesives and components of coatings (where Subpart B covers substances for use only as components of adhesives and lists permitted chemicals and includes limitations in packaging legislation).

21 CFR 189 lists food ingredients that have been prohibited from use in human food by the FDA. Use of any of the substances in violation of Section 21 CFR 189 causes the food involved to be adulterated in violation of the act (21 CFR 189 provides a list of substances prohibited from use in human food, but it is not a complete list of substances that may not lawfully be used in human food as no substance may be used in human food unless it meets all applicable requirements of the Act).

For the case of adulterated food, the prohibition of the movement and supply of contaminated food appears in several laws; for example, the Food, Drug and Cosmetic Act (21 U.S.C. §331), Meat Inspection Act (21 U.S.C. §610; 9 CFR 301.2), Poultry Products Inspection Act (21 U.S.C. §458; 9 CFR 381.1), and Egg Products Inspection Act (21 U.S.C. §1037; 9 CFR 590.5).

All of these articles and laws can be accessed from the Web site of law departments (such as Cornell University Law School) [26]. However, generally speaking, food in the U.S. is not as heavily regulated as in Europe.

From the introduction above, it can be seen that deliberate chemical contamination or processing contamination covers a wide area. As such, many techniques to rapidly identify and quantify contaminants have been used and one of the latest techniques is LC–MS/MS. When the first commercial LC–MS/MS instruments came to market in the 1980s, they were predominantly used by academic scientific research institutions and for quantitative analysis in the drug industry. However, over the last three decades, instrumentation and software have come a long way, making instruments cheaper, more sensitive, and easier to use. As a result, LC–MS/MS is no longer the domain of well-funded institutions and mass spectrometry experts and is now used across many application areas and industrial sectors and has revolutionized food testing in many laboratories.

Initially used in traditional food safety (e.g., pesticide detection), LC–MS/MS has since moved into areas such as packaging testing, colorant detection, food authenticity and functionality, allergen detection, and many more. LC–MS/MS is now particularly widespread in contaminant analysis. The following sections include examples of its use and describe how it has replaced more traditional techniques to both speed up analysis and improve detection limits, as well as its use in many of the food contamination cases listed above.

4.2 HEAT-INDUCED FOOD PROCESSING CONTAMINANTS

Of all the heat-induced food processing contaminants, acrylamide is probably one of the most famous. Acrylamide forms in food due to the reaction between the amino

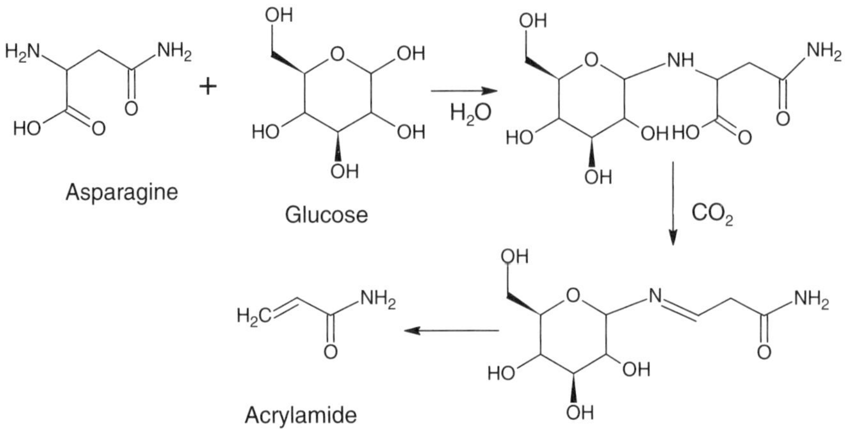

Figure 4.1. Formation of acrylamide in food processing.

acid asparagine and reducing sugars such as glucose and fructose at elevated temperatures and low-moisture conditions—a process known as the Maillard reaction (Figure 4.1) [27,28]. Acrylamide is therefore formed in the production of chips and crisps and is among a series of contaminants, including furan, which can be produced during food manufacturing.

There are two main techniques used for the analysis of acrylamide by laboratories all over the world: LC–MS/MS and GC–MS. In GC–MS, there are two approaches: the use of complicated multiple solvent extractions [29] producing detection limits that are above 10 μg/kg and can be affected by matrix, and the use of bromination [30] that offers adequate sensitivity with multiple ion confirmation. However, in the use of GC, there is a risk of a false positive result because of acrylamide formation in the hot GC injector if the reducing sugars and asparagine are present in the food extract. Determination of acrylamide using LC–MS/MS, therefore, offers an alternative approach where potential injector formation of acrylamide does not occur while providing lower detection limits and avoiding a time-consuming derivatization step that is sometimes needed for GC–MS.

LC–MS/MS methods often use a simple solvent extraction followed by dilution to minimize possible matrix effects [31]. The sample is then analyzed using a fast LC separation with accurate and reproducible MS/MS detection allowing quantitation down to low μg/kg (ppb) levels in food, as shown in Figure 4.2. LC–MS/MS methods tend to use electrospray ionization (ESI) in positive polarity mode with multiple reaction monitoring (MRM) and typical Q1/Q3 mass transitions of 72/55 and 72/44. A reversed-phase column using a polar endcapped stationary phase [32] or a Hypercarb phase [31] designed to retain small polar compounds is used in order to retain the acrylamide. To increase sensitivity, volatile acids such as formic acid [31] or acetic acid [32] are added to the mobile phase and acrylamide is eluted using a gradient from aqueous to higher levels of methanol. When the more retentive Hypercarb phase [31] is used, gradient elution with methanol is also used. When

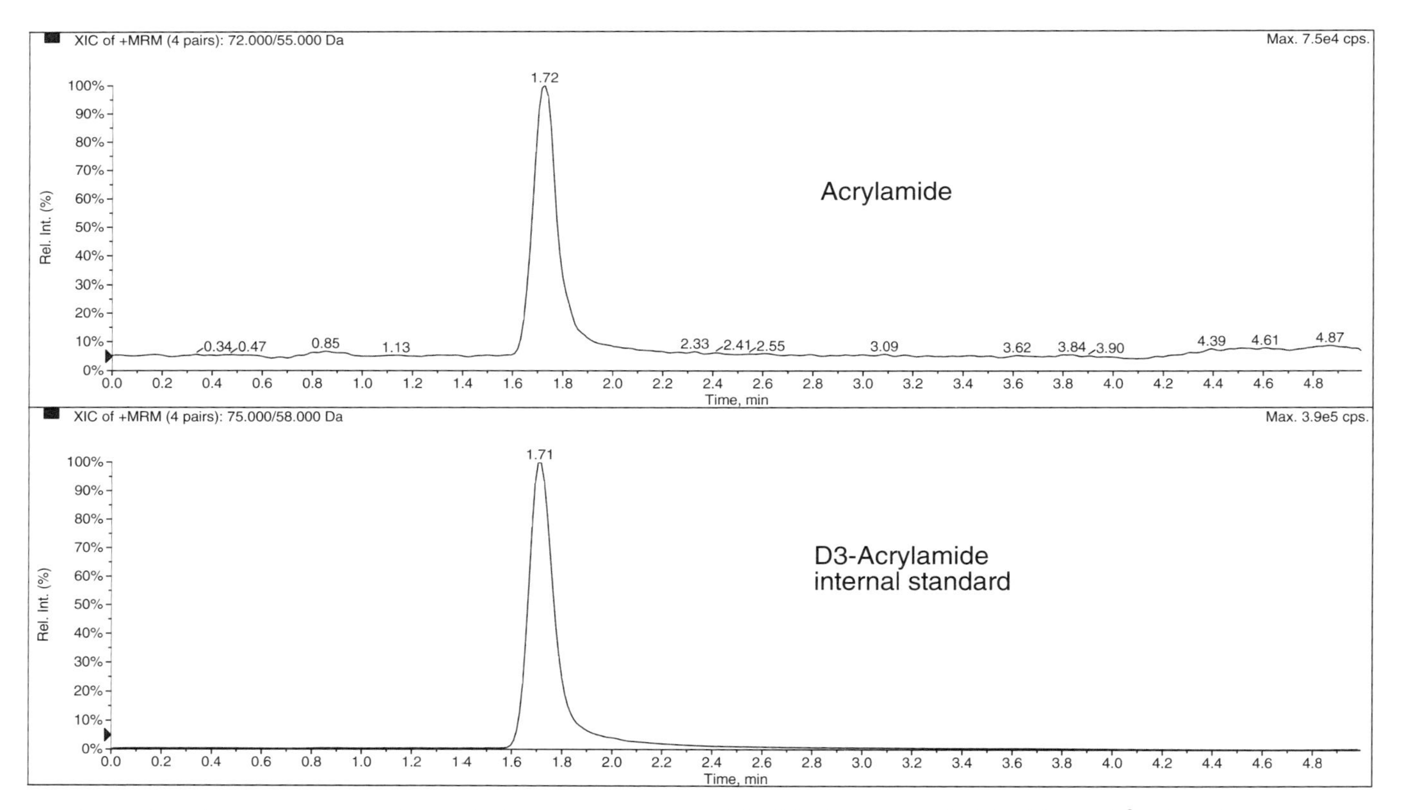

Figure 4.2. Chromatogram of a 5 ng/ml acrylamide standard analyzed by LC–MS/MS using an ABSCIEX 4500QTRAP® LC–MS/MS system.

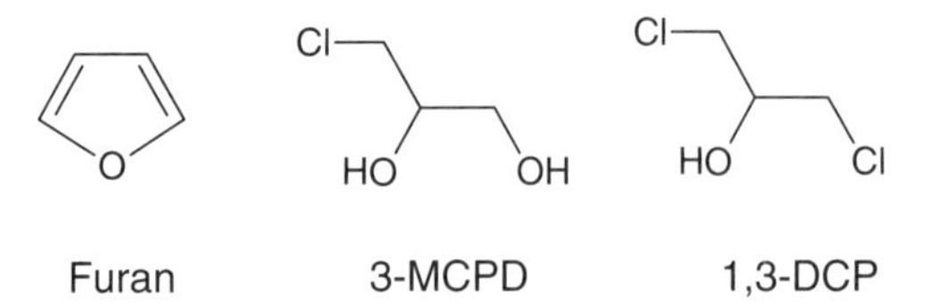

Figure 4.3. Chemical structures of furan, 3-MPCD, and 1,3-DCP.

a less retentive polar endcapped phase [32] is used, the mobile phase contains <1% methanol. If additional sensitivity is needed, either because a less sensitive LC–MS/MS detector is used or lower limits of detection are needed, then solid-phase extraction (SPE) has been used as a means to improve detection limits and remove matrix effects [32].

Apart from acrylamide, there are other contaminants that are produced as a result of chemical reactions caused by heat-treated food, including chloropropanols and furan (Figure 4.3). Chloropropanols, especially 3-monochloropropane-1,2-diol (3-MCPD) and 1,3-dichloropropan-2-ol (1,3-DCP), are recognized by-products from heat treatment of food [33]. 3-MCPD has been recognized by the European Commission's Scientific Committee on Food as a carcinogen and has been classified as an undesirable contaminant in food [34]. Chloropropanols are formed by the reaction of glycerol or acylglycerol (present in fats) with chloride ions (e.g., sodium chloride) in the heat processing of foods containing low levels of water. 3-MCPD can therefore be produced by acid hydrolysis, used, for example, in soy sauce production [33]. 3-MCPD is usually extracted from samples using SPE based on diatomaceous earth (Extrelut) together with solvent partitioning [33]. Due to the lack of a chromophore, HPLC–UV is not suitable and because of the small polar nature of 3-MPCD and 1,3-DCP, direct GC analysis is difficult—so derivatization is needed. Of the reagents available, heptafluorobutyrylimidazole (HFBI) is one of the most commonly used, and methods using HFBI derivatization followed by GC–MS analysis [33] have been recognized as fit for purpose by the AOAC (AOAC Official Method 2000.01) as well as by the European standardization body (EN 14573) [35]. LC–MS/MS has not been used to detect 3-MCPD directly as 3-MCPD ionization is not very efficient, resulting in low sensitivity. Recently, LC–MS/MS has been used to detect intermediates of 3-MPCD (3-MCPD esters) and presentations at the 2012 annual AOAC meeting [36,37] showed that LC–MS/MS can directly detect the individual glycidyl ester species. In one of these papers, a normal phase separation on an ODS column using a gradient from methanol (containing 10 mM ammonium formate and 0.1% formic acid) to isopropanol was used to separate the esters, followed by atmospheric pressure chemical ionization (APCI) to ionize them [36]. Samples were prepared by solvent extraction followed by SPE on a silica-based media [36]. In contrast, GC–MS uses acid hydrolysis to cleave these esters before detecting the total amount of 3-MCPD in the sample, preventing detection of the native esters. Such 3-MCPD esters have been found in margarine and baby food. There is ongoing research into this topic [38] as the toxic properties of these contaminants are still unknown and they can act as a further source of 3-MCPD contamination in food. Like 3-MPCD, furan is another small, polar, but also very volatile hazardous contaminant produced during heat treatment of foods. Furan can be produced in several pathways, including the pyrolysis of sugars and the breakdown of ascorbic acid [33]. Due to its polar and volatile nature, furan lends itself to headspace GC analysis and this is the most common technique used for its detection. In this area, solid-phase microextraction (SPME) or headspace analysis as an extraction technique followed by GC–MS detection is commonly used [39–41].

4.3 PACKAGING MIGRANTS

As mentioned in Section 4.1, chemicals that migrate into food are covered in EU legislation. This class of contaminants is quite extensive. This section will discuss the ones that have been highlighted in the press and are covered in current legislation regarding compounds that migrate from packaging into food.

Of this class of chemicals, one that has been in the news recently is bisphenol A (BPA). BPA together with bisphenol A diglycidylether (BADGE) and its derivatives are typical migrants from epoxy resins. Traditional sample preparation for these migrants involved solid–liquid–liquid extraction followed by SPE cleanup, but recent publications have shown that these migrants can be analyzed using a faster QuEChERS method [42]. QuEChERS extraction of a sample normally takes about 20 min, whereas solid–liquid–liquid extraction followed by SPE can take well over 1 h, especially if the SPE extract needs to be blown down and reconstituted prior to analysis. Analysis is by HPLC separation using a reversed-phase gradient on a C18 column where both the aqueous and organic phases contain formic acid followed by ESI in positive mode. Detection levels are at the μg/kg level using a midrange LC-MS system. In Ref [42], BPA and BADGE had typical Q1/Q3 mass transitions of 229.2/107 and 341.2/135, respectively. BADGE can actually be present as several derivatives (BADGE-2 H_2O with Q1/Q3 mass transition of 377.3/135; BADGE-H_2O with Q1/Q3 mass transition of 359.2/191; BADGE-H_2O-HCl with Q1/Q3 mass transition of 395.2/209.1; BADGE-2HCl with Q1/Q3 mass transition of 414.3/229.2; and BADGE-HCl with Q1/Q3 mass transition of 377.2.3/209) and these are normally all monitored.

Another group of compounds that made headlines in 2005 for contaminating infant formula are compounds that are used in inks on packaging labels. This class includes a number of substances, including ITX, Irgacure, and TRP—ITX is a mixture of 2-isopropylthioxanthone and 4-isopropylthioxanthone; Irgacure contains Irgacure 819 (phenylbis-2,4,6-trimethylbenzoyl-phosphine oxide); and TRP is tri(propylene glycol) diacrylate (Figure 4.4). They are specifically used as photoinitiators in UV cured inks. Again, as with BPA, LC–MS/MS methods have been developed [43] for quantitation of these migrants and they use ESI in positive mode with ITX having Q1/Q3 mass transitions of 255.1/213.1 and 255.1/184.1; Igracure having Q1/Q3 mass transitions of 419.2/147.2 and 419.2/119.2; and TRP having Q1/Q3 mass transitions of 301.2/113.3 and 301.2/55. HPLC separation of the ITX isomers is not easy (due to their structure) (Figure 4.4) and typically they are combined into one HPLC peak

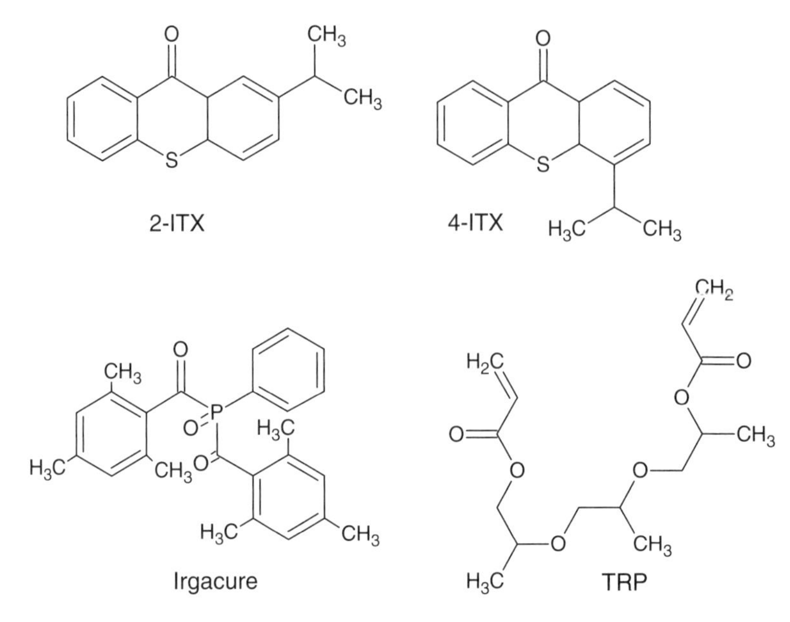

Figure 4.4. Chemical structures of photoinitiators found in food.

using reversed-phase chromatography on a C18 column (e.g., Hypersil BDS C18 column (5 μm, 100 × 2 mm)) with a gradient of water to acetonitrile containing formic acid. The sensitivity levels of the ABSCIEX 3200 QTRAP® LC–MS/MS system (a midrange system) were high enough to detect migrants at 0.01 mg/kg in extracts from packaging material by direct injection of the extracts [43].

Contamination of infant formula not only occurs via labeling but also comes from chemical migration from baby bottle teats. The contaminants of this group include *N*-nitrosamines that originate from the various dialkyl amines that are used as accelerators and stabilizers in the vulcanization process of the rubber used for the teats. These compounds have been shown to have significant health effects on infant ingestion. Traditional methodology used detection by GC analysis [44], but this approach suffers from several drawbacks, the most significant being the inability to identify the peak of interest due to coeluting or masking matrix peaks. Therefore, methods have been developed to detect these contaminants by LC–MS/MS. Several different *N*-nitrosamines have been reportedly found in rubber teats and they include *N*-nitrosomethylethylamine and *N*-nitrosodimethylamine (Table 4.1). In this case, detection is done by APCI in negative mode and the HPLC separation is again by reversed-phase chromatography. Because APCI was used, no HPLC modifiers are needed to boost sensitivity and a simple methanol gradient can be employed [45]. The MRM transitions used are shown in Table 4.1 and detection levels at low μg/kg can be easily achieved.

Probably the largest group of migrants is phthalates. The issue of phthalates in food was brought to the attention of the Directorate General for Health and Consumers (DG SANCO) in 2007 [46]. Phthalates (1,2-benzenedicarboxylic acid esters) are a group of

Table 4.1. List of Common *N*-Nitrosamines That Can Be Present in Rubber Used for Baby Teats with Their Corresponding Q1 and Q3 Transitions Used in LC–MS/MS Analysis

PFC	Formula	Abbreviation	Q1 Mass (amu)	Q3 Mass (amu)
N-Nitrosomethylethylamine	$C_3H_8N_2O$	NMEA	89.0	61
N-Nitrosodimethylamine	$C_2H_6N_2O$	NDMA	75.0	43
N-Nitrosodiethylamine	$C_4H_{10}N_2O$	NDEA	103.1	75
N-Nitrosomorpholine	$C_4H_8N_2O_2$	NMOR	117.0	87
N-Nitrosopiperidine	$C_5H_{80}N_2O$	NPIP	115.1	69
N-Nitrosopyrrolidine	$C_4H_8N_2O$	NPYR	101.0	55

compounds that are mainly used as plasticizers for polymers such as polyvinylchloride (PVC), but can also be used in adhesives, paints, films, glues, and cosmetics and so their potential sources are quite diverse. As such, they can contaminate food via packaging migration. In a 2007 survey [47], most laboratories used GC-based techniques with either electron capture detection, MS, or flame ionization detection, but sample preparation was lengthy or used harmful solvents and included liquid–liquid partitioning (using dichloromethane) or gel permeation chromatography (GPC). More recently, LC–MS/MS has been used to simplify sample extraction and speed up sample analysis [48]. Ionization depends on the phthalate class, but over 20 phthalate esters (Table 4.2) can be detected using ESI in positive mode. When phthalate analysis is moved over to LC–MS/MS, a simple methanol extraction followed by dilution is all that is needed to extract over 20 phthalates and the analysis time is then <10 min [49] compared with over 20 min using capillary GC–MS. Figure 4.5 is an example of 18 phthalates, detected in one run, and includes some of the more prevalent contaminants (e.g., bis(2-ethylhexyl) phthalate and benzyl butyl phthalate) (Table 4.2). In this example, the separation of compounds was accomplished by reversed phase using a Phenomenex Kinetex C18 column (100×4.6 mm; 2.6 μm) and a fast gradient of water containing 10 mM ammonium acetate to methanol at a flow rate of 500 μl/min.

Another group of packaging migrants are perfluorinated contaminants (PFCs). This class of compounds is used for a variety of industrial applications, including flame retardants and stain removers. This class does not normally enter food as a result of deliberate contamination or food processing, but rather enters the food chain via bioaccumulation, for example, in fish [50], so they are primarily environmental contaminants. However, they can leach into food from coated food contact materials, for example, nonstick cookware and food paper packaging that is oil and moisture resistant (e.g., microwave popcorn paper bags) [51–54]. The two most prevalent compounds in this class are perfluorooctanesulfonic acid (PFOA) and perfluorooctane sulfonate (PFOS). The current method of choice for this class is LC–MS/MS. There are numerous papers that have used LC–MS/MS to detect PFOS and PFOA and in all cases this class of compounds is detected in negative ion mode normally using ESI

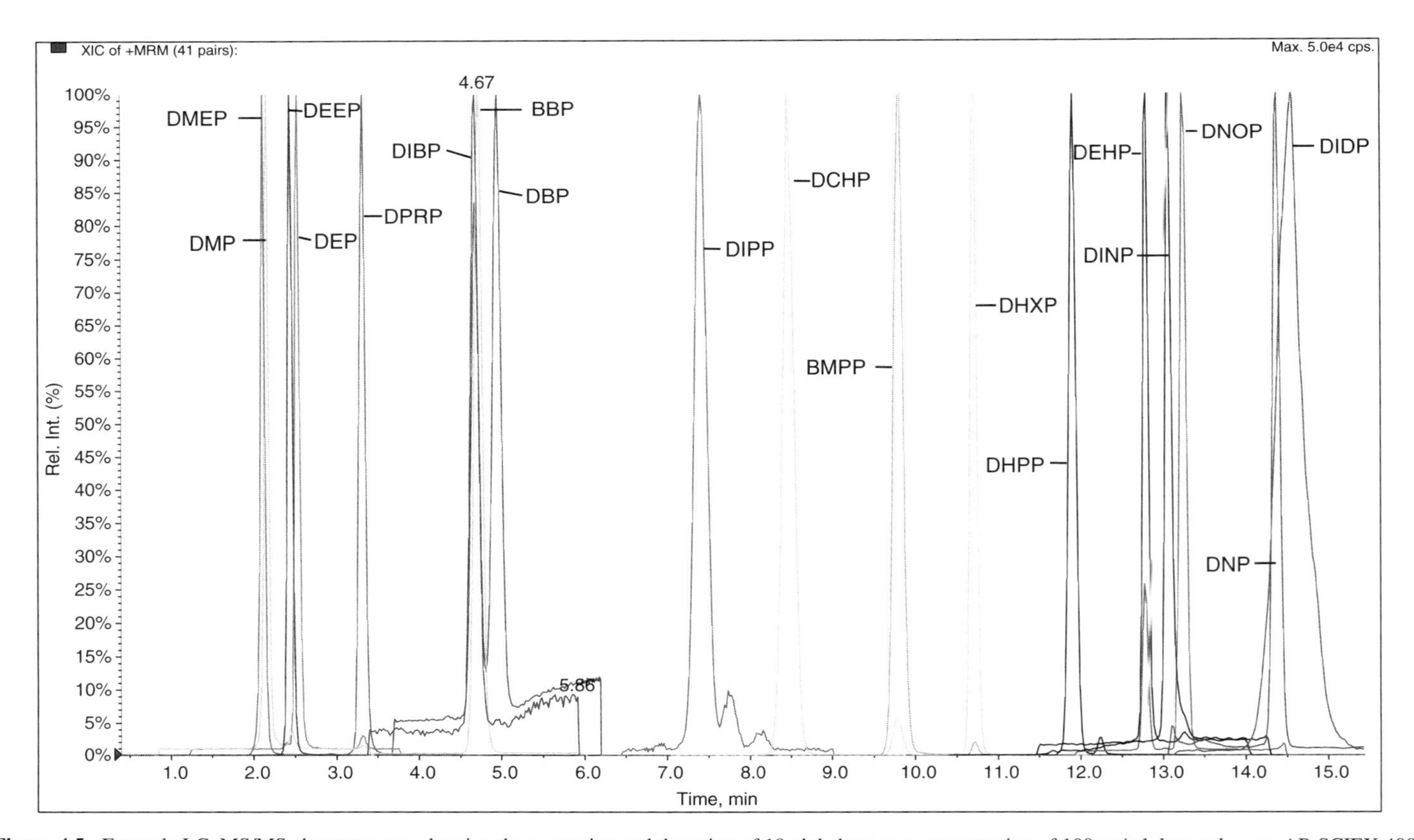

Figure 4.5. Example LC–MS/MS chromatogram showing the separation and detection of 18 phthalates at a concentration of 100 ng/ml detected on an AB SCIEX 4000 QTRAP® LC–MS/MS system.

Table 4.2. Examples of Phthalates, with Their Corresponding Q1 and Q3 Transitions Used in LC–MS/MS Analysis

Phthalate	Formula	Abbreviation	Q1 Mass (amu)	Q3 Mass (amu)
Dimethyl phthalate	$C_{10}H_{10}O_4$	DMP	195	163/133
Diethyl phthalate	$C_{12}H_{14}O_4$	DEP	223	149/177
Diallyl phthalate	$C_{14}H_{14}O_4$	DAP	247	189/149
Dipropyl phthalate	$C_{14}H_{18}O_4$	DPrP	251	149/191
Diisopropyl phthalate	$C_{14}H_{18}O_4$	DIPrP	251	149/191
Dibutyl phthalate[a,b]	$C_{16}H_{22}O_4$	DBP	279	149/205
Diisobutyl phthalate[b]	$C_{16}H_{22}O_4$	DIBP	279	149/205
Bis(2-methoxyethyl) phthalate	$C_{14}H_{18}O_6$	DMEP	283	207/59
Dipentyl phthalate[b]	$C_{18}H_{26}O_4$	DPP	306	219/149
Diisopentyl phthalate	$C_{18}H_{26}O_4$	DIPP	306	219/149
Bis(2-ethoxyethyl) phthalate	$C_{16}H_{22}O_6$	DEEP	311	221/149
Benzyl butyl phthalate[a,b]	$C_{19}H_{20}O_4$	BBP	313	149/205
Diphenyl phthalate	$C_{20}H_{14}O_4$	DPhP	319	225/77
Dicyclohexyl phthalate	$C_{20}H_{26}O_4$	DCHP	331	167/249
Bis(4-methyl–2-pentyl) phthalate	$C_{20}H_{30}O_4$	BMPP	335	167/251
Dihexyl phthalate	$C_{20}H_{30}O_4$	DHXP	335	149/233
Di-*n*-heptyl phthalate	$C_{22}H_{34}O_4$	DHP	363	149/233
Bis(2-*n*-butoxyethyl) phthalate	$C_{20}H_{30}O_6$	DBEP	367	101/249
Bis(2-ethylhexyl) phthalate[a,b]	$C_{24}H_{38}O_4$	DEHP	391	167/279
Di-*n*-octyl phthalate[a,b]	$C_{24}H_{38}O_4$	DNOP	391	261/149
Diisononyl *ortho*-phthalate[a,b]	$C_{26}H_{42}O_4$	DINP	419	275/149
Diisodecyl *ortho*-phthalate[a,b]	$C_{28}H_{46}O_4$	DIDP	447	149/289

[a]Restricted use in toys and childcare articles in Europe.
[b]Addressed in the phthalates action plan of the U.S. Environmental Protection Agency.

with the compounds separated by reversed-phase gradient of water to methanol with ammonium acetate buffering on a C18 column. Table 4.3 lists common PFC compounds with their corresponding MRM transitions. The main factor that can affect PFC analysis is that background interferences can come from the HPLC system used for the analysis. In order to get around these issues, plastic tubing can be replaced by polyetheretherketone (PEEK) or stainless steel tubing and the PTFE frits should be replaced by stainless steel. Finally, a trap column can be used at the most down gradient point before the solvent mixer [52].

4.4 MALICIOUS CONTAMINATION OF FOOD

A relatively recent example of malicious contamination of food for monetary gain is the addition of Sudan dyes to chili powder to enhance the physical characteristics of the spice and gain a larger profit. Sudan dyes are azo-dyes (Figure 4.6) and the

Table 4.3. List of Common PFC Compounds That Can Migrate into Food Samples with Their Corresponding Q1 and Q3 Transitions Used in LC–MS/MS Analysis

PFC	Formula	Abbreviation	Q1 Mass (amu)	Q3 Mass (amu)
Perfluorobutanoic acid	$CF_3(CF_2)_2COOH$	PFBA	213	169
Perfluoropentanoic acid	$CF_3(CF_2)_3COOH$	PFPA	263	219
Perfluoro-*n*-hexanoic acid	$CF_3(CF_2)_4COOH$	PFHxA (C6)	313	269/119
Perfluoro-*n*-heptanoic acid	$CF_3(CF_2)_5COOH$	PFHpA (C7)	363	319/169
Perfluoro-*n*-octanoic acid	$CF_3(CF_2)_6COOH$	PFOA (C8)	413	369/169
Perfluoro-*n*-nonanoic acid	$CF_3(CF_2)_7COOH$	PFNA (C9)	463	419/219
Perfluoro-*n*-decanoic acid	$CF_3(CF_2)_8COOH$	PFDA (C10)	513	469/219
Perfluoro-*n*-undecanoic acid	$CF_3(CF_2)_9COOH$	PFUnDA (C11)	563	519/269
Perfluoro-*n*-dodecanoic acid	$CF_3(CF_2)_{10}COOH$	PFDoDA (C12)	613	569/319
Perfluoro-*n*-tridecanoic acid	$CF_3(CF_2)_{11}COOH$	PFTriDA (C13)	663	619/319
Perfluoro-*n*-tetradecanoic acid	$CF_3(CF_2)_{12}COOH$	PFTeDA (C14)	713	669/319
Perfluorobutanesulfonic acid	$CF_3(CF_2)_3SO_3H$	PFBS (S4)	299	80/99
Perfluorohexanesulfonic acid	$CF_3(CF_2)_5SO_3H$	PFHxS (S6)	399	80/99
Perfluoro-*n*-heptanesulfonic acid	$CF_3(CF_2)_6SO_3H$	PFHpS (S7)	449	80/99
Perfluorooctanesulfonic acid	$CF_3(CF_2)_7SO_3H$	PFOS (S8)	499	80/99
Perfluorooctane sulfonamide	$CF_3(CF_2)_6SO_2NH_2$	PFOSA (S8)	498	78/478
N-Methylperfluoro-1-octanesulfonamide	$CF_3(CF_2)_7SO_2NHCH_3$	*N*-MeFOSA	512	169
N-Ethylperfluoro-1-octanesulfonamide	$CF_3(CF_2)_7SO_2NHC_2H_5$	*N*-EtFOSA	526	169
2-(*N*-Methylperfluoro-1-octanesulfonamido)-ethanol	$CF_3(CF_2)_7SO_2N$ $CH_3(C_2H_4OH)$	*N*-MeFOSE	616	59
2-(*N*-Ethylperfluoro-1-octa-nesulfonamido)-ethanol	$CF_3(CF_2)_7SO_2NC_2H_5$ (C_2H_4OH)	*N*-EtFOSE	630	59

International Agency for Research on Cancer (IARC) classified azo-dyes as potential carcinogenic substances. One of the first methods used to detect Sudan dyes was reversed-phase chromatography with APCI ionization in positive mode [55,56]. Most methods used to detect Sudan dyes are based on LC separation, with MS detection recently replacing UV detection due to the decrease in the analysis time. Initial LC–MS/MS methods were based on APCI and used a simple solvent extraction followed by filtration to detect Sudan dyes I–IV in chili powder and tomato-based products and were capable of detecting these dye contaminants at low part per billion levels [55]. These methods screened for only five or fewer dyes by either APCI [55–57] or ESI [58]. However, Sudan dyes are among a series of dyes that are used in textile

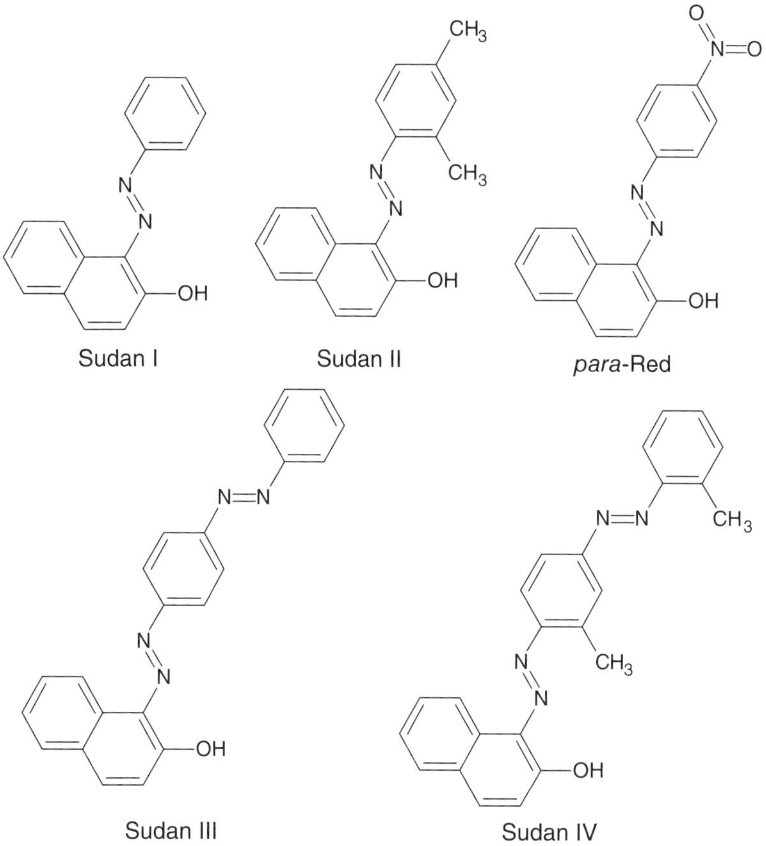

Figure 4.6. Structures for some common Sudan dyes found in food as contaminants.

manufacturing and could illegally be used to improve a food product's appearance, so methods have been expanded to include an increasing number of azo dyes. Either APCI ionization or ESI is used. Table 4.4 shows the LC-MS conditions used for a screen for 13 different dyes by a reversed-phase gradient separation on a C8 column with positive ESI [59].

As previously discussed, melamine became an issue in 2007 due to its illicit use for artificially increasing the nitrogen content in food and feed. The structures of melamine and cyanuric acid are shown in Figure 4.7. Melamine's nitrogen content is higher than that of a typical amino acid, valine, and even higher than one of the nitrogen-rich amino acids, asparagine. Due to this high nitrogen content and wide industrial use, it was ideally suited for this fraudulent activity. Methods for its detection are mainly based around LC–MS/MS, although GC–MS methods have been developed for the analysis of melamine in wheat, rice, and other gluten products. These GC–MS methods require extensive sample cleanup with hazardous solvents and derivatization is needed to give limits of detection typically in the mg/kg

Table 4.4. Examples of Sudan Dyes Found in Food with Their Corresponding Q1 and Q3 Transitions Used in LC–MS/MS Analysis

Sudan Dye	Formula	CAS No.	Q1 Mass (amu)	Q3 Mass (amu)
Dimethyl yellow	$C_{14}H_{15}N_3$	60-11-7	226.1	120.1/105.1
Fast Garner GBC	$C_{14}H_{15}N_3$	97-56-3	226.1	91.1/107.1
Orange II (positive)	$C_{16}H_{12}N_2O_4S$	633-96-5	329.1	156/128
Orange II (positive)	$C_{16}H_{12}N_2O_4S$	633-96-5	327.0	171/80
Para Red	$C_{16}H_{11}N_3O_3$	6410-10-2	294.1	156.1/128.1
Rhodamine B	$C_{28}H_{31}N_2O_3^+$	81-88-9	443.2	399.1/355.1
Sudan I	$C_{16}H_{12}N_2O$	842-07-9	249.1	93/156.1
Sudan II	$C_{18}H_{16}N_2O$	3118-97-6	277.1	121.1/106.1
Sudan III	$C_{22}H_{16}N_4O$	85-86-9	353.1	197.1/128.1
Sudan IV	$C_{24}H_{20}N_4O$	85-83-6	381.1	224.1/225.1
Sudan Orange G	$C_{12}H_{10}N_2O_2$	2051-85-6	215.1	93.1/122.1
Sudan Red 7B	$C_{24}H_{21}N_5$	6368-72-5	380.2	183.1/115.1
Sudan Red B	$C_{24}H_{20}N_4O$	3176-79-2	381.2	224.1/156.1
Sudan Red G	$C_{17}H_{14}N_2O_2$	1229-55-6	279.1	123.1/108.1
D5–Sudan I (internal standard)			254.1	156
D6–Sudan IV (internal standard)			387.1	106

range [60]. In comparison with GC–MS, LC–MS/MS methods have the benefit of reduced sample preparation and run time and still provide lower limits of detection. For example, when analyzing pet food and infant formula by LC–MS/MS, the samples can be easily extracted under acidic conditions before the sample is centrifuged, filtered, and diluted before analysis [61–63].

Melamine and cyanuric acid are small polar compounds and have been separated using a normal phase gradient on a hydrophilic interaction liquid chromatography (HILIC) column [61,63] or by adding heptafluorobutyric acid (HFBA) or trideca-fluoroheptanoic acid (TFHA) and analyzing samples on a C18 column (both ion pair reagents are added to both the sample vial and the mobile phase). Ion pair reagents suppress the response in negative mode, so typically just melamine is detected in this type of method and this approach has been used successfully to test for melamine in milk-based products [62]. When a HILIC separation is used, a period method can be set up, where half the analysis run is dedicated to the detection of cyanuric acid in negative mode and the polarity of the system is changed to positive mode to allow detection of melamine. An example of a chromatogram obtained using HILIC is shown in Figure 4.8. In this figure, both compounds have been detected by ESI, cyanuric acid in negative mode (Q1/Q3 mass transitions of 128/42 and 128/85) and melamine in positive mode (Q1/Q3 mass transitions of 127/60, 127/68, and 127/85) [63].

The case of melamine contamination for commercial gain has started a worrying trend and recently another nitrogen-rich compound, dicyandiamide, has been found in

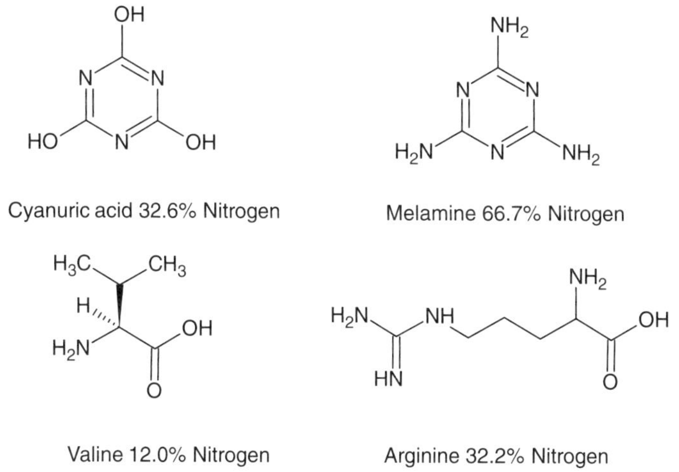

Cyanuric acid 32.6% Nitrogen

Melamine 66.7% Nitrogen

Valine 12.0% Nitrogen

Arginine 32.2% Nitrogen

Figure 4.7. Chemical structures of melamine and cyanuric acid compared with two common amino acids, one of which has the highest nitrogen content available for an amino acid.

milk [64]. In response to the recent issue, methods have been developed again using ESI, but the need for screening for unknown contaminants outside the scope of regulations is becoming a more pressing requirement for authorities. To this end, screening techniques are starting to be used by research organizations and government laboratories to look for new contaminants. When the structures of compounds are totally unknown, multiple analytical techniques are required and these include GC–MS, NMR, IR, and LC–MS/MS. Again, with no prior knowledge of structure, it is difficult to identify compounds. However, accurate mass measurements are important to provide molecular weight information, which together with the structural information of MS/MS spectra and statistical approaches (where control groups are compared with suspect samples) can enable the detection of unknown contaminants. Although this area is relatively new, it has been applied already to pesticide screening and it is starting to be applied to other classes of contaminants in food. Recent publications [65–68] have shown how this approach can help authorities maintain food safety in the future.

In summary, food testing over the last 10–15 years has seen the emergence of LC–MS as a routine technique. Although its use started in the detection of veterinary drug residues and pesticide residues, it is rapidly moving into the field of other chemical contamination detection. The increase in sensitivity of the newer LC–MS instruments means that sample preparations can be simplified, speeding up analyses. The ability to detect compounds that are thermally unstable means that LC–MS is replacing GC–MS in some areas due to speed improvements as a result of removing the need for a derivatization required previously for GC–MS.

Although LC–MS will not replace all GC–MS methods, as some compounds are not detectable by LC–MS at present, the use of accurate mass systems in the future will mean that more contaminants will be grouped together and analyzed simultaneously. An increasing amount of screening for unknown compounds is envisaged to

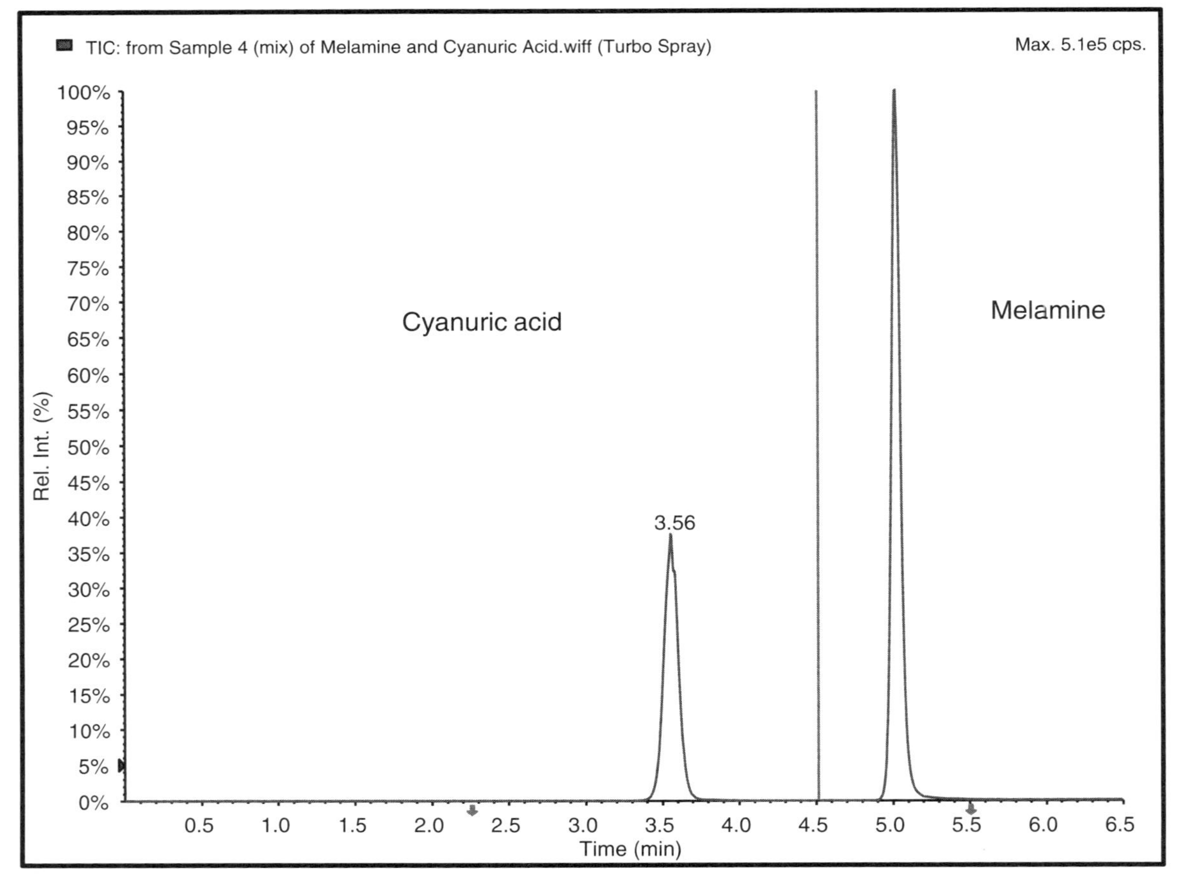

Figure 4.8. Example of a chromatogram obtained from the LC–MS/MS analysis of cyanuric acid and melamine standard.

earlier identify new contamination threats. This will require sensitive systems as sample preparation will have to be simple to enable multiple compound class detection simultaneously (an SPE step may remove the contaminant you are trying to detect), for example, using a simple solvent extraction followed by sample dilution before injection. In the future, sensitive and fast scanning systems with software solutions to enable fast data processing will help generate an increase in the use of accurate mass analysis for food contaminant screening.

REFERENCES

1. Food Standard Agency. Committee on toxicity of chemicals in food, consumer products and the environment. Available at http://cot.food.gov.uk/pdfs/TOX-2003-41.PDF (accessed September 2003).
2. Food Standard Agency. Contaminated spices and palm oils. Available at http://food.gov.uk/business-industry/imports/banned_restricted/spices (accessed December 28, 2011).
3. BBC. Food recalled in cancer dye scare. Available at http://news.bbc.co.uk/1/hi/health/4277677.stm (accessed February 18, 2005).
4. Spice Board India. Circular on sampling & testing of chili/chili products/turmeric powder regarding benefit of reduced tariff for testing. Available at http://www.indianspices.com/html/trade_notifc.html (accessed December 29, 2005).
5. Association of Official Analytical Chemists. Protein (crude) determination in animal feed. Copper Catalyst Kjeldahl Method (984.13), Official Methods of Analysis, 15th edition, **1990**.
6. U.S. Food and Drug Administration (FDA). Melamine Pet Food Recall of 2007. Available at http://www.fda.gov/animalveterinary/safetyhealth/recallswithdrawals/ucm129575.htm (accessed November 29, 2010).
7. USA Today. Melamine found in baby formula made in China. Available at http://usatoday30.usatoday.com/news/health/2008-09-11-tainted-formula_N.htm (accessed September 15, 2008).
8. China Daily. Two get death in tainted milk case. Available at http://www.chinadaily.com.cn/china/2009-01/23/content_7422983.htm (accessed January 23, 2009).
9. The Guardian. Baby milk fears spread across Europe. Available at http://www.guardian.co.uk/world/2005/nov/23/foodanddrink (accessed November 23, 2005).
10. European Food Safety Authority. Opinion of the scientific panel on food additives, flavourings, processing aids and materials in contact with food (AFC) related to 2-isopropyl thioxanthone (ITX) and 2-ethylhexyl-4-dimethylaminobenzoate (EHDAB) in food contact materials. Available at http://www.efsa.europa.eu/en/efsajournal/pub/293.htm (accessed December 9, 2005).
11. Tareke, E.; Rydberg, P.; Karlsson, P.; Eriksson, S.; Törnqvist, M. Analysis of acrylamide, a carcinogen formed in heated foodstuffs. *J. Agric. Food Chem.* **2002**, 50(17), 4998–5006.
12. Friedman, M. Chemistry, biochemistry, and safety of acrylamide: a review. *J. Agric. Food Chem.* **2003**, 51(16), 4504–4526.
13. Andrzejewski, D.; Roach, J. A. G.; Gay, M. L.; Musser, S. M. Analysis of coffee for the presence of acrylamide by LC–MS/MS. *J. Agric. Food Chem.* **2004**, 52, 1996–2002.

14. Hoenicke K.; Gatermann R.; Harder W.; Hartig L. Analysis of acrylamide in different foodstuffs using liquid chromatography–tandem mass spectrometry and gas chromatography–tandem mass spectrometry. *Anal. Chim. Acta* **2004**, 520(1–2), 207–215.
15. World Health Organization. Monitoring of contaminants in food. Available at http://www.who.int/foodsafety/about/Flyer_GEMS.pdf (accessed October 10, 2011).
16. U.S. Food and Drug Administration (FDA). Federal Food, Drug, and Cosmetic Act (FD&C Act), Chapter IV: Food. Available at http://www.fda.gov/RegulatoryInformation/Legislation/FederalFoodDrugandCosmeticActFDCAct/FDCActChapterIVFood/default.htm (accessed August 6, 2012).
17. U.S. Food and Drug Administration (FDA). The New FDA Food Safety Modernization Act (FSMA). Available at http://www.fda.gov/Food/GuidanceRegulation/FSMA/default.htm (accessed August 13, 2013).
18. EUR-Lex Access to European Union Law. Council Regulation (EEC) No. 315/93 of 8 February 1993 laying down community procedures for contaminants in food. Available at http://eur-lex.europa.eu/LexUriServ/LexUriServ.do?uri=OJ:L:1993:037:0001:0003:EN:PDF (accessed February 8, 1993).
19. EUR-Lex Access to European Union Law. REGULATION (EC) No. 1935/2004 of the European Parliament and of the Council of 27 October 2004 on materials and articles intended to come into contact with food and repealing Directives 80/590/EEC and 89/109/EEC. Available at http://eur-lex.europa.eu/LexUriServ/LexUriServ.do?uri=OJ:L:2004:338:0004:0017:en:PDF (accessed November 13, 2004).
20. EUR-Lex Access to European Union Law. Commission Regulation (EU) No. 10/2011 of 14 January 2011. Available at http://eur-lex.europa.eu/LexUriServ/LexUriServ.do?uri=OJ:L:2011:012:0001:0089:EN:PDF (accessed January 15, 2011).
21. EUR-Lex Access to European Union Law. Commission Regulation (EC) No. 450/2009 of 29 May 2009 on active and intelligent materials and articles intended to come into contact with food. Available at http://eur-lex.europa.eu/LexUriServ/LexUriServ.do?uri=OJ:L:2009:135:0003:0011:EN:PDF (accessed May 30, 2009).
22. EUR-Lex Access to European Union Law. Commission Regulation (EC) No. 1881/2006 of 19 December 2006 setting maximum levels for certain contaminants in foodstuffs. Available at http://eurlex.europa.eu/LexUriServ/LexUriServ.do?uri=OJ:L:2006:364:0005:0024:EN:PDF (accessed December 20, 2006).
23. EUR-Lex Access to European Union law. Commission Decision of 20 June 2003 on emergency measures regarding hot chilli and hot chilli products. Available at http://eur-lex.europa.eu/LexUriServ/LexUriServ.do?uri=CONSLEG:2003D0460:20030621:EN:PDF (accessed June 21, 2003).
24. U.S. Food and Drug Administration (FDA). Guidance for industry: action levels for poisonous or deleterious substances in human food and animal feed. Available at http://www.fda.gov/Food/GuidanceRegulation/GuidanceDocumentsRegulatoryInformation/ChemicalContaminantsMetalsNaturalToxinsPesticides/ucm077969.htm (accessed August 15, 2013).
25. U.S. Food and Drug Administration (FDA). Draft guidance for industry: assessing the effects of significant manufacturing process changes, including emerging technologies, on the safety and regulatory status of food ingredients and food contact substances, including food ingredients that are color additives. Program for a Substance Used in a Food Contact Article. Available at http://www.fda.gov/Food/GuidanceRegulation/GuidanceDocuments-RegulatoryInformation/IngredientsAdditivesGRASPackaging/ucm300661.htm (accessed April 29, 2012).

26. Cornell University Law School. 21 CFR Chapter I, Subchapter B: Food for Human Consumption. Available at http://www.law.cornell.edu/cfr/text/21/chapter-I/subchapter-B.
27. Yarnell, A. Acrylamide mystery solved. *Chem. Eng. News* **2002**, 80, 40, 70.
28. Becalski, A.; Lau, B.P.; Lewis, D.; Seaman, S.W. Acrylamide in foods: occurrence, sources and modelling. *J. Agric. Food Chem.* **2003**, 51, 801–808.
29. Biedermann, M.; Biedermann-Brem, S.; Noti, A.; Grob, K.; Egli, P.; Mändli, H. Two GC–MS methods for the analysis of acrylamide in foods. *Mitt. Lebensmittelunters. Hyg.* **2002**, 93(6), 638–652.
30. Rothweiler, B.; Kuhn, E.; Prest, H. GC–MS approaches to the analysis of acrylamide. PittCon 2003 Poster, Agilent Technologies, Publication No. B-0467, **2003**.
31. Schreiber, A. LC–MS/MS analysis of emerging food contaminants: quantitation and identification acrylamide in starch-rich food. ABSCIEX Application Note, Publication No. 5470412-01, **2012**.
32. U.S. Food and Drug Administration (FDA). Detection and quantitation of acrylamide in foods. Available at http://www.fda.gov/Food/FoodborneIllnessContaminants/Chemical-Contaminants/ucm053537.htm (accessed June 18, 2013).
33. Wenzl, T.; Lachenmeier, D.; W; Gökmen, V. Analysis of heat-induced contaminants (acrylamide, chloropropanols and furan) in carbohydrate-rich food. *Anal. Bioanal. Chem.* **2007**, 389, 119–137.
34. Scientific Committee on Food. Opinion of the Scientific Committee on Food on 3-monochloropropane-1, 2-diol (3-MCPD) updating the SCF opinion of 1994. Available at http://ec.europa.eu/food/fs/sc/scf/out91_en.pdf (accessed May 30, 2001).
35. European Committee for Standardization. Foodstuffs: determination of 3-monochloro-propane-1,2-diol by GC/MS. Ref. No. EN 14573:2004: E, **2004**.
36. MacMahon, S.; Begley, T.; Diachenko, G. LC–MS/MS detection of glycidyl esters and 3-MCPD esters in edible oils. AOAC Annual Meeting, Cincinnati, May 3, **2011**.
37. Pinkston, J.D.; Iannelli, D.P.; Mertens, T.R. Indirect method for the determination of 3-MCPD esters: hydrolysis time and recovery considerations for the acid hydrolysis method. AOAC Annual Meeting, Cincinnati, May **2012**.
38. Bordajandi, L. R. 3-Monochloropropane-1,2 diol esters (3-MCPD). Available at http://www.efsa.europa.eu/en/topics/topic/monochloropropane.htm (accessed September 6, 2011).
39. Kim, T.-K.; Kim, S.; Lee, K-G. Analysis of furan in heat-processed foods consumed in Korea using solid phase microextraction–gas chromatography/mass spectrometry (SPME–GC/MS). *Food Chem.* **2010**, 123(4), 1328–1333.
40. Bianchi, F.; Careri, M.; Mangia, A.; Musci, M. Development and validation of a solid phase micro-extraction–gas chromatography–mass spectrometry method for the determination of furan in baby-food. *J. Chromatogr. A* **2006**, 1102(1–2), 268–272.
41. Altaki, M.S.; Santos, F.J.; Galceran, M.T. Analysis of furan in foods by headspace solid-phase microextraction–gas chromatography–ion trap mass spectrometry. *J. Chromatogr. A* **2007**, 1146(1), 103–109.
42. Bonoral A.; Garbini, D.; Lorenzini, R.; Barbanera, M.; Mazzini, C.; Bonaga, G.; Rodriguez-Estrada, M.T. QuEChERS extraction and LC–MS/MS analysis for bisphenol A (BPA), bisphenol A diglycidil ether (BADGE) and its regulated derivatives: food simulants vs. food matrixes. AOAC Europe/ASFILAB International Workshop, Paris, November **2009**.

43. Busset. C.; Lock, S.J. Quantitative analysis and identification of migrants in food packaging using LC–MS/MS/MS. ABSCIEX Application Note, Publication No. 0460310-01, **2010**.
44. Hotchkiss, J.H.; Libbey, L.M.; Barbour, J.F.; Scanlan, R.A. Combination of a GC–TEA and a GC–MS-data system for the microgram/kg estimation and confirmation of volatile *N*-nitrosamines in foods. *IARC Sci. Publ.* **1980**, 31, 361–376.
45. Donegan, M.; Sawyers, W.; Stanislaw, E. Determination of *N*-nitrosamines in baby bottle rubber teats by liquid chromatography–atmospheric pressure chemical ionisation mass spectrometry. ABSCIEX Application Note, No. 8, **2003**.
46. Cao, X.-L. Phthalate esters in foods: sources occurrence, and analytical methods. *Compre. Rev. Food Sci. Food Saf.* **2010**, 9, 21–43.
47. Wenzl, T. *Methods for the Determination of Phthalates in Food: Outcome of a Survey Conducted among European Food Control Laboratories*, JRC Scientific and Technical Reports. Belgium: JRC European Commission; **2009**.
48. Chan, J. Rapid sensitive and robust detection of phthalates in food using GC/MS or LC–MS/MS. Agilent Technologies Application Note, Publication No. 5990–9510EN, **2012**.
49. Schreiber, A.; Fu, F.; Yang, O.; Eric W, E.; Gu, L.; LeBlanc, Y. Increasing selectivity and confidence in detection when analyzing phthalates by LC–MS/MS. ABSCIEX Application Note, Publication No. 3690411-01, **2011**.
50. Hrádková, P.; Poustka, J.; Hloušková, V.; Pulkrabová, J.; Tomaniová, M.; Hajšlová, J. Perfluorinated compounds: occurrence of emerging food contaminants in canned fish and seafood products. *Czech J. Food Sci.* **2010**, 28(4), 333–342.
51. Begley, T.H.; White, K.; Honigfort, P.; Twaroski, R.A.; Neches, R.; Walker, R.A. Perfluorochemicals: potential sources of and migration from food packaging. *Food Addit. Contam.* **2005**, 22(10), 1023–1031.
52. United States Office of Water Environmental Protection Agency (EPA). Draft procedure for analysis of perfluorinated carboxylic acids and sulfonic acids in sewage sludge and biosolids by HPLC–MS/MS/MS, **2011**.
53. United States Department of Agriculture Food Safety and Inspection Service, Office of Public Health Science. Determination and confirmation of PFOA and PFOS by UPLC–MS/MS/MS. Method CLG-PFC1.01, **2009**.
54. Payne, T. Rapid identification and quantitation of perfluorinated alkyl acids on the Varian 320-MS triple quadrupole LC–MS/MS. Varian Application Note 01692, **2008**.
55. Baynham, M.; Lock S. J. Detection of Sudan dyes in tomato paste and chili powder by LC–MS/MS/MS. Applied Biosystems Application Note 15, **2004**.
56. Tatoe, F.; Bononi, M. Fast determination of Sudan I by HPLC/APCI–MS in hot chili, spices, and oven-baked foods. *J. Agric. Food Chem.* **2004**, 52, 655–658.
57. Fang, Y. A rapid and sensitive analysis method for Sudan reds in curry and chili powder using LC–MS/MS/MS. Agilent Technologies Application Note, **2008**.
58. Botek, P.; Poustka, J.; Hajšlová J. Determination of banned dyes in spices by liquid chromatography–mass spectrometry. *Czech J. Food Sci.* **2007**, 25, 17–24.
59. Schreiber, A.; von Czapiewski, K.; Lock, S.J. Quantitation and confirmation of 13 azo-dyes in spices using the 3200 QTRAP® LC–MS/MS system. ABSCIEX Application Note, Publication No. 114AP69-01, **2004**.

60. Tittlemier, S. Background paper on methods for the analysis of melamine and related compounds in foods and animal feeds. World Health Organization, **2009**.
61. Turnipseed, S.; Casey, C.; Nochetto, C.; Heller, D. N. LIB 4421 Melamine and cyanuric acid residues in infant formula. Laboratory Information Bulletin LIB No. 4421. Available at http://www.fda.gov/Food/FoodScienceResearch/LaboratoryMethods/ucm071637.htm (accessed May 8, 2013).
62. Ibáñez, M.; Sancho J. V.; Hernández, F. Determination of melamine in milk-based products and other food and beverage products by ion-pair liquid chromatography–tandem mass spectrometry. *Anal. Chim. Acta* **2009**, 649, 91–97.
63. Sakuma, T.; Taylor, A.S.; Schreiber, A. A new, fast and sensitive LC–MS/MS method for the accurate quantitation and identification of melamine and cyanuric acid in pet food samples. ABSCIEX Application Note, Publication No. 1283110-01, **2010**.
64. Fu, F.; Schreiber, A. LC–MS/MS analysis of emerging food contaminants. ABSCIEX Application Note, Publication No. 7170213-01, **2013**.
65. Schreiber, A; Zou, Y. Y. LC–MS/MS based strategy for the non-targeted screening of an unlimited number of contaminants in food using the ABSCIEX TripleTOF™ 5600 System and advanced software tools. ABSCIEX Application Note, Publication No. 3690611-01, **2011**.
66. Filigenzia, M.S.; Ehrkea, N.; Astona, L.S.; Poppengaa, R.H. Evaluation of a rapid screening method for chemical contaminants of concern in four food-related matrices using QuEChERS extraction, UHPLC and high resolution mass spectrometry. *Food Addit. Contam. A* **2011**, 28(10), 1324–1339.
67. Oosterink, E.; Driessen, W.; Zuidema, T.; Pikkemaat, M.; Stolker, L. Identification of unknown microbial growth inhibitors in animal feed by LC–TOF–MS with accurate mass database searching. 5th International Symposium on Recent Advances in Food Analysis (RAFA 2011), Prague, Czech Republic, Europe, November 1–4, **2011**.
68. Driffield, M.; Lloyd, A.; Morphet, J.; Gay, M.; Gledhill, A. Food packaging migration: direct injection (ASAP) and LC analyses using QTOF MS. 5th International Symposium on Recent Advances in Food Analysis (RAFA 2011), Prague, Czech Republic, Europe, November 1–4, **2011**.

CHAPTER

5

MULTIRESIDUAL DETERMINATION OF 295 PESTICIDES AND CHEMICAL POLLUTANTS IN ANIMAL FAT BY GEL PERMEATION CHROMATOGRAPHY (GPC) CLEANUP COUPLED WITH GC–MS/MS, GC–NCI-MS, AND LC–MS/MS

YAN-ZHONG CAO, YONG-MING LIU, NA WANG, XIN-XIN JI, CUI-CUI YAO, XIANG LI, LI-LI SHI, QIAO-YING CHANG, CHUN-LIN FAN, and GUO-FANG PANG

5.1 INTRODUCTION

With the improvement in people's living standards, environmental protection and food safety have become a growing concern in the world. Environmental pollutant residues in cereals, vegetables, fruits, livestock products, aquatic products, soil, and water cause potential hazards to human health and safety, and they have directly affected economic development and social stability; thus, environmental pollutant residues are one of the important issues in food safety.

Environmental pollutants usually refer to the substances that can change the normal composition of the environment and can be directly or indirectly detrimental to growth, development, and reproduction of the living species. Long-term hazards of environmental pollutants to humans are mainly carcinogenic, teratogenic, and mutagenic effects. Data show that less than 5% of human cancers are caused by biological factors of the virus, less than 5% of human cancers are caused by physical factors of radiation, and about 90% of human cancers are caused by chemical substances—a considerable number of carcinogenic chemicals are environmental pollutants.

This chapter reports the determination of 295 pesticides and chemical pollutants (209 polychlorinated biphenyls (PCBs), 15 polycyclic aromatic hydrocarbons (PAHs), 3 phthalate esters (PAEs), and 68 pesticides) in animal fat by GC–MS/MS (gas chromatography–tandem mass spectrometry), GC–NCI-MS (gas chromatography–negative chemical ionization-mass spectrometry), and LC–MS/MS (liquid chromatography–tandem mass spectrometry). In the proposed method, fat samples are extracted with acetonitrile followed by gel permeation chromatography (GPC) cleanup, and 23–60 min fraction is collected and online concentrated to 1 ml for instrument analysis. Among the 295 compounds, 209 PCBs, 15 PAHs, 3 PAEs, and 57 pesticides are analyzed by GC–MS/MS, endosulfan I and endosulfan II are

High-Throughput Analysis for Food Safety, First Edition.
Edited by Perry G. Wang, Mark F. Vitha, and Jack F. Kay.

determined by GC–NCI-MS, and 9 pesticides, including trichlorphon, metsulfuron-methyl, chlortoluron, 2,4-D, bensulfuron-methyl, propanil, fipronil, phoxim, and hexythiazox, are determined by LC–MS/MS. The limits of detection (LODs) are between 0.1 and 233.0 μg/kg; the LODs of 275 compounds are lower than 10 μg/kg, accounting for 93.2% of the total; the recoveries of 259 compounds are between 60 and 120%, accounting for 87.8% of the total. The relative standard deviations (RSDs) of 278 compounds are below 20%, accounting for 94.2% of the total. The proposed method is applied to 633 human adipose tissue samples, in which 70 compounds (80 compounds if isomers are included) are detected. The results show that the proposed method is suitable for the qualitative and quantitative determination of 295 pesticides and chemical pollutants in animal fat.

5.1.1 Persistent Organic Pollutants

In May 1995, the United Nations Environment Programme (UNEP) Council proposed the first batch of 12 persistent organic pollutants (POPs), including 8 insecticides (aldrin, chlordane, DDT, dieldrin, endrin, heptachlor, mirex, and toxaphene), PCBs, hexachlorobenzene (HCB), polychlorinated dibenzo-*p*-dioxins, and polychlorinated dibenzofurans. The POPs were defined as belonging to a group of organic compounds that are toxic, persistent, easily gathered within the organism, easily transported over long distances, and damaging to the environment and human beings on or near the source [1]. Environmental study showed that [2] some species of wildlife were adversely affected by POPs in their food sources, including effects such as cancer, bone disease, and reproductive failure. GC, LC, GC–MS, GC–MS/MS, and LC–MS/MS were mainly applied in the detection of POPs and other pesticides [3–6]. Many extraction methods can be applied for POPs and other pesticides, including liquid–liquid extraction (LLE), solid–liquid extraction (SLE), Soxhlet extraction, accelerated solvent extraction (ASE), microwave-assisted extraction (MAE), QuEChERS, and matrix solid-phase dispersion (MSPD) [7–12]. GPC and solid-phase extraction (SPE) were most frequently applied in the cleanup of POPs and other pesticides [3,9,13]. Botella et al. [5] analyzed and determined 15 organochlorine pesticides in adipose tissue and blood from 200 Southern Spain women by LC and GC. The research found that *p,p*′-DDE was detected in all of the adipose tissue and blood samples, and DDTs and *p,p*′-DDT were the most commonly detected compounds. The paper suggested that women of reproductive age from Southern Spain had been exposed to organochlorine pesticides. Garrido Frenich et al. [7] established the method for simultaneous analysis of 47 organochlorine and organophosphorus pesticides in muscle of chicken, pork, and lamb. The samples were homogeneously extracted, cleaned up with GPC, and then analyzed by gas chromatography–triple quadrupole mass spectrometry. For most pesticides, recoveries were 70.0–90.0% and RSDs were 15%. LODs were below 2.0 μg/kg for all of the pesticides, except acephate. The extraction efficiencies of Soxhlet extraction and ASE were also compared with homogeneous extraction. Pérez et al. [12] developed the detection method for organophosphates, organochlorines, and pyrethrins in lanolin based on MSPD. Three analytical methods—that is, GC–FPD (flame photometric detection),

GC–ECD (electron capture detection), GC–MS/SIM (selected ion monitoring)—were used for the analysis of different pesticides. For most pesticides, recoveries were 83–118% with RSDs <20%, which were in accord with the requirements of European and U.S. pharmacopeias.

5.1.2 Polycyclic Aromatic Hydrocarbons

The most prominent characteristics of PAHs are carcinogenic, teratogenic, and mutagenic effects, and the carcinogenic effects increase with the increase in the number of benzene rings [14]. When PAHs react with $-NO_2$, $-OH$, and $-NH_2$, more carcinogenic PAH derivatives are generated [15]. Currently, PAHs have been the important content of environment monitoring in most nations, and the U.S. Environmental Department defines 16 PAHs as priority monitoring pollutants (EPA-PAHs for short): naphthalene, acenaphthene, acenaphthylene, fluorene, phenanthrene, anthracene, fluoranthene, pyrene, benzo(*a*)anthracene, chrysene, benzo(*b*)fluoranthene, benzo(*k*)fluoranthene, benzo(*a*)pyrene, indeno(1,2,3-*cd*)pyrene, dibenzo(*a*,*h*)anthracene, and benzo(*g*,*h*,*i*)perylene [16]. In 1933, it was confirmed that the carcinogens in soot were PAHs, especially benzo(*a*)pyrene [17]. Soxhlet extraction [18], ASE [19,20], ultrasonic extraction [21], MAE [22], solid-phase microextraction (SPME) [23], and other techniques had been applied in the extraction of PAHs in meat and fat. GPC [19,20,24] and SPE [18,22] were most frequently used for the cleanup, and GC–MS [20,23] and LC (combined with different detection methods) [22,25] were the analysis methods. Purcaroa et al. [22] developed a rapid extraction method based on MAE to analyze PAHs in smoked meat. After extraction, PAHs were cleaned up by SPE and analyzed by RP-HPLC and fluorescence detection. The extraction efficiency of MAE was better than that of solvent extraction assisted by sonication. LODs of the method were <0.4 μg/kg and recoveries were 77–103%. Jira et al. [20] determined 15 PAHs in smoked meat and edible oils by GC–MS. The sample preparation approaches included ASE extraction, GPC, and SPE cleanup. GC–HRMS (high-resolution mass spectroscopy) and GC–MSD (mass-selective detection) had comparable results in the research. The results showed that the concentrations of 16 PAHs in 22 smoked meats were in the range of 0.01–19 μg/kg.

5.1.3 Polychlorinated Biphenyls

PCBs are a group of chlorinated aromatic hydrocarbons with extremely stable chemical properties. As they resist degradation, they can be directly harmful to human beings by enrichment in the food chain, and they have become a part of global important pollutants [26]. PCBs can be found from the Arctic seals to the Antarctic seabird eggs [27]. Their toxicity can cause body acne, liver damage, and carcinogenic effects [28]. They are also “environmental hormones” that interfere with the body’s endocrine systems, and can bring serious problems and diseases in the reproductive system of humans and animals [29]. PCBs have been included in the “blacklist” by the United States Environmental Protection Agency (U.S. EPA), in which organic pollutants have priority to be detected. The congeners of PCBs constantly migrate in soil, water, air, and other environmental matrices and eventually bioaccumulate [30]. The analyses of PCBs in fat and tissues were mainly based on GC–ECD [31–34], GC–MS [35,36], and GC–MS/MS [37], and immunoassay had also been applied to assay PCBs in animal fat [38]. The sample preparation methods were SPE [39], ASE [40], SFE (supercritical fluid extraction) [41], MSPD [36], and other techniques. Bordet et al. [39] reported an interlaboratory study on the determination of 21 organochlorines, 6 pyrethroid pesticides, and 7 PCBs in milk, fish, eggs, and beef fat by GC. SPE with C18 and Florisil cartridges was used for cleanup of the samples after cryogenic extraction. The results showed the method had acceptable intra- and interlaboratory precision. Zhang et al. [40] developed a simultaneous extraction and cleanup method for the determination of polybrominated diphenyl ethers (PBDEs) and PCBs in sheep liver tissue by selective pressurized liquid extraction (SPLE). GC–MS was used to analyze PBDEs and PCBs. Related factors in extraction efficiency were optimized, for example, extraction solvent, temperature, pressure, and so on. The method developed compared favorably with traditional extraction methods (e.g., Soxhlet extraction, off-line pressurized liquid extraction (PLE), and ultrasonic and heating extraction methods).

5.1.4 Phthalate Esters

PAEs are a class of environmental estrogens. In 1995, the World Health Organization (WHO) promulgated chemicals that can disrupt human endocrine function and must be controlled, and PAEs were in the list. Six PAEs are included in the “list of priority monitoring pollutants” by the U.S. EPA: dimethyl phthalate (DMP), diethyl phthalate (DEP), dibutyl phthalate (DBP), phthalate bis-2-ethylhexyl ester (DEHP), dioctyl phthalate (DOP), and butyl benzyl phthalate (BBP) [42]. Many studies have been reported on the analysis of PAEs in water and other environmental samples. The most frequently applied methods were GC and GC-MS [43–46], and the sample preparation methods were LLE, SPE, SPME, and others [47–50]. Prieto et al. [49] established a method based on stir bar sorptive extraction (SBSE) and GC–MS to analyze several environmental pollutants in water samples. Sixteen PAHs, 12 PCBs, 6 PAEs, and 3 nonylphenols (NPs) could be determined simultaneously with LODs of 0.1–10 ng/l. Lin et al. [47] assayed 14 PAEs in six types of animal viscus by ultrasonic extraction and Florisil SPE cleanup. GC–EI-MS–SIM was used to determine the PAE residues. LODs of 12 PAEs were <1.74 μg/kg, and LODs of dimethyl glycol phthalate and bis(2-ethoxyethyl) phthalate were 3.30 and 2.25 μg/kg, respectively.

5.1.5 Multiclass and Multiresidue Analyses

We have been continuously committed to multiclass, multiresidue, and high-throughput analyses of pesticides and chemical pollutants in complex matrices, including animal tissues. As early as 1994, we established the multiresidue analytical method for nine pyrethroids in chicken, beef, mutton, and pork. The samples were cleaned up by two Florisil columns after solvent extraction, and nine pyrethroids were analyzed by GC. LODs were 5 μg/kg for all insecticides (except permethrin, for which LOD was 10 μg/kg), and the recoveries were between 76.9 and 88.0% [51]. In 2006, we established the multiresidue analysis method for 437 pesticides in animal tissues. The analytes included different types of pesticides and chemical pollutants, for example, organochlorines, organophosphorus, pyrethroids, carbamates, herbicides, PCBs, PAEs, and others. In the method, 10 g animal samples (beef, mutton, pork, and chicken) were mixed with 20 g sodium sulfate and extracted with 35 ml of cyclohexane + ethyl acetate (1 + 1) twice by blender homogenization, centrifugation, and filtration. Evaporation was conducted and an equivalent of 5 g sample was injected into a 400 mm × 25 mm S-X3 GPC column, with cyclohexane + ethyl acetate (1 + 1) as the mobile phase at a flow rate of 5 ml/min. The 22–40 min fraction was collected for subsequent analysis. For the 368 pesticides determined by GC–MS, the portions collected from GPC were concentrated to 0.5 ml and exchanged with 5 ml hexane twice. For the 69 pesticides determined by LC–MS/MS, the portions collected from GPC were dissolved with acetonitrile + water (60 + 40) after taking the extract to dryness with nitrogen gas. In the linear range of each pesticide, the correlation coefficient was $r \geq 0.98$, exceptions being dinobuton, linuron, and fenamiphos sulfoxide. At the three (low, medium, and high) fortification levels of 0.2–4800 g/kg, recoveries fell within 40–120%. The RSDs were below 28% for all 437 pesticides. The LODs for the method were 0.2–600 g/kg, depending on each pesticide. The LOD and LOQ (limit of quantification) of the method were obtained with fortified pesticides of different concentrations, and a S/N ratio of ≥5 was the criterion for the LOD, whereas a S/N ratio of ≥10 was the criterion for the LOQ [52]. In 2009, we further studied different types of pesticides and chemical pollutants, and the number of analytes reached 839. We also established the database of chemical pollutants, including a GPC database for 744 pesticides, a GC–MS database for 541 pesticides, and a LC–MS/MS database for 464 pesticides. The LODs of the analytes were between 0.1 and 1600 μg/kg [53]. We have also established high-throughput analysis methods in other complex matrices, including honey, cereal, fruit, vegetables, tea, Chinese herbal medicine, milk and milk powder, and others. Hundreds of pesticides and other chemical pollutants can be analyzed by high-throughput methods with simple and convenient sample preparation and rapid instrument analysis.

For animal tissues, although the varieties of the pesticides and chemical pollutants in our former research had reached nearly a thousand, they were mainly pesticides, and the matrices were animal muscles. But multiclass environmental pollutants with different chemical and physical properties may exist in animal fats, and comprehensive and universal analytical methods for detecting them have not been established heretofore. Therefore, it is important to establish a rapid and multiresidue monitoring technology for representative environmental pollutants, that is, POPs and chemical pesticides, PAHs, PCBs, and PAEs. This study established the qualitative and quantitative determination methods for 209 PCBs, 15 PAHs, 3 PAEs, and 68 pesticides in animal fat. The samples were extracted with acetonitrile followed by GPC cleanup and online concentration. Among the 295 compounds, 209 PCBs,

15 PAHs, 3 PAEs, and 57 pesticides are analyzed by GC–MS/MS, endosulfan I and endosulfan II are determined by GC–NCI-MS, and 9 pesticides, including trichlorphon, metsulfuron-methyl, chlortoluron, 2,4-D, bensulfuron-methyl, propanil, fipronil, phoxim, and hexythiazox, are determined by LC–MS/MS. The LODs of the analytes were between 0.1 and 233.0 μg/kg; LODs of 275 compounds were lower than 10 μg/kg, accounting for 93.2% of the total; recoveries of 259 compounds were between 60 and 120%, accounting for 87.8% of the total. The RSDs of 278 compounds were below 20%, accounting for 94.2% of the total. The research had also analyzed 633 human adipose tissue samples, in which 81 compounds were detected. The results showed that the proposed method was suitable for the qualitative and quantitative determination of 295 pesticides and chemical pollutants in animal fat.

5.2 EXPERIMENT

5.2.1 Instruments

In this work, we used GC–MS (Quattro micro GC), equipped with an electron impact (EI) source (Waters, USA); GC (6890N)–MS (5973N), equipped with an NCI source (Agilent, USA); LC–MS/MS (3200 Q TRAP), equipped with an electrospray ionization (ESI) source (Applied Biosystems, USA); ASE 300 (Dionex, USA); GPC (AccuPrep MPS), equipped with BIO-Beads S-X3, 360 mm × 25 mm column (J2 Scientific, USA); rotary evaporator (R-205, Büchi, Switzerland), equipped with a BP-51 vacuum cooling system (Yamato, Japan); T25 homogenizer (IKA, Germany); N-EVAP112 nitrogen concentrator (Orgamonation Associates, USA); SA 300 oscillator (Yamato, Japan); and KDC-40 low-speed centrifuge (USTC Chuangxin Co. Ltd.).

5.2.2 Reagents

We used acetonitrile, acetone, *n*-hexane, cyclohexane, isooctane, ethyl acetate, dichloromethane, toluene, methanol (pesticide residue grade) (Dikma, China), and anhydrous sodium sulfate (analytical grade), burned at 650 °C for 4 h and stored in a desiccator. Deionized water was obtained from a Milli-Q system (Millipore, USA); Sep-Pak® NH_2-Carb cartridge (6 ml, 500 mg; Waters, USA); Sep-Pak® Alumina N cartridge (12 ml, 2 g; Water, USA); ENVI™-18 cartridge (12 ml, 2 g; Supelco, USA); membrane (0.45 and 0.2 μm); standards of individual environmental pollutant (purities ≥95%, LGC Promochem GmbH, Germany).

5.2.3 Preparation of Standard Solutions

Individual standard stock solutions: Weigh 5–10 mg (accurate to 0.1 mg) standard into a 10 ml volumetric flask and dissolve with toluene, *n*-hexane, acetone, methanol, isooctane, and so on to the volume based on the solubility of the standard. The standard solutions are stored in the dark at 0–4 °C.

Mixed standard solutions: The concentration of mixed standard solutions depends upon the sensitivity of the method for each compound. Pipette an adequate amount of individual stock standard solution into a 50 ml volumetric flask and dilute to volume with toluene or *n*-hexane. Mixed standard solutions should be stored in dark below 4 °C and can be used for 1 month.

The internal standard solution was prepared by accurately weighing 3.5 mg heptachlor epoxide into a 10 ml volumetric flask and dissolving and diluting with toluene.

Matrix mixed standard solutions were prepared by diluting 40 µl internal standard solution and an appropriate amount of mixed standard solution to 1.0 ml with blank extract, which had been taken through the method with the rest of the samples, and mixing thoroughly. These solutions were used to construct calibration plots.

5.2.4 Sample Preparation

Extraction: Weigh 5 g fat sample (from pork or human body; accurate to 0.01 g) into an 80 ml centrifuge tube containing 15 g anhydrous sodium sulfate (to absorb the moisture in fat) and add 35 ml acetonitrile, then homogenize at 15,000 rpm for 1 min, and centrifuge at 4200 rpm for 5 min.

The supernatants are made to pass through a glass funnel containing a glass wool plug and ~15 g anhydrous sodium sulfate and collected in a 100 ml pear-shaped flask. Repeat extracting the dregs with 35 ml acetonitrile one time, centrifuge, consolidate the extractions of over two times, and rotary evaporate in water bath at 45 °C until about 2 ml remain. Then two separate additions of 7 ml ethyl acetate–cyclohexane (1 + 1) are made for solvent exchange and the residue is evaporated to 1 ml for cleanup.

Cleanup: 40 µl internal standard solution is added to the 1 ml concentrate and mixed. The mixture is transferred to a 10 ml colorimetric tube. The 100 ml pear-shaped flask is rinsed with 8 ml ethyl acetate–cyclohexane (1 + 1, v/v) (8 ml solvent is added in three separate times), and the solvent is transferred to the colorimetric tube. The colorimetric tube is diluted with ethyl acetate to the volume. After membrane filtration (0.45 µm) to a 10 ml cuvette, the solution is cleaned by GPC, and the 23–60 min elution fraction is collected. The eluate is treated in two different ways for analyzing by different methods: (a) The eluate is automatically concentrated to 1 ml with acetate–cyclohexane (1 + 1) for analyzing by GC–MS/MS and GC–MS. (b) Two separate additions of 5 ml acetonitrile are made for solvent exchange of the eluate until nearly dry. Add 1 ml acetonitrile–water (3 + 2) to the residue and filter with 0.2 µm membrane for analyzing by LC–MS/MS.

Simultaneously, a blank fat sample is extracted by the extraction and cleanup steps above to prepare blank extraction for matrix mixed standard solution preparation.

5.2.5 Analytical Methods

5.2.5.1 *GPC Cleanup*

GPC column: 360 mm × 25 mm, filled with BIO-Beads S-X3; detection wavelength: 254 nm; mobile phase: ethyl acetate–cyclohexane (1 + 1, v/v); flow rate: 5.0 ml/min; injection volume: 5 ml; start collect time: 23 min; stop collect time: 60 min; online concentration temperature and degree of vacuum: zone 1: 45 °C, 33.3 kPa; zone 2: 49 °C, 29.3 kPa; zone 3: 52 °C, 26.6 kPa; termination mode: liquid-level sensor mode; termination temperature and degree of vacuum: zone 1: 51 °C, 26.60 kPa; zone 2: 50 °C, 23.94 kPa.

5.2.5.2 *GC–MS/MS with EI Source*

Column: DB-1701, 30 m × 0.25 mm × 0.25 μm silica capillary column; the oven temperature is held isothermally at 40 °C for 1 min, then increased to 130 °C at 30 °C/min, then increased to 250 °C at 5 °C/min, and finally increased to 300 °C at 10 °C/min for 5 min; carrier gas: helium, purity ≥99.999%; inlet temperature: 200 °C; injection volume: 1 μl; injection mode: splitless injection, purge on after 1.5 min; ionization energy: 70 eV; ion source temperature: 200 °C; interface temperature: 250 °C; solvent delay: 5 min; data acquisition mode: multiple reaction monitoring (MRM); quantifying and qualifying ions and collision energies (CEs) of 284 environmental pollutants are given in Table 5.1.

5.2.5.3 *GC–MS with NCI Source*

Column: DB-1701, 30 m × 0.25 mm × 0.25 μm silica capillary column; the oven temperature is held isothermally at 70 °C for 1 min, then increased to 260 °C at 20 °C/min, held for 5 min; carrier gas: helium, purity ≥99.999%; inlet temperature: 280 °C; injection volume: 1 μl; injection mode: splitless injection, purge on after 0.75 min; ionization mode: NCI, 30 eV; ion source temperature: 230 °C; interface temperature: 280 °C; reaction gas: methane (CH_4); solvent delay: 10 min; data acquisition mode: selective ion monitoring; selective ions and relative abundances of endosulfans are given in Table 5.2.

5.2.5.4 *LC–MS/MS*

Column: Atlantis®T3, 3 μm, 150 mm × 2.1 mm; column temperature: 35 °C; mobile phase: 0.1% formic acid (A), 5 mmol/l ammonium acetate (B), and acetonitrile (C); flow rate: 0.2 ml/min; injection volume: 20 μl; the elution steps of positive ion mode are as follows: 0–3 min isocratic at 90% A, 3–4 min linear gradient from 90 to 20% A, 4–12 min isocratic at 20% A, 12–12.1 min linear gradient from 20 to 90% A, 12.1–20 min isocratic at 90% A; the elution steps of negative ion mode are 0–10 min isocratic at 90% A; ion source: ESI; scan mode: positive/negative ion scan; ion spray voltage: 5500 V; nebulizer gas pressure: 0.076 MPa; curtain gas pressure: 0.069 MPa;

Table 5.1. Names, Retention Times, Quantifying and Qualifying Ions, Collision Energies, Linear Ranges, Linear Equations, and Correlation Coefficients of 284 Environmental Pollutants Determined by GC–MS/MS

No.	Abbreviation	Name	Retention Time (min)	Quantifying Ion	Qualifying Ion	Collision Energy (eV)	Linear Range (μg/l)	Linear Equation	Correlation Coefficient
	–	Heptachlor (ISTD)	22.04	353/263	353/263; 353/282	17; 17			
Group A									
1	PCB 001	2-Chlorobiphenyl	11.04	188/152	188/152; 188/153	20; 10	1.5–120	$y=2.4459x+0.0321$	0.9916
2	PCB 004	2,2′-Dichlorobiphenyl	13.39	152/151	152/151; 152/150	20; 40	1.5–120	$y=0.6702x-0.3194$	0.9968
3	PCB 008	2,4′-Dichlorobiphenyl	14.91	224/152	224/152; 224/151	30; 50	1.5–120	$y=1.3927x-0.7712$	0.9980
4	PCB 019	2,2′,6-Trichlorobiphenyl	15.77	256/221	256/221; 256/186	10; 20	1.5–120	$y=1.2178x-0.5426$	0.9979
5	PCB 012	3,4-Dichlorobiphenyl	16.55	222/152	222/152; 222/151	30; 50	1.5–120	$y=2.1949x-0.6531$	0.9980
6	PCB 027	2,3′,6-Trichlorobiphenyl	16.94	186/151	186/151; 186/150	20; 30	1.5–120	$y=1.0444x-0.0164$	0.9960
7	PCB 016	2,2′,3-Trichlorobiphenyl	17.40	256/186	256/186; 256/221	20; 10	1.5–120	$y=1.2102x-0.1932$	0.9952
8	PCB 025	2,3′,4-Trichlorobiphenyl	17.93	256/186	256/186; 256/151	30; 40	1.5–120	$y=2.2189x-1.1486$	0.9985
9	PCB 021	2,3,4-Trichlorobiphenyl	18.60	256/186	256/186; 186/151	20; 20	1.5–120	$y=1.7531x-0.3355$	0.9996
10	PCB 020	2,3,3′-Trichlorobiphenyl	18.81	186/151	186/151; 186/150	20; 30	1.5–120	$y=1.0537x-0.3765$	0.9985
11	PCB 036	3,3′,5-Trichlorobiphenyl	19.24	186/151	186/151; 186/150	20; 30	1.5–120	$y=0.8907x+0.0022$	0.9932
12	PCB 043	2,2′,3,5-Tetrachlorobiphenyl	19.57	294/222	294/222; 294/150	30; 50	4.5–120	$y=0.3876x-0.4500$	0.9999
13	PCB 065	2,3,5,6-Tetrachlorobiphenyl	19.67	292/222	292/222; 292/220	20; 20	1.5–120	$y=0.7969x-0.2484$	0.9982
14	PCB 104	2,2′,4,6,6′-Pentachlorobiphenyl	19.84	254/184	254/184; 254/219	30; 20	1.5–120	$y=0.5316x-0.2535$	0.9958
15	PCB 072	2,3′,5,5′-Tetrachlorobiphenyl	20.52	292/220	292/220; 292/150	30; 50	1.5–120	$y=0.8374x-0.3825$	0.9996
16	PCB 103	2,2′,4,5′,6-Pentachlorobiphenyl	20.73	326/256	326/256; 326/184	40; 50	1.5–120	$y=0.5428x+0.1439$	0.9951
17	PCB 041	2,2′,3,4-Tetrachlorobiphenyl	20.93	292/220	292/220; 292/150	30; 50	1.5–120	$y=0.6036x-0.1385$	0.9994
18	PCB 067	2,3′,4,5-Tetrachlorobiphenyl	21.22	292/220	292/220; 292/185	30; 40	1.5–120	$y=0.8405x-0.3066$	0.9990
19	PCB 040	2,2′,3,3′-Tetrachlorobiphenyl	21.43	292/220	292/220; 292/150	30; 50	1.5–120	$y=0.4695x-0.1937$	0.9979
20	PCB 074	2,4,4′,5-Tetrachlorobiphenyl	21.57	290/220	290/220; 290/150	20; 50	1.5–120	$y=0.9181x-0.0497$	0.9989
21	PCB 102	2,2′,4,5,6′-Pentachlorobiphenyl	21.72	254/184	254/184; 254/219	30; 20	1.5–120	$y=0.5123x+0.0418$	0.9982
22	PCB 095	2,2′,3,5′,6-Pentachlorobiphenyl	21.97	254/184	254/184; 254/219	30; 20	1.5–120	$y=0.6398x-0.2194$	0.9963
23	PCB 092	2,2′,3,5,5′-Pentachlorobiphenyl	22.46	184/149	184/149; 328/256	20; 40	2.5–200	$y=0.2195x+0.0848$	0.9953
24	PCB 099	2,2′,4,4′,5-Pentachlorobiphenyl	22.77	326/184	326/184; 326/256	50; 50	1.5–120	$y=0.3347x-0.0855$	0.9994
25	PCB 084	2,2′,3,3′,6-Pentachlorobiphenyl	22.92	254/184	254/184; 254/219	30; 20	1.5–120	$y=0.4903x+0.1542$	0.9904
26	PCB 109	2,3,3′,4,6-Pentachlorobiphenyl	23.24	326/184	326/184; 326/256	50; 40	1.5–120	$y=0.4238x-0.0125$	0.9946
27	PCB 083	2,2′,3,3′,5-Pentachlorobiphenyl	23.42	184/149	184/149; 184/123	20; 30	2.5–200	$y=0.2220x-0.0859$	0.9989

(continued)

Table 5.1 (*Continued*)

No.	Abbreviation	Name	Retention Time (min)	Quantifying Ion	Qualifying Ion	Collision Energy (eV)	Linear Range (μg/l)	Linear Equation	Correlation Coefficient
28	PCB 086	2,2′,3,4,5-Pentachlorobiphenyl	23.53	326/291	326/291; 326/256	10; 20	1.5–120	$y = 0.5546x - 0.2389$	0.9998
29	PCB 125	2′,3,4,5,6′-Pentachlorobiphenyl	23.66	254/184	254/184; 254/219	20; 20	1.5–120	$y = 0.5407x - 0.0427$	0.9997
30	PCB 087	2,2′,3,4,5′-Pentachlorobiphenyl	23.87	328/256	328/256; 328/258	30; 30	1.5–120	$y = 0.3698x - 0.2884$	0.9998
31	PCB 110	2,3,3′,4′,6-Pentachlorobiphenyl	24.29	324/254	324/254; 324/184	30; 50	1.5–120	$y = 0.7662x - 0.9562$	0.9930
32	PCB 135	2,2′,3,3′,5,6′-Hexachlorobiphenyl	24.58	325/290	325/290; 325/288	10; 10	1.5–120	$y = 0.4681x - 0.3688$	0.9974
33	PCB 124	2′,3,4,5,5′-Pentachlorobiphenyl	24.83	326/256	326/256; 326/254	30; 30	1.5–120	$y = 0.7963x - 0.2467$	0.9993
34	PCB 123	2′,3,4,4′,5-Pentachlorobiphenyl	24.97	328/256	328/256; 328/258	35; 35	1.5–120	$y = 0.3608x - 0.3634$	0.9988
35	PCB 118	2,3′,4,4′,5-Pentachlorobiphenyl	25.13	326/256	326/256; 326/254	30; 30	1.5–120	$y = 0.6858x - 0.2333$	0.9978
36	PCB 134	2,2′,3,3′,5,6-Hexachlorobiphenyl	25.31	325/290	325/290; 325/288	10; 10	1.5–120	$y = 0.4407x - 0.2854$	0.9990
37	PCB 114	2,3,4,4′,5-Pentachlorobiphenyl	25.45	254/184	254/184; 254/219	20; 20	2.5–200	$y = 0.3100x - 0.2190$	0.9993
38	PCB 168	2,3′,4,4′,5′,6-Hexachlorobiphenyl	25.69	358/218	358/218; 358/288	50; 40	1.5–120	$y = 0.1898x - 0.0113$	0.9978
39	PCB 127	3,3′,4,5,5′-Pentachlorobiphenyl	26.27	326/256	326/256; 326/254	20; 20	1.5–120	$y = 0.7606x - 0.0661$	0.9987
40	PCB 137	2,2′,3,4,4′,5-Hexachlorobiphenyl	26.43	362/290	362/290; 362/292	25; 25	1.5–120	$y = 0.4780x - 0.2686$	0.9987
41	PCB 163	2,3,3′,4′,5,6-Hexachlorobiphenyl	26.86	360/290	360/290; 360/288	30; 30	1.5–120	$y = 0.7916x - 0.1005$	0.9991
42	PCB 178	2,2′,3,3′,5,5′,6-Heptachlorobiphenyl	26.93	396/326	396/326; 396/324	30; 30	1.5–120	$y = 0.3163x - 0.0915$	0.9996
43	PCB 187	2,2′,3,4′,5,5′,6-Heptachlorobiphenyl	27.25	396/361	396/361; 396/359	10; 10	1.5–120	$y = 0.3358x - 0.1223$	0.9973
44	PCB 162	2,3,3′,4′,5,5′-Hexachlorobiphenyl	27.70	358/288	358/288; 358/218	30; 50	1.5–120	$y = 0.6615x - 0.2135$	0.9990
45	PCB 202	2,2′,3,3′,5,5′,6,6′-Octachlorobiphenyl	28.07	432/360	432/360; 432/362	30; 30	1.5–120	$y = 0.3761x - 0.1960$	0.9992
46	PCB 204	2,2′,3,4,4′,5,6,6′-Octachlorobiphenyl	28.30	432/360	432/360; 432/362	30; 30	1.5–120	$y = 0.3587x - 0.0924$	0.9990
47	PCB 197	2,2′,3,3′,4,4′,6,6′-Octachlorobiphenyl	28.58	428/358	428/358; 428/356	30; 30	1.5–120	$y = 0.7242x - 0.1286$	0.9988
48	PCB 192	2,3,3′,4,5,5′,6-Heptachlorobiphenyl	28.96	396/324	396/324; 396/326	40; 40	1.5–120	$y = 0.3291x - 0.0827$	0.9991
49	PCB 193	2,3,3′,4′,5,5′,6-Heptachlorobiphenyl	29.31	324/254	324/254; 324/252	30; 30	1.5–120	$y = 0.2592x - 0.0653$	0.9967
50	PCB 190	2,3,3′,4,4′,5,6-Heptachlorobiphenyl	30.19	394/324	394/324; 394/322	20; 20	1.5–120	$y = 0.6880x - 0.0956$	0.9989

51	PCB 169	3,3′,4,4′,5,5′-Hexachlorobiphenyl	30.48	358/288	358/288; 362/290	20; 20	1.5–120	$y = 0.5396x + 0.0405$	0.9964
52	PCB 195	2,2′,3,3′,4,4′,5,6-Octachlorobiphenyl	31.18	428/358	428/358; 428/356	30; 30	1.5–120	$y = 0.5381x - 0.1630$	0.9984
53	PCB 206	2,2′,3,3′,4,4′,5,5′,6-Nonachlorobiphenyl	32.44	466/394	466/394; 466/396	40; 40	1.5–120	$y = 0.2142x - 0.0274$	0.9980
54	PCB 209	2,2′,3,3′,4,4′,5,5′,6,6′-Decachlorobiphenyl	32.78	500/429	500/429; 500/428	30; 30	1.5–120	$y = 0.3092x - 0.1418$	0.9999
Group B									
55	PCB 002	3-Chlorobiphenyl	12.46	188/152	188/152; 188/151	30; 50	1.5–120	$y = 1.8077x - 0.1286$	0.9986
56	PCB 007	2,4-Dichlorobiphenyl	14.07	224/152	224/152; 152/151	10; 20	1.5–120	$y = 1.4773x + 0.4931$	0.9974
57	PCB 005	2,3-Dichlorobiphenyl	14.97	222/152	222/152; 152/151	10; 20	1.5–120	$y = 1.7668x - 0.0261$	0.9981
58	PCB 011	3,3′-Dichlorobiphenyl	16.34	224/152	224/152; 224/151	20; 50	1.5–120	$y = 1.3893x + 0.1723$	0.9990
59	PCB 013	3,4′-Dichlorobiphenyl	16.63	152/151	152/151; 152/150	20; 40	2.5–200	$y = 0.5225x + 3.6859$	0.9879
60	PCB 032	2,4′,6-Trichlorobiphenyl	17.23	256/186	256/186; 256/151	30; 40	1.5–120	$y = 1.4701x + 0.2391$	0.9987
61	PCB 029	2,4,5-Trichlorobiphenyl	17.42	256/151	256/151; 256/150	40; 50	1.5–120	$y = 0.4704x - 0.0694$	0.9962
62	PCB 050	2,2′,4,6-Tetrachlorobiphenyl	17.94	292/220	292/220; 292/222	40; 40	1.5–120	$y = 0.4179x + 0.0127$	0.9995
63	PCB 053	2,2′,5,6′-Tetrachlorobiphenyl	18.68	292/150	292/150; 292/220	50; 50	1.5–120	$y = 0.3821x + 0.0380$	0.9972
64	PCB 022	2,3,4′-Trichlorobiphenyl	19.11	256/186	256/186; 256/151	20; 40	1.5–120	$y = 1.8027x + 0.3811$	0.9973
65	PCB 073	2,3′,5′,6-Tetrachlorobiphenyl	19.44	290/220	290/220; 290/150	30; 50	1.5–120	$y = 1.0349x + 0.2342$	0.9975
66	PCB 039	3,4′,5-Trichlorobiphenyl	19.61	258/151	258/151; 258/186	50; 40	1.5–120	$y = 0.5034x + 0.0375$	0.9977
67	PCB 062	2,3,4,6-Tetrachlorobiphenyl	19.68	292/222	292/222; 292/150	20; 50	1.5–120	$y = 0.8622x + 0.0859$	0.9989
68	PCB 038	3,4,5-Trichlorobiphenyl	19.93	258/151	258/151; 258/186	30; 50	2.5–200	$y = 0.4391x + 0.2602$	0.9963
69	PCB 035	3,3′,4-Trichlorobiphenyl	20.55	186/151	186/151; 186/150	20; 30	1.5–120	$y = 0.8346x - 0.3293$	0.9992
70	PCB 064	2,3,4′,6-Tetrachlorobiphenyl	20.83	220/150	220/150; 294/222	30; 20	1.5–120	$y = 0.6745x + 0.4506$	0.9833
71	PCB 037	3,4,4′-Trichlorobiphenyl	20.97	186/151	186/151; 186/150	20; 30	1.5–120	$y = 0.8905x + 0.0762$	0.9985
72	PCB 080	3,3′,5,5′-Tetrachlorobiphenyl	21.39	292/220	292/220; 292/150	30; 50	1.5–120	$y = 0.8679x + 0.0574$	0.9991
73	PCB 058	2,3,3′,5′-Tetrachlorobiphenyl	21.45	292/222	292/222; 292/220	20; 20	1.5–120	$y = 0.8447x - 0.0013$	0.9988
74	PCB 121	2,3′,4,5′,6-Pentachlorobiphenyl	21.57	328/256	328/256; 328/258	40; 40	2.5–200	$y = 0.3361x - 0.0584$	0.9993
75	PCB 093	2,2′,3,5,6-Pentachlorobiphenyl	21.76	326/291	326/291; 326/289	10; 10	1.5–120	$y = 0.6106x + 0.0649$	0.9994
76	PCB 066	2,3′,4,4′-Tetrachlorobiphenyl	22.00	292/220	292/220; 292/222	20; 20	1.5–120	$y = 0.8585x + 0.1412$	0.9971
77	PCB 090	2,2′,3,4′,5-Pentachlorobiphenyl	22.58	324/254	324/254; 326/291	30; 10	1.5–120	$y = 0.6510x + 0.4744$	0.9971
78	PCB 113	2,3,3′,5′,6-Pentachlorobiphenyl	22.82	324/254	324/254; 326/256	30; 20	1.5–120	$y = 0.7591x + 0.3328$	0.9864
79	PCB 089	2,2′,3,4,6′-Pentachlorobiphenyl	22.93	254/184	254/184; 254/219	30; 20	1.5–120	$y = 0.5683x + 0.1734$	0.9984
80	PCB 152	2,2′,3,5,6,6′-Hexachlorobiphenyl	23.26	358/288	358/288; 358/218	30; 50	1.5–120	$y = 0.4630x + 0.1136$	0.9960
81	PCB 145	2,2′,3,4,6,6′-Hexachlorobiphenyl	23.45	290/218	290/218; 290/220	30; 30	1.5–120	$y = 0.2597x + 0.0437$	0.9993

(continued)

Table 5.1 (*Continued*)

No.	Abbreviation	Name	Retention Time (min)	Quantifying Ion	Qualifying Ion	Collision Energy (eV)	Linear Range (μg/l)	Linear Equation	Correlation Coefficient
82	PCB 115	2,3,4,4′,6-Pentachlorobiphenyl	23.59	326/256	326/256; 326/254	35; 35	1.5–120	$y = 0.7787x + 0.1503$	0.9965
83	PCB 154	2,2′,4,4′,5,6′-Hexachlorobiphenyl	23.66	358/288	358/288; 358/218	40; 50	1.5–120	$y = 0.4710x + 0.1900$	0.9957
84	PCB 085	2,2′,3,4,4′-Pentachlorobiphenyl	23.97	326/256	326/256; 326/254	40; 40	1.5–120	$y = 0.5159x - 0.2218$	0.9985
85	PCB 151	2,2′,3,5,5′,6-Hexachlorobiphenyl	24.36	358/288	358/288; 358/323	30; 10	1.5–120	$y = 0.3866x - 0.1639$	0.9973
86	PCB 139	2,2′,3,4,4′,6-Hexachlorobiphenyl	24.67	358/288	358/288; 358/218	30; 50	1.5–120	$y = 0.4978x - 0.1412$	0.9944
87	PCB 140	2,2′,3,4,4′,6′-Hexachlorobiphenyl	24.86	360/325	360/325; 360/290	10; 20	1.5–120	$y = 0.4286x - 0.0733$	0.9966
88	PCB 107	2,3,3′,4′,5-Pentachlorobiphenyl	25.04	254/184	254/184; 254/219	30; 20	1.5–120	$y = 0.4346x + 0.2180$	0.9944
89	PCB 143	2,2′,3,4,5,6′-Hexachlorobiphenyl	25.20	290/218	290/218; 290/220	30; 30	1.5–120	$y = 0.5288x - 0.1023$	0.9967
90	PCB 142	2,2′,3,4,5,6-Hexachlorobiphenyl	25.32	362/237	362/290; 362/237	10; 20	1.5–120	$y = 0.2089x - 0.0063$	0.9975
91	PCB 146	2,2′,3,4′,5,5′-Hexachlorobiphenyl	25.45	358/288	358/288; 358/323	20; 10	1.5–120	$y = 0.5088x + 0.3019$	0.9957
92	PCB 122	2′,3,3′,4,5-Pentachlorobiphenyl	25.76	326/256	326/256; 326/254	40; 41	1.5–120	$y = 0.4733x - 0.1304$	0.9996
93	PCB 141	2,2′,3,4,5,5′-Hexachlorobiphenyl	26.28	290/218	290/218; 290/220	30; 30	1.5–120	$y = 0.2364x + 0.0591$	0.9991
94	PCB 130	2,2′,3,3′,4,5′-Hexachlorobiphenyl	26.67	358/288	358/288; 358/218	50; 50	1.5–120	$y = 0.1488x + 0.0030$	0.9944
95	PCB 138	2,2′,3,4,4′,5′-Hexachlorobiphenyl	26.87	360/290	360/290; 360/288	30; 30	1.5–120	$y = 0.6538x + 0.4058$	0.9964
96	PCB 175	2,2′,3,3′,4,5′,6-Heptachlorobiphenyl	26.97	359/324	359/324; 359/322	10; 10	1.5–120	$y = 0.3863x - 0.0539$	0.9957
97	PCB 183	2,2′,3,4,4′,5′,6-Heptachlorobiphenyl	27.42	396/361	396/361; 396/359	10; 10	1.5–120	$y = 0.2910x + 0.0142$	0.9968
98	PCB 166	2,3,4,4′,5,6-Hexachlorobiphenyl	27.80	362/290	362/290; 362/292	20; 20	1.5–120	$y = 0.2723x + 0.2410$	0.9956
99	PCB 128	2,2′,3,3′,4,4′-Hexachlorobiphenyl	28.09	325/290	325/290; 325/288	10; 10	1.5–120	$y = 0.3715x + 0.1531$	0.9957
100	PCB 200	2,2′,3,3′,4,5′,6,6′-Octachlorobiphenyl	28.34	428/358	428/358; 428/356	30; 30	1.5–120	$y = 0.8084x - 0.1319$	0.9997
101	PCB 173	2,2′,3,3′,4,5,6-Heptachlorobiphenyl	28.74	396/361	396/361; 396/359	10; 10	1.5–120	$y = 0.4234x - 0.0281$	0.9987
102	PCB 157	2,3,3′,4,4′,5′-Hexachlorobiphenyl	29.05	360/290	360/290; 362/290	30; 30	1.5–120	$y = 0.7321x + 0.6309$	0.9980
103	PCB 191	2,3,3′,4,4′,5′,6-Heptachlorobiphenyl	29.40	396/326	396/326; 396/324	30; 30	1.5–120	$y = 0.5642x + 0.1188$	0.9969
104	PCB 203	2,2′,3,4,4′,5,5′,6-Octachlorobiphenyl	30.26	358/288	358/288; 358/286	40; 40	1.5–120	$y = 0.2451x + 0.0263$	0.9965
105	PCB 208	2,2′,3,3′,4,5,5′,6,6′-Nonachlorobiphenyl	30.73	462/392	462/392; 462/390	30; 30	1.5–120	$y = 0.5763x + 0.1776$	0.9967
106	PCB 194	2,2′,3,3′,4,4′,5,5′-Octachlorobiphenyl	31.77	358/288	358/288; 358/286	30; 30	2.5–200	$y = 0.1471x + 0.1204$	0.9989

Group C									
107	PCB 003	4-Chlorobiphenyl	12.66	188/152	188/152; 188/153	20; 10	1.5–120	$y = 2.5852x - 0.7325$	0.9992
108	PCB 009	2,5-Dichlorobiphenyl	14.09	224/152	224/152; 224/151	20; 40	1.5–120	$y = 1.7036x - 0.9649$	0.9953
109	PCB 014	3,5-Dichlorobiphenyl	15.27	222/152	222/152; 222/151	20; 50	1.5–120	$y = 3.0308x - 2.0712$	0.9973
110	PCB 018	2,2′,5-Trichlorobiphenyl	16.47	186/151	186/151; 186/150	20; 30	1.5–120	$y = 1.0880x - 1.0872$	0.9938
111	PCB 024	2,3,6-Trichlorobiphenyl	16.82	258/151	258/151; 258/150	50; 50	1.5–120	$y = 0.3710x - 0.4487$	0.9916
112	PCB 023	2,3,5-Trichlorobiphenyl	17.29	186/151	186/151; 186/150	20; 30	1.5–120	$y = 1.0118x - 0.4367$	0.9962
113	PCB 054	2,2′,6,6′-Tetrachlorobiphenyl	17.87	292/222	292/222; 292/220	30; 30	1.5–120	$y = 0.7109x - 0.1322$	0.9971
114	PCB 031	2,4′,5-Trichlorobiphenyl	18.19	258/151	258/151; 258/166	50; 50	1.5–120	$y = 0.4591x + 0.0668$	0.9990
115	PCB 033	2′,3,4-Trichlorobiphenyl	18.73	258/186	258/186; 258/188	20; 20	1.5–120	$y = 1.2614x - 0.5768$	0.9917
116	PCB 069	2,3′,4,6-Tetrachlorobiphenyl	19.18	294/222	294/222; 220/150	20; 40	1.5–120	$y = 0.5131x - 0.3672$	0.9945
117	PCB 075	2,4,4′,6-Tetrachlorobiphenyl	19.52	292/220	292/220; 292/150	30; 50	1.5–120	$y = 0.7469x - 0.2298$	0.9978
118	PCB 046	2,2′,3,6′-Tetrachlorobiphenyl	19.62	292/220	292/220; 292/222	30; 30	1.5–120	$y = 1.1978x - 0.5608$	0.9969
119	PCB 047	2,2′,4,4′-Tetrachlorobiphenyl	19.70	290/220	290/220; 290/255	20; 20	1.5–120	$y = 1.9345x - 0.7181$	0.9983
120	PCB 044	2,2′,3,5′-Tetrachlorobiphenyl	20.48	292/150	292/150; 292/220	50; 40	1.5–120	$y = 0.3093x - 0.0989$	0.9941
121	PCB 042	2,2′,3,4′-Tetrachlorobiphenyl	20.56	294/222	294/222; 294/150	30; 50	1.5–120	$y = 0.4026x - 0.1862$	0.9923
122	PCB 071	2,3′,4′,6-Tetrachlorobiphenyl	20.84	294/220	294/220; 220/150	20; 40	1.5–120	$y = 0.1522x - 0.0493$	0.9988
123	PCB 096	2,2′,3,6,6′-Pentachlorobiphenyl	21.02	324/254	324/254; 328/256	20; 20	1.5–120	$y = 0.8585x - 0.2699$	0.9995
124	PCB 088	2,2′,3,4,6-Pentachlorobiphenyl	21.39	328/256	328/256; 328/258	40; 40	1.5–120	$y = 0.3834x - 0.2149$	0.9902
125	PCB 094	2,2′,3,5,6′-Pentachlorobiphenyl	21.46	254/184	254/184; 254/219	30; 20	1.5–120	$y = 0.5321x - 0.2199$	0.9983
126	PCB 098	2,2′,3′,4,6-Pentachlorobiphenyl	21.67	254/184	254/184; 254/219	20; 20	1.5–120	$y = 0.5715x - 0.3194$	0.9990
127	PCB 076	2′,3,4,5-Tetrachlorobiphenyl	21.85	220/150	220/150; 294/222	30; 20	1.5–120	$y = 0.6350x - 0.3119$	0.9989
128	PCB 091	2,2′,3,4′,6-Pentachlorobiphenyl	22.13	328/256	328/256; 328/258	30; 30	1.5–120	$y = 0.3782x - 0.1951$	0.9955
129	PCB 101	2,2′,4,5,5′-Pentachlorobiphenyl	22.66	328/256	328/256; 328/293	30; 10	1.5–120	$y = 0.4283x - 0.1392$	0.9959
130	PCB 056	2,3,3′,4′-Tetrachlorobiphenyl	22.84	290/220	290/220; 290/150	30; 50	1.5–120	$y = 1.4684x - 0.4890$	0.9973
131	PCB 119	2,3′,4,4′,6-Pentachlorobiphenyl	23.01	326/256	326/256; 326/254	40; 40	2.5–200	$y = 0.5558x - 0.6791$	0.9992
132	PCB 079	3,3′,4,5′-Tetrachlorobiphenyl	23.31	220/150	220/150; 294/222	30; 20	1.5–120	$y = 0.6847x - 0.4840$	0.9916
133	PCB 116	2,3,4,5,6-Pentachlorobiphenyl	23.48	328/256	328/256; 328/258	30; 30	1.5–120	$y = 0.4426x - 0.3617$	0.9936
134	PCB 117	2,3,4′,5,6-Pentachlorobiphenyl	23.59	328/256	328/256; 328/258	40; 40	1.5–120	$y = 0.3923x - 0.2829$	0.9970
135	PCB 078	3,3′,4,5-Tetrachlorobiphenyl	23.72	294/222	294/222; 294/150	30; 50	1.5–120	$y = 0.6307x - 0.4254$	0.9968
136	PCB 136	2,2′,3,3′,6,6′-Hexachlorobiphenyl	23.99	362/290	362/290; 362/292	20; 20	1.5–120	$y = 0.4211x - 0.4115$	0.9956
137	PCB 144	2,2′,3,4,5′,6-Hexachlorobiphenyl	24.48	360/290	360/290; 360/288	30; 30	1.5–120	$y = 0.6040x - 1.2141$	0.9695
138	PCB 077	3,3′,4,4′-Tetrachlorobiphenyl	24.77	290/220	290/220; 290/150	30; 50	1.5–120	$y = 1.1559x - 1.7888$	0.9868
139	PCB 149	2,2′,3,4′,5′,6-Hexachlorobiphenyl	24.86	360/325	360/325; 360/290	10; 20	1.5–120	$y = 0.4670x - 0.4010$	0.9908
140	PCB 188	2,2′,3,4′,5,6,6′-Heptachlorobiphenyl	25.06	324/254	324/254; 324/252	30; 30	1.5–120	$y = 0.3181x - 0.3664$	0.9884

(*continued*)

Table 5.1 (*Continued*)

No.	Abbreviation	Name	Retention Time (min)	Quantifying Ion	Qualifying Ion	Collision Energy (eV)	Linear Range (μg/l)	Linear Equation	Correlation Coefficient
141	PCB 133	2,2′,3,3′,5,5′-Hexachlorobiphenyl	25.22	358/288	358/288; 358/323	30; 20	1.5–120	$y = 0.3672x - 0.2899$	0.9952
142	PCB 165	2,3,3′,5,5′,6-Hexachlorobiphenyl	25.39	358/218	358/218; 358/288	50; 40	1.5–120	$y = 0.1395x - 0.0626$	0.9963
143	PCB 161	2,3,3′,4,5′,6-Hexachlorobiphenyl	25.47	362/290	362/290; 362/292	20; 20	1.5–120	$y = 0.5669x - 0.7815$	0.9906
144	PCB 132	2,2′,3,3′,4,6′-Hexachlorobiphenyl	26.08	360/290	360/290; 360/288	25; 25	1.5–120	$y = 0.6102x - 0.5038$	0.9904
145	PCB 105	2,3,3′,4,4′-Pentachlorobiphenyl	26.34	254/184	254/184; 254/219	30; 20	1.5–120	$y = 0.5021x - 0.3422$	0.9939
146	PCB 186	2,2′,3,4,5,6,6′-Heptachlorobiphenyl	26.69	324/254	324/254; 324/252	30; 30	1.5–120	$y = 0.3276x - 0.0352$	0.9996
147	PCB 158	2,3,3′,4,4′,6-Hexachlorobiphenyl	26.92	290/218	290/218; 290/220	30; 30	1.5–120	$y = 0.2539x - 0.0976$	0.9987
148	PCB 182	2,2′,3,4,4′,5,6′-Heptachlorobiphenyl	27.16	398/326	398/326; 398/328	30; 30	1.5–120	$y = 0.3004x - 0.1752$	0.9982
149	PCB 159	2,3,3′,4,5,5′-Hexachlorobiphenyl	27.44	358/288	358/288; 362/290	20; 20	1.5–120	$y = 0.6772x - 0.4259$	0.9975
150	PCB 167	2,3′,4,4′,5,5′-Hexachlorobiphenyl	27.91	358/218	358/218; 358/288	50; 40	2.5–200	$y = 0.2552x - 0.4003$	0.9961
151	PCB 174	2,2′,3,3′,4,5,6′-Heptachlorobiphenyl	28.17	324/254	324/254; 324/252	30; 30	1.5–120	$y = 0.3061x - 0.3686$	0.9933
152	PCB 177	2,2′,3,3′,4′,5,6-Heptachlorobiphenyl	28.44	394/324	394/324; 394/322	30; 30	1.5–120	$y = 0.6042x - 0.3641$	0.9936
153	PCB 156	2,3,3′,4,4′,5-Hexachlorobiphenyl	28.82	362/290	362/290; 360/290	30; 40	1.5–120	$y = 0.5292x - 0.084$	0.9979
154	PCB 180	2,2′,3,4,4′,5,5′-Heptachlorobiphenyl	29.23	396/324	396/324; 396/326	30; 30	1.5–120	$y = 0.4919x - 0.1138$	0.9983
155	PCB 198	2,2′,3,3′,4,5,5′,6-Octachlorobiphenyl	30.01	430/360	430/360; 430/358	30; 30	1.5–120	$y = 0.4285x - 0.2733$	0.9987
156	PCB 196	2,2′,3,3′,4,4′,5,6′-Octachlorobiphenyl	30.27	358/288	358/288; 358/286	30; 30	1.5–120	$y = 0.2514x - 0.1195$	0.9965
157	PCB 207	2,2′,3,3′,4,4′,5,6,6′-Nonachlorobiphenyl	30.94	464/463	464/463; 464/394	10; 30	1.5–120	$y = 1.2273x - 0.6215$	0.9965
158	PCB 205	2,3,3′,4,4′,5,5′,6-Octachlorobiphenyl	31.90	430/360	430/360; 430/358	20; 20	1.5–120	$y = 0.7079x - 0.2265$	0.9990
Group D									
159	PCB 010	2,6-Dichlorobiphenyl	13.27	152/151	152/151; 152/150	20; 40	1.5–120	$y = 0.5098x + 0.6470$	0.9948
160	PCB 006	2,3′-Dichlorobiphenyl	14.67	152/151	152/151; 152/150	20; 40	2.5–200	$y = 0.5777x + 0.7616$	0.9949

161	PCB 030	2,4,6-Trichlorobiphenyl	15.50	186/151	186/151; 186/150	20; 30	1.5–120	$y = 0.9428x - 0.6238$	0.9988
162	PCB 017	2,2′,4-Trichlorobiphenyl	16.48	221/186	221/186; 221/151	20; 40	1.5–120	$y = 0.6599x - 0.0527$	0.9994
163	PCB 015	4,4′-Dichlorobiphenyl	16.91	222/152	222/152; 222/151	20; 40	1.5–120	$y = 2.6365x - 0.7447$	0.9992
164	PCB 034	2′,3,5-Trichlorobiphenyl	17.39	258/186	258/186; 258/188	20; 20	1.5–120	$y = 1.1703x - 0.3340$	0.9994
165	PCB 026	2,3′,5-Trichlorobiphenyl	17.89	258/186	258/186; 258/151	20; 40	1.5–120	$y = 1.1810x - 0.4684$	0.9996
166	PCB 028	2,4,4′-Trichlorobiphenyl	18.22	256/151	256/151; 256/150	50; 50	1.5–120	$y = 0.4340x - 0.0469$	0.9989
167	PCB 051	2,2′,4,6′-Tetrachlorobiphenyl	18.80	294/222	294/222; 294/224	30; 30	1.5–120	$y = 0.3543x - 0.1073$	0.9991
168	PCB 045	2,2′,3,6-Tetrachlorobiphenyl	19.20	220/150	220/150; 220/185	30; 20	1.5–120	$y = 0.4630x - 0.1403$	0.9995
169	PCB 052	2,2′,5,5′-Tetrachlorobiphenyl	19.56	220/150	220/150; 220/185	30; 20	1.5–120	$y = 0.5072x - 0.2072$	0.9984
170	PCB 049	2,2′,4,5′-Tetrachlorobiphenyl	19.64	290/220	290/220; 290/185	40; 40	1.5–120	$y = 0.5634x - 0.2119$	0.9996
171	PCB 048	2,2′,4,5-Tetrachlorobiphenyl	19.75	220/150	220/150; 220/185	10; 10	1.5–120	$y = 0.5424x - 0.1651$	0.9999
172	PCB 059	2,3,3′,6-Tetrachlorobiphenyl	20.49	220/150	220/150; 220/185	30; 20	1.5–120	$y = 0.5976x - 0.0754$	0.9992
173	PCB 068	2,3′,4,5′-Tetrachlorobiphenyl	20.62	294/222	294/222; 294/220	30; 30	1.5–120	$y = 0.4237x - 0.1662$	0.9984
174	PCB 100	2,2′,4,4′,6-Pentachlorobiphenyl	20.87	328/256	328/256; 328/184	30; 50	1.5–120	$y = 0.4722x + 0.0052$	0.9977
175	PCB 057	2,3,3′,5-Tetrachlorobiphenyl	21.05	220/150	220/150; 220/185	30; 20	1.5–120	$y = 0.6239x + 0.1033$	0.9966
176	PCB 063	2,3,4′,5-Tetrachlorobiphenyl	21.41	292/220	292/220; 292/222	30; 30	1.5–120	$y = 0.8219x + 0.2707$	0.9974
177	PCB 061	2,3,4,5-Tetrachlorobiphenyl	21.47	294/222	294/222; 294/150	30; 50	1.5–120	$y = 0.5322x - 0.2370$	0.9990
178	PCB 155	2,2′,4,4′,6,6′-Hexachlorobiphenyl	21.70	360/290	360/290; 360/288	30; 30	1.5–120	$y = 0.8670x - 0.2959$	0.9980
179	PCB 070	2,3′,4′,5-Tetrachlorobiphenyl	21.91	294/222	294/222; 220/150	20; 40	1.5–120	$y = 0.6048x + 0.0786$	0.9992
180	PCB 055	2,3,3′,4-Tetrachlorobiphenyl	22.42	292/222	292/222; 292/220	20; 20	1.5–120	$y = 0.8607x - 0.3357$	0.9990
181	PCB 060	2,3,4,4′-Tetrachlorobiphenyl	22.76	294/222	294/222; 294/224	20; 20	1.5–120	$y = 0.5273x - 0.0476$	0.9980
182	PCB 150	2,2′,3,4′,6,6′-Hexachlorobiphenyl	22.85	325/290	325/290; 325/288	10; 10	1.5–120	$y = 0.1093x + 0.0513$	0.9966
183	PCB 112	2,3,3′,5,6-Pentachlorobiphenyl	23.15	326/256	326/256; 326/254	30; 30	1.5–120	$y = 0.8518x - 0.2858$	0.9993
184	PCB 148	2,2′,3,4′,5,6′-Hexachlorobiphenyl	23.39	362/327	362/327; 362/290	10; 20	1.5–120	$y = 0.3181x + 0.0355$	0.9968
185	PCB 111	2,3,3′,5,5′-Pentachlorobiphenyl	23.49	324/254	324/254; 328/256	20; 20	1.5–120	$y = 0.8594x - 0.4715$	0.9991
186	PCB 097	2,2′,3′,4,5-Pentachlorobiphenyl	23.65	328/256	328/256; 328/293	20; 20	1.5–120	$y = 0.3223x + 0.3961$	0.9791
187	PCB 120	2,3′,4,5,5′-Pentachlorobiphenyl	23.73	254/184	254/184; 254/219	20; 20	1.5–120	$y = 0.3360x - 0.1284$	0.9991
188	PCB 081	3,4,4′,5-Tetrachlorobiphenyl	24.21	290/220	290/220; 290/150	30; 50	1.5–120	$y = 0.7896x - 0.2314$	0.9825
189	PCB 147	2,2′,3,4′,5,6-Hexachlorobiphenyl	24.57	290/218	290/218; 290/220	30; 30	1.5–120	$y = 0.2069x - 0.1182$	0.9991
190	PCB 082	2,2′,3,3′,4-Pentachlorobiphenyl	24.82	328/256	328/256; 328/258	40; 40	1.5–120	$y = 0.2137x + 0.0537$	0.9880
191	PCB 108	2,3,3′,4,5′-Pentachlorobiphenyl	24.95	254/184	254/184; 254/219	30; 20	1.5–120	$y = 0.4724x - 0.2849$	0.9998
192	PCB 106	2,3,3′,4,5-Pentachlorobiphenyl	25.10	328/256	328/256; 328/184	40; 50	2.5–200	$y = 0.2961x - 0.0953$	0.9984
193	PCB 184	2,2′,3,4,4′,6,6′-Heptachlorobiphenyl	25.31	396/326	396/326; 396/324	30; 30	1.5–120	$y = 0.5236x - 0.1083$	0.9983
194	PCB 131	2,2′,3,3′,4,6-Hexachlorobiphenyl	25.44	360/290	360/290; 360/288	30; 30	1.5–120	$y = 0.5262x - 0.3005$	0.9994

(continued)

Table 5.1 (*Continued*)

No.	Abbreviation	Name	Retention Time (min)	Quantifying Ion	Qualifying Ion	Collision Energy (eV)	Linear Range (μg/l)	Linear Equation	Correlation Coefficient
195	PCB 153	2,2′,4,4′,5,5′-Hexachlorobiphenyl	25.67	290/218	290/218; 290/220	20; 20	2.5–200	$y = 0.2334x - 0.1276$	0.9945
196	PCB 179	2,2′,3,3′,5,6,6′-Heptachlorobiphenyl	26.14	398/326	398/326; 398/328	20; 20	1.5–120	$y = 0.3483x - 0.0831$	0.9911
197	PCB 176	2,2′,3,3′,4,6,6′-Heptachlorobiphenyl	26.39	324/254	324/254; 324/252	40; 40	1.5–120	$y = 0.2166x + 0.0822$	0.9965
198	PCB 160	2,3,3′,4,5,6-Hexachlorobiphenyl	26.85	360/290	360/290; 360/288	20; 20	1.5–120	$y = 0.7255x - 0.2075$	0.9973
199	PCB 164	2,3,3′,4′,5′,6-Hexachlorobiphenyl	26.92	360/290	360/290; 360/288	20; 20	1.5–120	$y = 0.6470x - 0.1528$	0.9994
200	PCB 129	2,2′,3,3′,4,5-Hexachlorobiphenyl	27.23	325/290	325/290; 325/218	10; 40	1.5–120	$y = 0.4375x - 0.1344$	0.9991
201	PCB 126	3,3′,4,4′,5-Pentachlorobiphenyl	27.69	254/184	254/184; 254/220	30; 20	1.5–120	$y = 0.4254x - 0.0264$	0.9979
202	PCB 185	2,2′,3,4,5,5′,6-Heptachlorobiphenyl	27.92	394/320	394/322; 394/320	30; 30	1.5–120	$y = 0.2588x + 0.0625$	0.9998
203	PCB 181	2,2′,3,4,4′,5,6-Heptachlorobiphenyl	28.17	394/324	394/324; 394/322	30; 30	1.5–120	$y = 0.5657x - 0.2639$	0.9990
204	PCB 171	2,2′,3,3′,4,4′,6-Heptachlorobiphenyl	28.49	398/326	398/326; 398/328	30; 30	1.5–120	$y = 0.2589x + 0.0282$	0.9996
205	PCB 172	2,2′,3,3′,4,5,5′-Heptachlorobiphenyl	28.93	394/324	394/324; 394/322	30; 30	1.5–120	$y = 0.5493x - 0.2220$	0.9993
206	PCB 199	2,2′,3,3′,4,5,6,6′-Octachlorobiphenyl	29.27	358/288	358/288; 358/286	40; 40	1.5–120	$y = 0.2126x + 0.0138$	0.9985
207	PCB 201	2,2′,3,3′,4,5,5′,6′-Octachlorobiphenyl	30.13	432/360	432/360; 432/361	30; 30	1.5–120	$y = 0.2837x - 0.0349$	0.9992
208	PCB 170	2,2′,3,3′,4,4′,5-Heptachlorobiphenyl	30.36	359/324	359/324; 359/322	20; 20	1.5–120	$y = 0.2592x + 0.0453$	0.9983
209	PCB 189	2,3,3′,4,4′,5,5′-Heptachlorobiphenyl	31.02	394/324	394/324; 394/322	20; 20	1.5–120	$y = 0.8290x - 0.0279$	0.9997
Group E									
210	–	Naphthalene	6.41	128/101	128/101; 128/77	15; 15	1.0–95.6	$y = 0.1955x - 0.5254$	0.9956
211	–	Isoproturon	6.58	146/128	146/128; 146/91	15; 15	126.7–10135.2	$y = 0.1309x + 39.3624$	0.9920
212	–	Dichlorvos	7.88	185/93	185/93; 185/109	15; 10	5.9–588.5	$y = 0.0816x - 0.9668$	0.9876

213	–	Carbofuran	8.36	164/149	164/149; 164/103	15; 25	4.3–432.0	$y = 0.1310x + 0.1248$	0.9986
214	–	Methamidophos	9.35	141/95	141/95; 141/80	10; 15	85.4–8541.5	$y = 0.0240x - 7.7487$	0.9684
215	–	Acenaphthylene	10.55	152/126	152/126; 151/99	15; 25	2.2–224.7	$y = 0.1953x + 1.2657$	0.9875
216	–	Acenaphthene	10.85	152/126	152/126; 151/99	15; 25	4.6–455.4	$y = 0.0593x + 0.0296$	0.9957
217	–	Fluorene	12.94	165/164	165/164; 165/163	25; 25	6.6–658.8	$y = 0.7383x - 3.5999$	0.9970
218	–	Hexachlorobenzene	14.36	284/249	284/249; 284/214	18; 25	2.9–288.8	$y = 0.2330x + 0.0854$	0.9982
219	–	Ethoprophos	14.40	158/97	158/97; 158/114	12; 7	1.3–129.9	$y = 0.1949x - 0.0944$	0.9967
220	–	Chlordimeform	14.91	196/181	196/181; 196/152	5; 25	5.8–580.0	$y = 0.3243x + 0.0058$	0.9969
221	–	Trifluralin	15.37	306/264	306/264; 306/206	12; 15	4.5–451.2	$y = 0.2438x - 1.6645$	0.9942
222	–	α-HCH	16.14	219/183	219/183; 219/147	5; 15	4.4–437.0	$y = 0.1961x - 1.0487$	0.9944
223	–	Omethoate	16.82	156/110	156/110; 156/80	5; 10	18.2–1094.2	$y = 0.0198x + 0.7319$	0.9969
224	–	Anthracene	17.03	176/150	176/150; 178/152	20; 12	0.1–14.7	$y = 20.9615x - 1.1090$	1.0000
225	–	Clomazone	17.04	204/107	204/107; 204/78	25; 25	2.9–288.6	$y = 0.2356x - 0.2193$	0.9969
226	–	Diazinon	17.09	304/179	304/179; 304/162	8; 8	2.7–270.3	$y = 0.3229x - 0.2373$	0.9983
227	–	Phenanthrene	17.13	178/150	178/150; 178/151	45; 40	5.3–525.0	$y = 0.1750x - 0.6505$	0.9998
228	–	γ-HCH	17.72	219/183	219/183; 219/147	5; 15	4.3–431.8	$y = 0.1740x - 1.0058$	0.9932
229	–	Atrazine	17.95	215/173	215/173; 215/200	5; 5	4.0–396.4	$y = 0.2420x - 0.0580$	0.9975
230	–	Simazine	18.03	201/173	201/173; 201/138	5; 15	11.8–1177.2	$y = 0.0621x - 0.2475$	0.9975
231	–	Heptachlor	18.40	272/237	272/237; 272/235	25; 25	6.5–649.8	$y = 0.1095x - 0.8773$	0.9960
232	–	Pirimicarb	18.98	238/166	238/166; 238/96	15; 25	3.1–306.0	$y = 0.4814x - 1.0088$	0.9962
233	–	Dimethoate	19.32	125/79	125/79; 143/111	8; 12	6.5–654.9	$y = 0.0286x - 0.6293$	0.9767
234	–	Aldrin	19.41	263/193	263/193; 263/191	25; 35	4.6–455.4	$y = 0.0467x + 0.0240$	0.9983
235	–	Alachlor	20.16	237/160	237/160; 237/146	8; 20	2.9–289.1	$y = 0.1279x - 0.3246$	0.9949
236	–	Prometryne	20.19	241/199	241/199; 241/184	5; 5	2.7–271.9	$y = 0.4000x - 0.1287$	0.9985
237	–	Chlorothalonil	20.35	266/231	266/231; 266/170	20; 35	13977.6–23296.0	$y = 0.0160x - 180.1627$	0.9957
238	–	Phthalic acid bis-butyl ester	20.69	149/121	149/121; 149/93	10; 10	6.6–664.4	$y = 7.3006x - 52.6826$	0.9986
239	–	β-HCH	20.72	219/183	219/183; 219/147	10; 20	4.5–448.0	$y = 0.1844x - 0.9595$	0.9934
240	–	Chlorpyrifos	20.92	314/286	314/286; 314/258	5; 5	4.5–451.5	$y = 0.1324x - 0.5356$	0.9960
241	–	Parathion-methyl	21.05	263/109	263/109; 263/246	12; 5	4.3–430.9	$y = 0.1515x - 1.3505$	0.9901
242	–	Dicofol	21.34	250/139	250/139; 250/215	15; 10	3.4–336.4	$y = 0.2491x - 0.1258$	0.9986
243	–	Metolachlor	21.44	238/162	238/162; 238/133	15; 25	1.1–108.0	$y = 1.3801x - 1.2331$	0.9962
244	–	δ-HCH	21.50	219/183	219/183; 219/147	10; 20	4.3–434.0	$y = 0.0680x - 0.7771$	0.9862
245	–	Triadimefon	22.42	210/183	210/183; 210/129	5; 10	4.8–481.8	$y = 0.1295x - 0.3226$	0.9972
246	–	Fluoranthene	22.58	202/152	202/152; 202/176	30; 30	3.0–302.6	$y = 0.1309x - 0.4220$	0.9989

(continued)

Table 5.1 (*Continued*)

No.	Abbreviation	Name	Retention Time (min)	Quantifying Ion	Qualifying Ion	Collision Energy (eV)	Linear Range (μg/l)	Linear Equation	Correlation Coefficient
247	–	2,4′-DDE	22.70	246/176	246/176; 246/211	25; 25	6.7–666.0	$y = 0.5787x + 0.2554$	0.9985
248	–	*cis*-Chlordane	23.21	373/266	373/266; 373/301	12; 12	6.3–625.1	$y = 0.0595x - 0.1824$	0.9983
249	–	Phenthoate	23.38	274/246	274/246; 274/121	5; 25	0.4–43.4	$y = 2.1028x - 0.8472$	0.9921
250	–	*trans*-Chlordane	23.50	373/266	373/266; 373/301	12; 12	6.2–618.8	$y = 0.0539x - 0.1086$	0.9975
251	–	Pyrene	23.62	202/199	202/199; 202/200	45; 40	5.2–524.6	$y = 0.9615x - 92.2317$	0.9917
252	–	4,4′-DDE	23.90	246/176	246/176; 246/211	25; 25	6.6–657.2	$y = 0.5374x + 1.0065$	0.9990
253	–	Butachlor	23.98	176/150	176/150; 176/126	25; 25	0.7–74.4	$y = 0.1201x + 0.2435$	0.9900
254	–	Dieldrin	24.47	277/241	277/241; 277/207	12; 12	13.2–1315.6	$y = 0.0244x + 0.0047$	0.9984
255	–	2,4′-DDD	25.04	235/165	235/165; 235/199	15; 15	2.7–268.8	$y = 1.1094x + 1.8793$	0.9994
256	–	Buprofezin	25.05	105/77	105/77; 172/116	18; 7	14.5–1452.8	$y = 0.2211x + 3.1681$	0.9959
257	–	Endrin	25.06	263/191	263/191; 263/193	20; 12	30–2998.6	$y = 0.0156x + 0.1034$	0.9991
258	–	2,4′-DDT	25.47	235/165	235/165; 235/199	25; 25	2.1–208.3	$y = 0.5975x - 1.2217$	0.9955
259	–	Nithophen	26.27	283/162	283/162; 283/202	25; 25	8.1–809.2	$y = 0.0629x - 1.3208$	0.9875
260	–	Oxyfluorfen	26.45	300/223	300/223; 188/144	18; 17	14.6–1458	$y = 0.0775x - 1.6855$	0.9954
261	–	4,4′-DDD	26.73	235/165	235/165; 235/199	15; 15	6.6–664.2	$y = 1.0664x - 2.5332$	0.9978
262	–	4,4′-DDT	27.23	235/199	235/199; 235/165	25; 25	3.0–295.0	$y = 0.4644x - 5.2499$	0.9709
263	–	Phthalic acid benzyl butyl ester	27.84	206/149	206/149; 149/65	5; 25	6.6–656.0	$y = 0.3329x - 2.9928$	0.9984
264	–	Propargite	28.08	173/117	173/117; 173/145	10; 10	127.5–12750.0	$y = 0.0047x - 1.1361$	0.9958
265	–	Tricyclazole	28.39	189/162	189/162; 189/135	10; 15	0.7–73.8	$y = 1.3515x - 2.0591$	0.9879
266	–	Triazophos	28.54	161/134	161/134; 161/106	8; 15	11.4–1142.6	$y = 0.0074x - 0.0245$	0.9987
267	–	Mirex	28.70	272/237	272/237; 272/235	15; 15	2.8–275.0	$y = 0.4583x - 0.8473$	0.9976
268	–	Benzo(*a*)anthracene	29.27	228/226	228/226; 228/202	30; 30	2.0–200.0	$y = 1.0534x - 0.3460$	0.9999
269	–	Phthalic acid bis-2-ethylhexyl ester	29.47	167/149	167/149; 167/65	10; 25	1.3-133.2	$y = 1.7761x + 16.1890$	1.0000
270	–	Amitraz	30.37	293/162	293/162; 293/132	5; 15	1.6-155.9	$y = 0.0066x + 0.0371$	0.9704
271	–	Lambda-cyhalothrin	31.41	197/141	197/141; 197/161	15; 5	6.4-638.4	$y = 0.1814x - 0.9969$	0.9956
272	–	Pyridaben	32.07	147/117	147/117; 147/132	25; 15	6.7-667.0	$y = 0.8733x - 5.9209$	0.9954
273	–	Benzo(*b*)fluoranthene	32.94	252/250	252/250; 252/224	40; 50	0.1-5.4	$y = 8.4098x + 0.3645$	0.9952
274	–	Benzo(*k*)fluoranthene	32.94	252/250	252/250; 252/224	40; 50	0.2-21.8	$y = 0.3051x + 0.0760$	0.9962

275	–	Cyfluthrin	33.33	206/151	206/151; 206/177	15; 20	8.7-873.2	$y = 0.0272x - 0.3225$	0.9801
276	–	Cypermethrin	33.53	163/127	163/127; 163/91	5; 10	18.5-1854.6	$y = 0.1491x - 4.4244$	0.9939
277	–	Benzo(*a*)pyrene	33.70	252/250	252/250; 252/226	25	1.8-175.5	$y = 0.8605x + 1.5570$	1.0000
278	–	Acetamiprid	33.78	152/116	152/116; 166/139	20; 8	27.3-2728	$y = 0.0487x - 2.0732$	0.9893
279	–	Fenvalerate 1	34.61	419/225	419/225; 419/167	5; 5	29.0-2904.0	$y = 0.0832x - 2.4583$	0.9966
280	–	Fenvalerate 2	34.96	419/225	419/225; 419/167	5; 5	29.0-2904.0	$y = 0.0535x - 2.4768$	0.9949
281	–	Deltamethrin	35.98	181/152	181/152; 181/127	25; 25	24.2-2418.2	$y = 0.0314x - 2.5842$	0.9767
282	–	Indeno(1,2,3-*cd*)pyrene	37.63	276/274	276/274; 276/248	40; 50	2.2-220.4	$y = 1.2713x + 1.5120$	0.9992
283	–	Dibenzo[*a*,*h*]anthracene	37.83	278/276	278/276; 278/274	40; 55	2.2-220.4	$y = 1.2873x - 0.1164$	0.9815
284	–	Benzo(*g*,*h*,*i*)perylene	38.64	274/272	274/272; 274/248	25; 25	2.2-220.0	$y = 0.0709x - 0.1860$	0.9990

Table 5.2. Names, Retention Times, Selective Ions, Relative Abundances, Linear Ranges, Linear Equations, and Correlation Coefficients of Endosulfans Determined by GC–NCI–MS

No.	Name	Retention Time (min)	Selective Ion	Relative Abundance	Linear Range (μg/l)	Linear Equation	Correlation Coefficient
1	Endosulfan I	11.14	406, 408, 372	100:55:44.6	0.5–30.0	$y = 2022.5831x + 486.7318$	0.9973
2	Endosulfan II	12.56	406, 408, 372	100:78:17.5	0.5–30.0	$y = 822.2465x + 14.4991$	0.9995

flow rate of auxiliary gas: 6 l/min; ion source temperature: 350 °C; data acquisition mode: MRM; MRM transitions, CEs, declustering potentials (DPs), and collision cell exit potentials of nine environmental pollutants are given in Table 5.3.

5.2.6 Qualitative and Quantitative Determination

5.2.6.1 Qualitative Determination

When the samples are determined, if the retention times of peaks of the sample solution are the same as those of the peaks of the working standard mixed solution, the selected ions appeared in the background-subtracted mass spectrum, and the abundance ratios of the selected ions are within the expected limits (abundance ratios >50%, permitted tolerances are ±20%; abundance ratios >20–50%, permitted tolerances are ±25%; abundance ratios >10–20%, permitted tolerances are ±30%; abundance ratios ≤10%, permitted tolerances are ±50%), then the sample is confirmed to contain the environmental pollutants.

5.2.6.2 Quantitative Determination

An internal standard calibration curve is used for quantitative determination in GC–MS/MS, and the internal standard is heptachlor epoxide. External standard calibration curve is used for quantitative determination in GC–MS and LC–MS/MS. To decrease the influences of the matrix, the matrix mixed standard solutions are used as the standard solutions for quantitative determination and standard working curve. Furthermore, the responses of environmental pollutants in the samples are all in the linear ranges of the instrument.

5.3 RESULTS AND DISCUSSION

5.3.1 Selection of GPC Cleanup Conditions

Ninety-one pesticides and environmental pollutant monomers and mixed standard solution of 16 PCBs are used in this section, and they are called "environmental

Table 5.3. Names, Retention Times, Quantifying and Qualifying Ions, Declustering Potentials, Collision Energies, Collision Cell Exit Potentials, Linear Ranges, Linear Equations, and Correlation Coefficients of Nine Environmental Pollutants Detected by LC–MS/MS

No.	Name	Retention Time (min)	Quantifying Ion	Qualifying Ion	DP (V)	CE (V)	Collision Cell Exit Potential (V)	Linear Range (μg/l)	Linear Equation	Correlation Coefficient
1	Trichlorphon	9.25	257/109	257/109; 257/127.1	28	25; 23	2; 2	3.9–38.8	$y = 1703.5808x - 42.5213$	0.9991
2	Metsulfuron-methyl	11.18	382/167.1	382/167.1; 382/199.1	31	25; 30	3; 3	2.3–23.0	$y = 45166.5928x + 27852.7660$	0.9918
3	Chlortoluron	11.74	213.1/72.0	213.1/72.0; 213.1/140.1	29	38; 33	3; 3	2.6–25.9	$y = 16574.7308x - 12467.6596$	0.9989
4	2,4-D	12.12	219/160.8	219/124.8; 219/89	−35	−21; −38; −50	−2; −2; −2	12.5–124.8	$y = 940.7392x - 9211.9149$	0.9939
5	Bensulfuron-methyl	12.57	411.1/149.1	411.1/149.1; 411.1/182.1	29	31; 30	2; 2	0.6–6.0	$y = 45466.5694x - 14917.6596$	0.9891
6	Propanil	13.25	218/162.1	218/162.1; 218/127	45	23; 41	2; 2	4.7–47.0	$y = 4618.9752x + 15965.5319$	0.9883
7	Fipronil	13.59	436.9/368	436.9/368; 436.9/290	44	23; 35	4; 4	26.4–264.0	$y = 118.3954x + 206.6277$	0.9962
8	Phoxim	18.53	299/129	299/129; 299/97	36	18; 35	3; 2	6.6–66.2	$y = 180.0994x + 275.5957$	0.9842
9	Hexythiazox	22.05	353.1/228.1	353.1/228.1; 353.1/168.1	50	21; 35	2.5; 1.8	13–129.6	$y = 1377.9394x - 19979.0476$	0.9874

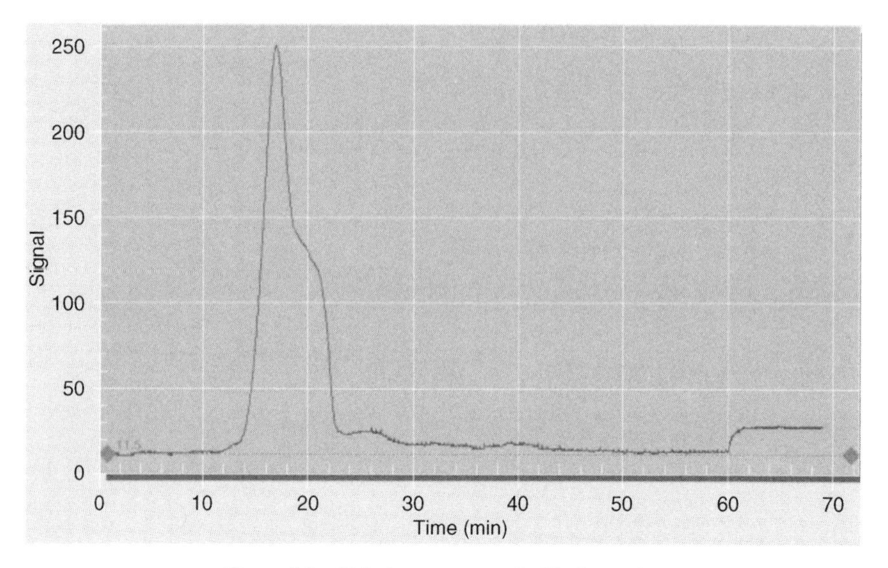

Figure 5.1. GPC chromatogram of a blank sample.

pollutants" hereinafter. Chromatographic behavior of these environmental pollutants is estimated by GPC within the experiment. One hundred and two environmental pollutants elute at 16–56 min, and 77 environmental pollutants elute after 23 min, accounting for 72.0% of the total. *cis*-Chlordane, benzo(*g*,*h*,*i*)pyrene, dimehypo, monosultap, and paraquat may not have UV absorption at 254 nm, and thus their collection times cannot be determined.

For multiresidue analysis, it is essential to meet the requirements of the detection methods for most analytes, to improve the removal of impurities (fat, etc.), and to ensure that the instrument is not contaminated. The blank samples are selected to compare the conditions of cleanup. The contents of samples collected at 21–60, 22–60, 23–60, 24–60, 25–60, and 26–60 min are compared. It is found that if the start collect time is set at 21 and 22 min, the eluate contains a little fat, and if the start collect time is set at 23 min, the interferences caused by fat can be mostly excluded. The GPC chromatograms of blank and standard fortified samples are shown in Figures 5.1 and 5.2.

In Figure 5.1, fat elutes between 14 and 23 min, and in Figure 5.2, most pesticides elute between 28 and 50 min. Thus, the condition of GPC is determined: 23–60 min is the start collect time. But the recovery of *cis*-chlordane is only about 90%, the recovery of benzo(*g*,*h*,*i*)pyrene is unstable (may be influenced by the matrix), and dimehypo, monosultap, and paraquat are not detected. The cleanup conditions determined by this method are suitable for *cis*-chlordane and most environmental pollutants and unsuitable for dimehypo, monosultap, and paraquat.

5.3.2 Selection of Extraction Solvent

Seven solvents—*n*-hexane, *n*-hexane–acetone (1 + 1, v/v), *n*-hexane–acetone (3 + 1, v/v), ethyl acetate–cyclohexane (1 + 1, v/v), *n*-hexane–dichloromethane

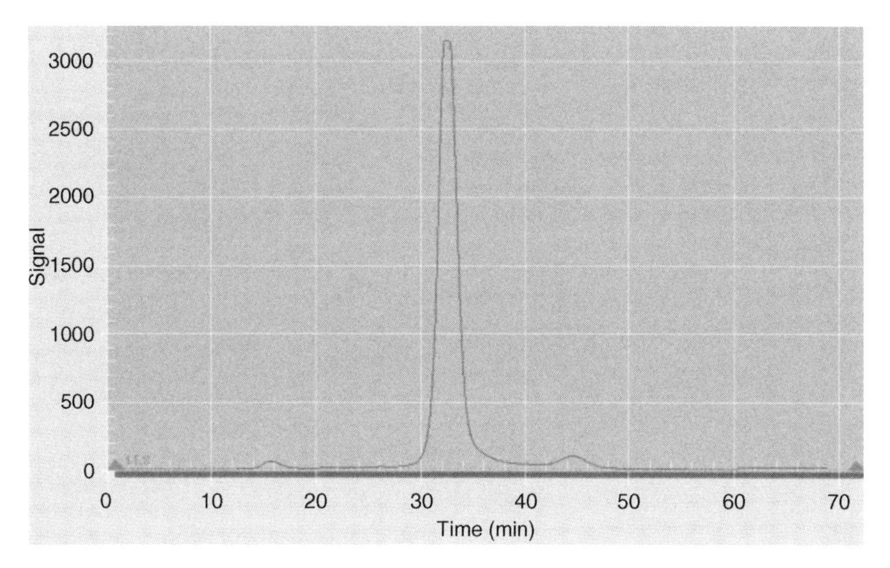

Figure 5.2. GPC chromatogram of a standard fortified sample.

(4 + 1, v/v), acetonitrile, and acetonitrile with 1% acetic acid (v/v)—are selected to compare their extraction efficiencies for 77 representative environmental pollutants.

For each solvent, three standard added samples and one matrix standard are prepared for calculating recoveries of 77 environmental pollutants. The results of three standard added samples are averaged before the calculation of recoveries. Accurately weigh 5 g fat sample, add standard solution of 2 LOQ (a S/N ratio of >10 is the criterion for LOD in the research; see Table 5.4) and 35 ml extraction solvent, the extraction method is homogeneous extraction, and repeat twice. After concentration and GPC cleanup, GC–MS/MS is used to detect the analytes, and internal standard is used for quantification. The results of comparison are shown in Figure 5.3.

On comparing the recoveries of seven solvents for 77 environmental pollutants, it is found that the recoveries obtained with acetonitrile, ethyl acetate–cyclohexane (1 + 1, v/v), *n*-hexane–acetone (1 + 1, v/v), and *n*-hexane–dichloromethane (4 + 1, v/v) are similar and these solvents are suitable, but the recovery with acetonitrile with 1% acetic acid is the worst. On comparing the RSDs, it is found that acetonitrile is the best, and acetonitrile with 1% acetic acid is the worst. Thus, these environmental pollutants are unsuitable to be extracted in acidic conditions. Furthermore, the fat is not easily dissolved in acetonitrile, and the matrix influence by acetonitrile extraction is less than the influence by other solvent extractions, thus facilitating the cleanup. Therefore, acetonitrile is selected as the extraction solvent in this study.

Table 5.4. LOD, LOQ, Recoveries, and Precision of 295 Environmental Pollutants

					LOQ		2 LOQ		4 LOQ	
No.	Abbreviation	Name	LOD (μg/kg)	LOQ (μg/kg)	Average (%)	RSD (%)	Average (%)	RSD (%)	Average (%)	RSD (%)
GC–MS/MS Group A										
1	PCB 001	2-Chlorobiphenyl	4.0	12.0	101.85	4.24	86.03	2.31	101.55	3.46
2	PCB 004	2,2′-Dichlorobiphenyl	4.0	12.0	123.65	12.07	99.03	8.82	109.25	6.81
3	PCB 008	2,4′-Dichlorobiphenyl	4.0	12.0	97.68	4.27	86.63	6.30	100.63	3.43
4	PCB 019	2,2′,6-Trichlorobiphenyl	4.0	12.0	95.87	5.80	91.62	2.29	98.98	3.59
5	PCB 012	3,4-Dichlorobiphenyl	4.0	12.0	93.78	3.96	86.68	5.39	92.77	4.17
6	PCB 027	2,3′,6-Trichlorobiphenyl	4.0	12.0	95.52	5.95	87.05	4.09	93.80	3.75
7	PCB 016	2,2′,3-Trichlorobiphenyl	4.0	12.0	100.67	3.07	88.65	3.15	98.25	3.44
8	PCB 025	2,3′,4-Trichlorobiphenyl	4.0	12.0	84.55	3.76	81.35	2.51	86.48	4.31
9	PCB 021	2,3,4-Trichlorobiphenyl	4.0	12.0	88.92	2.90	81.63	2.25	89.60	3.55
10	PCB 020	2,3,3′-Trichlorobiphenyl	4.0	12.0	94.28	7.66	85.12	2.37	92.40	4.03
11	PCB 036	3,3′,5-Trichlorobiphenyl	4.0	12.0	77.98	6.75	75.53	3.49	84.78	4.82
12	PCB 043	2,2′,3,5-Tetrachlorobiphenyl	4.0	12.0	87.67	4.29	85.12	1.69	89.35	4.27
13	PCB 065	2,3,5,6-Tetrachlorobiphenyl	4.0	12.0	90.03	5.94	81.13	2.52	94.05	3.95
14	PCB 104	2,2′,4,6,6′-Pentachlorobiphenyl	4.0	12.0	85.33	4.10	78.33	6.05	83.32	5.43
15	PCB 072	2,3′,5,5′-Tetrachlorobiphenyl	4.0	12.0	77.63	7.05	74.80	2.35	79.42	3.32
16	PCB 103	2,2′,4,5′,6-Pentachlorobiphenyl	4.0	12.0	86.28	6.15	74.98	3.88	81.72	5.37
17	PCB 041	2,2′,3,4-Tetrachlorobiphenyl	4.0	12.0	88.55	4.47	82.25	3.33	90.10	4.06
18	PCB 067	2,3′,4,5-Tetrachlorobiphenyl	4.0	12.0	77.90	5.16	74.38	2.36	79.70	3.00
19	PCB 040	2,2′,3,3′-Tetrachlorobiphenyl	4.0	12.0	80.58	6.84	70.00	3.41	97.73	4.81
20	PCB 074	2,4,4′,5-Tetrachlorobiphenyl	4.0	12.0	80.75	6.98	69.97	3.36	79.57	4.02
21	PCB 102	2,2′,4,5,6′-Pentachlorobiphenyl	4.0	12.0	80.68	6.00	74.05	5.95	80.15	3.11
22	PCB 095	2,2′,3,5′,6-Pentachlorobiphenyl	4.0	12.0	91.60	6.33	76.92	4.97	87.82	4.40
23	PCB 092	2,2′,3,5,5′-Pentachlorobiphenyl	6.7	20.0	82.47	8.36	77.42	3.13	78.60	5.54
24	PCB 099	2,2′,4,4′,5-Pentachlorobiphenyl	4.0	12.0	78.65	9.20	69.10	4.71	78.47	4.25
25	PCB 084	2,2′,3,3′,6-Pentachlorobiphenyl	4.0	12.0	96.62	4.17	81.20	3.48	90.60	2.82
26	PCB 109	2,3,3′,4,6-Pentachlorobiphenyl	4.0	12.0	76.82	4.02	78.55	3.58	82.27	3.39
27	PCB 083	2,2′,3,3′,5-Pentachlorobiphenyl	6.7	20.0	86.90	8.09	73.28	8.74	82.70	6.67
28	PCB 086	2,2′,3,4,5-Pentachlorobiphenyl	4.0	12.0	78.57	4.61	75.60	3.97	80.92	3.42
29	PCB 125	2′,3,4,5,6′-Pentachlorobiphenyl	4.0	12.0	78.20	4.33	71.20	4.44	82.27	2.58

30	PCB 087	2,2′,3,4,5′-Pentachlorobiphenyl	4.0	12.0	77.45	3.73	77.83	3.42	81.38	3.87
31	PCB 110	2,3,3′,4′,6-Pentachlorobiphenyl	4.0	12.0	79.72	6.40	73.82	2.33	84.62	2.60
32	PCB 135	2,2′,3,3′,5,6′-Hexachlorobiphenyl	4.0	12.0	79.37	5.97	70.13	4.96	78.80	3.21
33	PCB 124	2′,3,4,5,5′-Pentachlorobiphenyl	4.0	12.0	70.83	3.44	64.02	3.72	70.22	4.45
34	PCB 123	2′,3,4,4′,5-Pentachlorobiphenyl	4.0	12.0	67.37	6.85	61.38	4.34	63.65	4.42
35	PCB 118	2,3′,4,4′,5-Pentachlorobiphenyl	4.0	12.0	71.50	5.77	63.13	2.48	70.00	4.10
36	PCB 134	2,2′,3,3′,5,6-Hexachlorobiphenyl	4.0	12.0	84.03	3.23	73.33	4.26	83.20	3.25
37	PCB 114	2,3,4,4′,5-Pentachlorobiphenyl	6.7	20.0	71.40	5.61	66.88	4.87	73.68	4.39
38	PCB 168	2,3′,4,4′,5′,6-Hexachlorobiphenyl	4.0	12.0	69.43	5.10	60.85	8.24	66.90	5.49
39	PCB 127	3,3′,4,5,5′-Pentachlorobiphenyl	4.0	12.0	64.15	5.02	57.83	2.55	64.65	3.97
40	PCB 137	2,2′,3,4,4′,5-Hexachlorobiphenyl	4.0	12.0	65.98	5.25	62.07	4.03	68.03	5.25
41	PCB 163	2,3,3′,4′,5,6-Hexachlorobiphenyl	4.0	12.0	71.70	4.96	67.30	3.33	74.97	3.68
42	PCB 178	2,2′,3,3′,5,5′,6-Heptachlorobiphenyl	4.0	12.0	64.88	3.51	60.90	3.04	68.30	4.21
43	PCB 187	2,2′,3,4′,5,5′,6-Heptachlorobiphenyl	4.0	12.0	66.67	5.59	62.82	2.50	64.05	3.54
44	PCB 162	2,3,3′,4′,5,5′-Hexachlorobiphenyl	4.0	12.0	61.88	5.29	54.93	2.52	62.83	4.88
45	PCB 202	2,2′,3,3′,5,5′,6,6′-Octachlorobiphenyl	4.0	12.0	56.03	5.82	52.85	1.98	59.42	4.95
46	PCB 204	2,2′,3,4,4′,5,6,6′-Octachlorobiphenyl	4.0	12.0	46.88	5.09	40.58	3.73	46.78	4.97
47	PCB 197	2,2′,3,3′,4,4′,6,6′-Octachlorobiphenyl	4.0	12.0	49.57	3.87	45.77	2.01	49.62	3.57
48	PCB 192	2,3,3′,4,5,5′,6-Heptachlorobiphenyl	4.0	12.0	63.83	8.48	52.38	2.67	62.38	3.45
49	PCB 193	2,3,3′,4′,5,5′,6-Heptachlorobiphenyl	4.0	12.0	62.67	6.84	53.32	3.56	63.72	5.96
50	PCB 190	2,3,3′,4,4′,5,6-Heptachlorobiphenyl	4.0	12.0	61.78	6.06	57.73	4.58	63.47	3.75
51	PCB 169	3,3′,4,4′,5,5′-Hexachlorobiphenyl	4.0	12.0	57.30	6.84	52.53	4.49	61.30	4.08
52	PCB 195	2,2′,3,3′,4,4′,5,6-Octachlorobiphenyl	4.0	12.0	54.12	4.57	49.70	4.07	57.23	4.49
53	PCB 206	2,2′,3,3′,4,4′,5,5′,6-Nonachlorobiphenyl	4.0	12.0	42.77	5.05	41.47	7.12	44.93	5.56
54	PCB 209	2,2′,3,3′,4,4′,5,5′,6,6′-Decachlorobiphenyl	4.0	12.0	35.30	14.89	32.08	6.54	34.50	5.84
GC–MS/MS Group B										
55	PCB 002	3-Chlorobiphenyl	4.0	12.0	86.18	6.32	89.55	2.71	82.78	1.53
56	PCB 007	2,4-Dichlorobiphenyl	4.0	12.0	83.78	4.50	89.67	1.50	85.05	2.35
57	PCB 005	2,3-Dichlorobiphenyl	4.0	12.0	88.53	4.97	94.58	2.34	87.22	2.45
58	PCB 011	3,3′-Dichlorobiphenyl	4.0	12.0	90.00	4.78	91.02	3.40	84.25	2.55
59	PCB 013	3,4′-Dichlorobiphenyl	6.7	20.0	88.75	15.70	89.68	4.82	84.12	2.05
60	PCB 032	2,4′,6-Trichlorobiphenyl	4.0	12.0	89.93	4.16	93.90	1.37	93.07	2.12
61	PCB 029	2,4,5-Trichlorobiphenyl	4.0	12.0	80.42	9.71	88.82	1.16	88.18	2.79
62	PCB 050	2,2′,4,6-Tetrachlorobiphenyl	4.0	12.0	96.32	4.73	92.25	3.93	91.73	1.81
63	PCB 053	2,2′,5,6′-Tetrachlorobiphenyl	4.0	12.0	88.98	4.31	104.05	3.62	90.67	4.68

(*continued*)

Table 5.4 (*Continued*)

No.	Abbreviation	Name	LOD (μg/kg)	LOQ (μg/kg)	LOQ		2 LOQ		4 LOQ	
					Average (%)	RSD (%)	Average (%)	RSD (%)	Average (%)	RSD (%)
64	PCB 022	2,3,4′-Trichlorobiphenyl	4.0	12.0	87.42	4.95	94.12	2.55	90.17	1.98
65	PCB 073	2,3′,5′,6-Tetrachlorobiphenyl	4.0	12.0	82.20	4.68	88.42	3.77	85.53	2.56
66	PCB 039	3,4′,5-Trichlorobiphenyl	4.0	12.0	81.97	7.64	84.28	4.10	77.12	2.31
67	PCB 062	2,3,4,6-Tetrachlorobiphenyl	4.0	12.0	81.75	3.80	94.05	1.95	86.68	2.18
68	PCB 038	3,4,5-Trichlorobiphenyl	6.7	20.0	83.42	6.29	86.40	3.21	80.98	1.70
69	PCB 035	3,3′,4-Trichlorobiphenyl	4.0	12.0	84.05	6.14	87.73	2.59	82.70	3.39
70	PCB 064	2,3,4′,6-Tetrachlorobiphenyl	4.0	12.0	87.28	6.20	89.63	2.47	82.68	1.47
71	PCB 037	3,4,4′-Trichlorobiphenyl	4.0	12.0	82.17	4.24	87.48	2.76	79.95	1.31
72	PCB 080	3,3′,5,5′-Tetrachlorobiphenyl	4.0	12.0	76.12	5.75	81.87	2.32	77.80	1.41
73	PCB 058	2,3,3′,5′-Tetrachlorobiphenyl	4.0	12.0	81.45	4.52	81.83	3.70	76.52	2.50
74	PCB 121	2,3′,4,5′,6-Pentachlorobiphenyl	6.7	20.0	66.73	6.28	77.57	3.21	67.80	3.09
75	PCB 093	2,2′,3,5,6-Pentachlorobiphenyl	4.0	12.0	84.72	6.62	90.82	4.05	85.62	1.30
76	PCB 066	2,3′,4,4′-Tetrachlorobiphenyl	4.0	12.0	76.10	6.20	78.25	2.27	71.60	2.10
77	PCB 090	2,2′,3,4′,5-Pentachlorobiphenyl	4.0	12.0	71.98	8.86	78.75	1.79	71.73	2.20
78	PCB 113	2,3,3′,5′,6-Pentachlorobiphenyl	4.0	12.0	76.80	8.87	82.93	6.35	70.07	3.85
79	PCB 089	2,2′,3,4,6′-Pentachlorobiphenyl	4.0	12.0	80.32	6.60	96.20	3.26	88.35	1.66
80	PCB 152	2,2′,3,5,6,6′-Hexachlorobiphenyl	4.0	12.0	79.17	6.01	81.83	3.67	78.98	1.60
81	PCB 145	2,2′,3,4,6,6′-Hexachlorobiphenyl	4.0	12.0	79.75	8.71	83.40	4.23	75.32	2.70
82	PCB 115	2,3,4,4′,6-Pentachlorobiphenyl	4.0	12.0	77.27	4.14	82.32	3.08	75.97	2.68
83	PCB 154	2,2′,4,4′,5,6′-Hexachlorobiphenyl	4.0	12.0	62.80	6.13	69.43	3.30	64.60	3.22
84	PCB 085	2,2′,3,4,4′-Pentachlorobiphenyl	4.0	12.0	80.03	5.85	86.08	2.50	76.28	2.08
85	PCB 151	2,2′,3,5,5′,6-Hexachlorobiphenyl	4.0	12.0	75.77	8.55	83.60	3.31	75.08	3.87
86	PCB 139	2,2′,3,4,4′,6-Hexachlorobiphenyl	4.0	12.0	70.10	6.33	70.07	2.94	70.88	1.53
87	PCB 140	2,2′,3,4,4′,6′-Hexachlorobiphenyl	4.0	12.0	69.55	6.32	74.10	3.65	68.48	2.62
88	PCB 107	2,3,3′,4′,5-Pentachlorobiphenyl	4.0	12.0	70.83	4.65	66.98	5.14	70.35	3.27
89	PCB 143	2,2′,3,4,5,6′-Hexachlorobiphenyl	4.0	12.0	79.40	7.87	75.90	4.89	71.08	3.08
90	PCB 142	2,2′,3,4,5,6-Hexachlorobiphenyl	4.0	12.0	82.48	9.84	78.65	6.24	76.23	5.28
91	PCB 146	2,2′,3,4′,5,5′-Hexachlorobiphenyl	4.0	12.0	67.32	5.99	72.68	2.74	63.73	2.73
92	PCB 122	2′,3,3′,4,5-Pentachlorobiphenyl	4.0	12.0	76.05	9.67	73.72	3.42	68.38	1.97
93	PCB 141	2,2′,3,4,5,5′-Hexachlorobiphenyl	4.0	12.0	66.75	6.30	72.60	4.91	61.90	2.11

94	PCB 130	2,2′,3,3′,4,5′-Hexachlorobiphenyl	4.0	12.0	74.60	13.07	78.72	7.10	64.08	8.18
95	PCB 138	2,2′,3,4,4′,5′-Hexachlorobiphenyl	4.0	12.0	68.75	7.98	72.82	4.80	66.37	1.62
96	PCB 175	2,2′,3,3′,4,5′,6-Heptachlorobiphenyl	4.0	12.0	60.57	5.67	70.82	4.95	58.48	3.11
97	PCB 183	2,2′,3,4,4′,5′,6-Heptachlorobiphenyl	4.0	12.0	62.88	9.56	65.37	6.48	58.35	3.87
98	PCB 166	2,3,4,4′,5,6-Hexachlorobiphenyl	4.0	12.0	73.20	6.60	80.17	3.25	70.68	1.91
99	PCB 128	2,2′,3,3′,4,4′-Hexachlorobiphenyl	4.0	12.0	65.05	7.04	76.03	5.04	70.58	2.80
100	PCB 200	2,2′,3,3′,4,5′,6,6′-Octachlorobiphenyl	4.0	12.0	52.35	7.23	54.88	4.44	49.30	1.70
101	PCB 173	2,2′,3,3′,4,5,6-Heptachlorobiphenyl	4.0	12.0	73.90	6.98	76.93	5.42	69.17	1.59
102	PCB 157	2,3,3′,4,4′,5′-Hexachlorobiphenyl	4.0	12.0	60.90	7.37	65.27	1.97	58.57	2.34
103	PCB 191	2,3,3′,4,4′,5′,6-Heptachlorobiphenyl	4.0	12.0	55.98	7.42	58.25	4.14	53.47	2.59
104	PCB 203	2,2′,3,4,4′,5,5′,6-Octachlorobiphenyl	4.0	12.0	53.40	12.51	56.73	4.90	47.23	1.60
105	PCB 208	2,2′,3,3′,4,5,5′,6,6′-Nonachlorobiphenyl	4.0	12.0	41.83	7.08	46.53	2.22	41.83	2.88
106	PCB 194	2,2′,3,3′,4,4′,5,5′-Octachlorobiphenyl	6.7	20.0	50.17	9.19	50.08	6.01	46.68	1.89
GC–MS/MS Group C										
107	PCB 003	4-Chlorobiphenyl	4.0	12.0	97.45	2.56	89.68	3.95	97.68	2.24
108	PCB 009	2,5-Dichlorobiphenyl	4.0	12.0	91.77	1.88	96.85	4.88	97.75	2.54
109	PCB 014	3,5-Dichlorobiphenyl	4.0	12.0	91.70	3.92	90.33	5.61	95.43	1.99
110	PCB 018	2,2′,5-Trichlorobiphenyl	4.0	12.0	109.83	4.61	92.57	2.66	98.03	1.89
111	PCB 024	2,3,6-Trichlorobiphenyl	4.0	12.0	99.80	6.13	101.17	3.80	98.80	3.34
112	PCB 023	2,3,5-Trichlorobiphenyl	4.0	12.0	91.25	6.09	84.43	3.27	94.52	2.67
113	PCB 054	2,2′,6,6′-Tetrachlorobiphenyl	4.0	12.0	105.75	3.87	96.07	3.96	103.20	2.15
114	PCB 031	2,4′,5-Trichlorobiphenyl	4.0	12.0	88.03	8.27	90.05	9.87	92.05	2.31
115	PCB 033	2′,3,4-Trichlorobiphenyl	4.0	12.0	96.50	3.75	90.88	4.24	95.18	2.55
116	PCB 069	2,3′,4,6-Tetrachlorobiphenyl	4.0	12.0	98.38	6.14	89.87	2.37	92.17	1.45
117	PCB 075	2,4,4′,6-Tetrachlorobiphenyl	4.0	12.0	98.65	10.62	81.32	5.15	89.93	1.83
118	PCB 046	2,2′,3,6′-Tetrachlorobiphenyl	4.0	12.0	101.73	6.35	82.40	3.66	94.35	1.31
119	PCB 047	2,2′,4,4′-Tetrachlorobiphenyl	4.0	12.0	95.75	2.92	92.58	1.39	95.27	1.53
120	PCB 044	2,2′,3,5′-Tetrachlorobiphenyl	4.0	12.0	107.18	7.69	96.57	5.67	107.22	8.70
121	PCB 042	2,2′,3,4′-Tetrachlorobiphenyl	4.0	12.0	100.73	5.97	86.02	4.77	92.78	2.75
122	PCB 071	2,3′,4′,6-Tetrachlorobiphenyl	4.0	12.0	99.53	10.10	90.35	7.76	97.25	3.86
123	PCB 096	2,2′,3,6,6′-Pentachlorobiphenyl	4.0	12.0	104.67	5.04	93.73	4.42	98.15	2.33
124	PCB 088	2,2′,3,4,6-Pentachlorobiphenyl	4.0	12.0	100.42	9.60	88.92	3.41	96.22	3.27
125	PCB 094	2,2′,3,5,6′-Pentachlorobiphenyl	4.0	12.0	100.93	4.60	88.82	0.92	91.58	2.63
126	PCB 098	2,2′,3′,4,6-Pentachlorobiphenyl	4.0	12.0	97.45	5.10	85.75	4.54	89.57	2.64
127	PCB 076	2′,3,4,5-Tetrachlorobiphenyl	4.0	12.0	88.07	8.04	77.10	2.62	85.30	1.47

(*continued*)

Table 5.4 (*Continued*)

No.	Abbreviation	Name	LOD (μg/kg)	LOQ (μg/kg)	LOQ Average (%)	LOQ RSD (%)	2 LOQ Average (%)	2 LOQ RSD (%)	4 LOQ Average (%)	4 LOQ RSD (%)
128	PCB 091	2,2′,3,4′,6-Pentachlorobiphenyl	4.0	12.0	105.08	6.91	80.85	5.98	95.88	3.61
129	PCB 101	2,2′,4,5,5′-Pentachlorobiphenyl	4.0	12.0	85.00	9.40	75.58	6.01	82.83	1.57
130	PCB 056	2,3,3′,4′-Tetrachlorobiphenyl	4.0	12.0	97.77	9.44	76.85	6.44	89.95	1.16
131	PCB 119	2,3′,4,4′,6-Pentachlorobiphenyl	6.7	20.0	86.25	11.58	68.35	4.34	78.12	2.36
132	PCB 079	3,3′,4,5′-Tetrachlorobiphenyl	4.0	12.0	82.65	11.37	69.73	5.12	83.02	1.63
133	PCB 116	2,3,4,5,6-Pentachlorobiphenyl	4.0	12.0	93.05	11.77	74.65	5.46	88.48	3.23
134	PCB 117	2,3,4′,5,6-Pentachlorobiphenyl	4.0	12.0	91.72	6.79	76.07	4.11	87.38	1.48
135	PCB 078	3,3′,4,5-Tetrachlorobiphenyl	4.0	12.0	91.92	9.55	72.78	3.20	81.02	3.13
136	PCB 136	2,2′,3,3′,6,6′-Hexachlorobiphenyl	4.0	12.0	92.68	6.80	80.73	3.04	92.28	2.37
137	PCB 144	2,2′,3,4,5′,6-Hexachlorobiphenyl	4.0	12.0	80.63	7.59	75.93	5.84	81.00	1.94
138	PCB 077	3,3′,4,4′-Tetrachlorobiphenyl	4.0	12.0	91.17	9.19	71.70	5.92	86.80	3.19
139	PCB 149	2,2′,3,4′,5′,6-Hexachlorobiphenyl	4.0	12.0	86.02	5.49	74.18	2.58	80.67	3.13
140	PCB 188	2,2′,3,4′,5,6,6′-Heptachlorobiphenyl	4.0	12.0	71.43	10.20	60.95	2.81	66.22	3.91
141	PCB 133	2,2′,3,3′,5,5′-Hexachlorobiphenyl	4.0	12.0	78.32	8.88	67.67	8.11	77.23	3.88
142	PCB 165	2,3,3′,5,5′,6-Hexachlorobiphenyl	4.0	12.0	85.77	9.80	71.83	4.16	84.65	3.24
143	PCB 161	2,3,3′,4,5′,6-Hexachlorobiphenyl	4.0	12.0	74.58	8.42	64.30	3.04	73.05	3.85
144	PCB 132	2,2′,3,3′,4,6′-Hexachlorobiphenyl	4.0	12.0	90.17	7.39	72.30	3.37	87.05	2.35
145	PCB 105	2,3,3′,4,4′-Pentachlorobiphenyl	4.0	12.0	82.80	10.50	69.63	4.35	81.40	1.76
146	PCB 186	2,2′,3,4,5,6,6′-Heptachlorobiphenyl	4.0	12.0	77.70	8.01	67.82	3.48	76.03	2.87
147	PCB 158	2,3,3′,4,4′,6-Hexachlorobiphenyl	4.0	12.0	80.27	13.97	62.27	7.76	72.57	4.14
148	PCB 182	2,2′,3,4,4′,5,6′-Heptachlorobiphenyl	4.0	12.0	69.38	12.76	51.80	5.40	62.38	3.92
149	PCB 159	2,3,3′,4,5,5′-Hexachlorobiphenyl	4.0	12.0	69.08	12.58	60.57	3.87	67.20	3.17
150	PCB 167	2,3′,4,4′,5,5′-Hexachlorobiphenyl	6.7	20.0	66.28	8.78	57.85	3.91	62.62	3.40
151	PCB 174	2,2′,3,3′,4,5,6′-Heptachlorobiphenyl	4.0	12.0	69.93	9.74	61.23	5.82	71.35	3.39
152	PCB 177	2,2′,3,3′,4′,5,6-Heptachlorobiphenyl	4.0	12.0	81.25	10.65	59.53	6.16	72.57	3.58
153	PCB 156	2,3,3′,4,4′,5-Hexachlorobiphenyl	4.0	12.0	76.35	11.59	58.73	5.14	70.75	2.98
154	PCB 180	2,2′,3,4,4′,5,5′-Heptachlorobiphenyl	4.0	12.0	46.02	22.23	38.47	25.07	56.50	2.83
155	PCB 198	2,2′,3,3′,4,5,5′,6-Octachlorobiphenyl	4.0	12.0	63.65	11.30	48.98	8.12	57.57	4.22
156	PCB 196	2,2′,3,3′,4,4′,5,6′-Octachlorobiphenyl	4.0	12.0	56.90	10.70	46.70	6.81	52.02	3.13
157	PCB 207	2,2′,3,3′,4,4′,5,6,6′-Nonachlorobiphenyl	4.0	12.0	49.68	16.46	36.23	9.05	43.40	2.91
158	PCB 205	2,3,3′,4,4′,5,5′,6-Octachlorobiphenyl	4.0	12.0	63.67	14.66	44.85	8.84	53.43	2.44

GC–MS/MS Group D										
159	PCB 010	2,6-Dichlorobiphenyl	4.0	12.0	113.75	5.70	87.96	5.89	88.65	4.72
160	PCB 006	2,3′-Dichlorobiphenyl	6.7	20.0	108.72	3.60	87.13	5.69	83.75	4.01
161	PCB 030	2,4,6-Trichlorobiphenyl	4.0	12.0	98.55	5.12	82.47	6.31	81.00	5.84
162	PCB 017	2,2′,4-Trichlorobiphenyl	4.0	12.0	94.33	4.22	88.81	6.72	87.17	5.45
163	PCB 015	4,4′-Dichlorobiphenyl	4.0	12.0	97.80	6.20	83.71	7.18	81.78	5.74
164	PCB 034	2′,3,5-Trichlorobiphenyl	4.0	12.0	91.35	5.50	76.99	6.61	76.83	6.94
165	PCB 026	2,3′,5-Trichlorobiphenyl	4.0	12.0	94.73	5.42	78.91	6.21	82.38	5.08
166	PCB 028	2,4,4′-Trichlorobiphenyl	4.0	12.0	87.25	4.32	84.39	10.40	77.98	6.82
167	PCB 051	2,2′,4,6′-Tetrachlorobiphenyl	4.0	12.0	99.52	3.85	85.36	5.56	86.92	5.33
168	PCB 045	2,2′,3,6-Tetrachlorobiphenyl	4.0	12.0	102.43	4.17	92.70	6.84	90.93	4.61
169	PCB 052	2,2′,5,5′-Tetrachlorobiphenyl	4.0	12.0	93.53	4.91	80.41	7.57	87.05	4.37
170	PCB 049	2,2′,4,5′-Tetrachlorobiphenyl	4.0	12.0	89.28	8.81	78.50	6.39	88.77	4.86
171	PCB 048	2,2′,4,5-Tetrachlorobiphenyl	4.0	12.0	93.98	6.11	83.83	7.33	77.82	6.85
172	PCB 059	2,3,3′,6-Tetrachlorobiphenyl	4.0	12.0	100.47	5.13	89.07	7.48	79.45	4.61
173	PCB 068	2,3′,4,5′-Tetrachlorobiphenyl	4.0	12.0	79.72	5.34	73.89	5.48	69.33	6.10
174	PCB 100	2,2′,4,4′,6-Pentachlorobiphenyl	4.0	12.0	89.02	6.09	76.13	8.51	71.72	3.59
175	PCB 057	2,3,3′,5-Tetrachlorobiphenyl	4.0	12.0	87.32	4.08	76.03	4.02	71.62	5.18
176	PCB 063	2,3,4′,5-Tetrachlorobiphenyl	4.0	12.0	93.32	14.05	81.17	5.93	74.48	2.10
177	PCB 061	2,3,4,5-Tetrachlorobiphenyl	4.0	12.0	88.83	6.52	75.21	6.28	75.00	3.68
178	PCB 155	2,2′,4,4′,6,6′-Hexachlorobiphenyl	4.0	12.0	70.28	5.56	64.13	5.82	63.12	4.47
179	PCB 070	2,3′,4′,5-Tetrachlorobiphenyl	4.0	12.0	87.62	5.83	75.83	4.77	73.27	5.39
180	PCB 055	2,3,3′,4-Tetrachlorobiphenyl	4.0	12.0	87.58	9.01	81.16	5.38	75.82	5.23
181	PCB 060	2,3,4,4′-Tetrachlorobiphenyl	4.0	12.0	88.77	7.09	80.59	6.32	73.30	4.23
182	PCB 150	2,2′,3,4′,6,6′-Hexachlorobiphenyl	4.0	12.0	82.28	9.08	75.24	7.90	71.50	5.27
183	PCB 112	2,3,3′,5,6-Pentachlorobiphenyl	4.0	12.0	88.67	5.67	79.54	5.70	76.77	4.30
184	PCB 148	2,2′,3,4′,5,6′-Hexachlorobiphenyl	4.0	12.0	72.25	6.79	65.73	7.10	66.17	5.05
185	PCB 111	2,3,3′,5,5′-Pentachlorobiphenyl	4.0	12.0	76.95	7.12	68.66	8.60	60.18	4.01
186	PCB 097	2,2′,3′,4,5-Pentachlorobiphenyl	4.0	12.0	93.22	5.80	75.49	13.62	67.40	12.64
187	PCB 120	2,3′,4,5,5′-Pentachlorobiphenyl	4.0	12.0	76.02	6.91	62.14	8.94	57.05	4.39
188	PCB 081	3,4,4′,5-Tetrachlorobiphenyl	4.0	12.0	72.75	9.43	70.19	6.86	60.78	5.74
189	PCB 147	2,2′,3,4′,5,6-Hexachlorobiphenyl	4.0	12.0	74.03	8.48	64.00	10.44	51.23	10.70
190	PCB 082	2,2′,3,3′,4-Pentachlorobiphenyl	4.0	12.0	84.75	4.80	72.41	3.05	66.77	5.17
191	PCB 108	2,3,3′,4,5′-Pentachlorobiphenyl	4.0	12.0	74.02	2.12	60.34	7.42	53.40	5.00
192	PCB 106	2,3,3′,4,5-Pentachlorobiphenyl	6.7	20.0	70.43	3.86	64.24	6.12	60.32	4.90

(continued)

Table 5.4 (*Continued*)

No.	Abbreviation	Name	LOD (μg/kg)	LOQ (μg/kg)	LOQ Average (%)	LOQ RSD (%)	2 LOQ Average (%)	2 LOQ RSD (%)	4 LOQ Average (%)	4 LOQ RSD (%)
193	PCB 184	2,2′,3,4,4′,6,6′-Heptachlorobiphenyl	4.0	12.0	59.28	4.17	53.49	6.75	47.72	5.66
194	PCB 131	2,2′,3,3′,4,6-Hexachlorobiphenyl	4.0	12.0	77.57	2.09	70.81	6.03	67.00	5.33
195	PCB 153	2,2′,4,4′,5,5′-Hexachlorobiphenyl	6.7	20.0	74.78	4.16	57.54	10.46	53.97	5.23
196	PCB 179	2,2′,3,3′,5,6,6′-Heptachlorobiphenyl	4.0	12.0	79.27	5.24	70.16	7.41	63.73	4.08
197	PCB 176	2,2′,3,3′,4,6,6′-Heptachlorobiphenyl	4.0	12.0	69.93	6.82	65.20	5.24	58.93	4.61
198	PCB 160	2,3,3′,4,5,6-Hexachlorobiphenyl	4.0	12.0	81.55	9.14	79.90	5.74	65.35	6.16
199	PCB 164	2,3,3′,4′,5′,6-Hexachlorobiphenyl	4.0	12.0	76.22	6.50	64.87	7.58	67.47	13.76
200	PCB 129	2,2′,3,3′,4,5-Hexachlorobiphenyl	4.0	12.0	76.27	7.70	69.63	6.92	66.10	7.22
201	PCB 126	3,3′,4,4′,5-Pentachlorobiphenyl	4.0	12.0	72.10	8.32	67.46	7.33	58.82	7.19
202	PCB 185	2,2′,3,4,5,5′,6-Heptachlorobiphenyl	4.0	12.0	74.43	6.64	65.23	6.35	59.27	6.36
203	PCB 181	2,2′,3,4,4′,5,6-Heptachlorobiphenyl	4.0	12.0	64.07	7.38	59.74	7.96	56.57	5.15
204	PCB 171	2,2′,3,3′,4,4′,6-Heptachlorobiphenyl	4.0	12.0	71.38	9.30	62.50	6.96	58.88	7.06
205	PCB 172	2,2′,3,3′,4,5,5′-Heptachlorobiphenyl	4.0	12.0	65.62	8.27	57.91	7.66	53.87	6.26
206	PCB 199	2,2′,3,3′,4,5,6,6′-Octachlorobiphenyl	4.0	12.0	62.20	5.54	54.34	6.38	53.55	6.02
207	PCB 201	2,2′,3,3′,4,5,5′,6′-Octachlorobiphenyl	4.0	12.0	58.60	11.61	53.23	8.42	52.10	5.51
208	PCB 170	2,2′,3,3′,4,4′,5-Heptachlorobiphenyl	4.0	12.0	63.60	8.71	62.16	9.93	55.35	5.00
209	PCB 189	2,3,3′,4,4′,5,5′-Heptachlorobiphenyl	4.0	12.0	57.67	9.86	49.63	7.26	46.52	7.24
GC–MS/MS Group E										
210	–	Naphthalene	1.0	1.9	81.98	18.23	65.07	12.15	64.62	18.43
211	–	Isoproturon	126.7	253.4	85.90	6.60	102.92	7.24	87.60	3.91
212	–	Dichlorvos	5.9	11.8	82.68	10.24	86.83	6.48	88.85	6.55
213	–	Carbofuran	4.3	8.6	90.90	9.12	100.82	4.73	108.68	6.27
214	–	Methamidophos	85.4	170.8	124.15	10.78	101.42	6.28	94.80	5.59
215	–	Acenaphthylene	2.2	4.5	122.50	18.20	122.53	22.51	91.88	10.21
216	–	Acenaphthene	4.6	9.1	116.30	8.48	105.47	11.31	95.55	8.46
217	–	Fluorene	6.6	13.2	129.97	16.71	128.35	25.39	101.28	13.53
218	–	Hexachlorobenzene	2.9	5.8	90.65	4.70	89.93	8.27	91.23	3.91
219	–	Ethoprophos	1.3	2.6	110.37	25.11	99.05	10.16	114.95	12.76
220	–	Chlordimeform	5.8	11.6	110.32	12.58	108.90	11.56	105.38	6.50
221	–	Trifluralin	4.5	9.0	63.48	4.36	66.13	2.70	82.13	2.49

222	–	α-HCH	4.4	8.7	106.48	5.18	113.52	4.62	101.98	1.60
223	–	Omethoate	18.2	36.5	82.77	16.10	78.22	8.18	80.95	7.37
224	–	Anthracene	0.1	0.3	107.78	21.87	105.38	19.21	109.85	16.00
225	–	Clomazone	2.9	5.8	99.68	6.10	93.55	8.07	112.88	3.23
226	–	Diazinon	2.7	5.4	73.53	8.70	95.40	4.81	114.22	4.13
227	–	Phenanthrene	5.3	10.5	125.13	23.00	130.00	20.80	118.60	16.53
228	–	γ-HCH	4.3	8.6	94.38	4.02	102.67	3.82	101.75	3.86
229	–	Atrazine	4.0	7.9	82.52	7.40	91.35	5.60	98.62	2.22
230	–	Simazine	11.8	23.5	91.43	5.72	102.50	4.06	115.03	2.93
231	–	Heptachlor	6.5	13.0	96.65	3.00	112.33	4.34	114.07	4.39
232	–	Pirimicarb	3.1	6.1	93.10	2.78	105.93	3.31	111.97	1.79
233	–	Dimethoate	6.5	13.1	116.38	5.97	123.63	11.32	123.83	3.56
234	–	Aldrin	4.6	9.1	93.73	18.13	83.03	6.39	104.03	7.13
235	–	Alachlor	2.9	5.8	91.90	11.26	110.32	4.28	118.38	3.92
236	–	Prometryne	2.7	5.4	86.88	6.14	94.63	4.83	107.62	3.05
237	–	Chlorothalonil	233.0	466.0	69.02	26.85	101.28	22.14	105.40	22.30
238	–	Phthalic acid bis-butyl ester	6.6	13.3	133.97	18.76	88.30	8.52	109.35	5.00
239	–	β-HCH	4.5	9.0	90.60	5.27	104.45	6.59	105.58	5.60
240	–	Chlorpyrifos	4.5	9.0	88.75	10.05	89.50	7.00	103.75	5.98
241	–	Parathion-methyl	4.3	8.6	79.95	8.72	102.07	4.87	113.23	4.70
242	–	Dicofol	3.4	6.7	86.72	6.37	100.33	20.05	98.12	2.35
243	–	Metolachlor	1.1	2.2	94.50	1.52	103.77	2.03	108.88	1.43
244	–	δ-HCH	4.3	8.7	109.90	6.11	108.00	7.51	111.77	4.42
245	–	Triadimefon	4.8	9.6	85.22	4.48	100.75	4.29	111.48	2.78
246	–	Fluoranthene	3.0	6.1	123.87	22.11	124.25	60.71	112.07	18.42
247	–	2,4′-DDE	6.7	13.3	90.75	1.89	101.82	3.65	96.65	2.16
248	–	*cis*-Chlordane	6.3	12.5	81.47	5.65	100.63	5.16	96.28	3.61
249	–	Phenthoate	0.4	0.9	96.95	2.89	110.70	6.61	113.00	4.48
250	–	*trans*-Chlordane	6.2	12.4	102.28	6.36	87.87	6.36	100.52	3.72
251	–	Pyrene	5.2	10.5	126.38	22.61	119.25	21.09	109.03	18.31
252	–	4,4′-DDE	6.6	13.1	94.03	2.29	94.33	2.38	87.98	2.41
253	–	Butachlor	0.7	1.5	104.60	24.76	124.02	9.85	107.24	7.93
254	–	Dieldrin	13.2	26.3	95.57	5.93	98.27	9.23	104.02	3.91
255	–	2,4′-DDD	2.7	5.4	97.42	1.74	99.80	3.38	103.02	7.82
256	–	Buprofezin	14.5	29.1	112.02	12.47	105.58	5.42	103.57	2.81

(*continued*)

Table 5.4 (*Continued*)

					LOQ		2 LOQ		4 LOQ	
No.	Abbreviation	Name	LOD (μg/kg)	LOQ (μg/kg)	Average (%)	RSD (%)	Average (%)	RSD (%)	Average (%)	RSD (%)
257	–	Endrin	30.0	60.0	98.93	6.49	106.87	4.52	105.77	4.06
258	–	2,4′-DDT	2.1	4.2	87.27	8.19	101.38	5.48	124.77	28.85
259	–	Nithophen	8.1	16.2	103.67	9.19	97.92	2.55	109.10	5.80
260	–	Oxyfluorfen	14.6	29.2	96.33	5.74	90.15	4.73	109.98	3.88
261	–	4,4′-DDD	6.6	13.3	95.62	1.66	100.83	3.10	102.52	1.90
262	–	4,4′-DDT	3.0	5.9	98.57	9.99	117.25	7.90	109.78	8.00
263	–	Phthalic acid benzyl butyl ester	6.6	13.1	90.17	20.91	78.70	4.22	98.35	1.89
264	–	Propargite	127.5	255.0	94.60	13.72	106.52	7.95	103.25	4.30
265	–	Tricyclazole	0.7	1.5	92.25	7.38	109.22	5.79	117.98	7.81
266	–	Triazophos	11.4	22.9	105.97	9.76	106.88	10.55	101.82	6.62
267	–	Mirex	2.8	5.5	75.25	4.39	73.87	3.28	77.77	17.89
268	–	Benzo(*a*)anthracene	2.0	4.0	118.80	23.25	118.43	21.04	109.93	18.67
269	–	Phthalic acid bis-2-ethylhexyl ester	1.3	2.7	122.82	13.94	137.15	24.56	124.98	21.89
270	–	Amitraz	1.6	3.1	77.65	5.55	76.98	3.63	88.80	2.41
271	–	Lambda-cyhalothrin	6.4	12.8	25.67	12.89	25.30	6.84	24.75	5.70
272	–	Pyridaben	6.7	13.3	96.12	1.12	100.82	2.77	100.92	2.71
273	–	Benzo(*b*)fluoranthene	0.1	0.2	120.73	22.85	112.00	23.32	109.35	20.62
274	–	Benzo(*k*)fluoranthene	0.2	0.4	127.72	19.38	111.38	25.22	109.55	22.39
275	–	Cyfluthrin	8.7	17.5	92.43	8.89	125.48	14.23	117.03	5.99
276	–	Cypermethrin	18.5	37.1	101.45	3.57	115.70	5.50	103.90	4.90
277	–	Benzo(*a*)pyrene	1.8	3.5	109.10	20.69	116.98	19.38	108.23	13.30
278	–	Acetamiprid	27.3	54.6	80.85	7.75	91.52	4.81	102.90	4.50
279	–	Fenvalerate 1	29.0	58.1	100.05	6.60	101.65	4.28	113.33	7.11
280	–	Fenvalerate 2	29.0	58.1	103.05	9.05	106.95	7.18	112.77	7.82
281	–	Deltamethrin	24.2	48.4	128.15	27.73	120.08	8.63	125.00	16.08
282	–	Indeno(1,2,3-*cd*)pyrene	2.2	4.4	128.95	29.01	121.92	28.98	110.15	19.67
283	–	Dibenzo(*a*,*h*)anthracene	2.2	4.4	126.72	23.70	111.72	20.01	85.62	7.02
284	–	Benzo(*g*,*h*,*i*)perylene	2.2	4.4	121.85	26.13	125.85	28.65	120.27	25.22
GC–MS										
285	–	Endosulfan I	0.1	0.2	98.32	5.05	97.80	2.94	97.88	4.43

286	–	Endosulfan II	0.1	0.2	105.78	9.36	110.23	9.39	85.92	7.95
LC–MS/MS										
287	–	Trichlorphon	3.6	7.1	96.97	15.04	115.00	7.17	99.93	10.66
288	–	Metsulfuron-methyl	3.5	7.1	100.52	21.54	87.08	10.44	85.07	19.82
289	–	Chlortoluron	13.2	26.4	69.73	10.18	81.20	5.63	50.93	11.90
290	–	2,4-D	55.3	110.5	37.08	12.09	70.88	7.11	42.37	3.22
291	–	Bensulfuron-methyl	9.4	18.8	95.25	15.63	115.33	6.07	75.97	18.87
292	–	Propanil	8.0	16.0	90.12	11.02	101.00	22.71	89.78	20.40
293	–	Fipronil	14.5	29.0	25.98	7.32	18.02	28.66	16.37	14.40
294	–	Phoxim	4.1	8.2	99.87	8.51	126.50	8.95	109.22	11.70
295	–	Hexythiazox	31.5	63.0	69.53	8.90	97.12	25.17	74.35	4.28

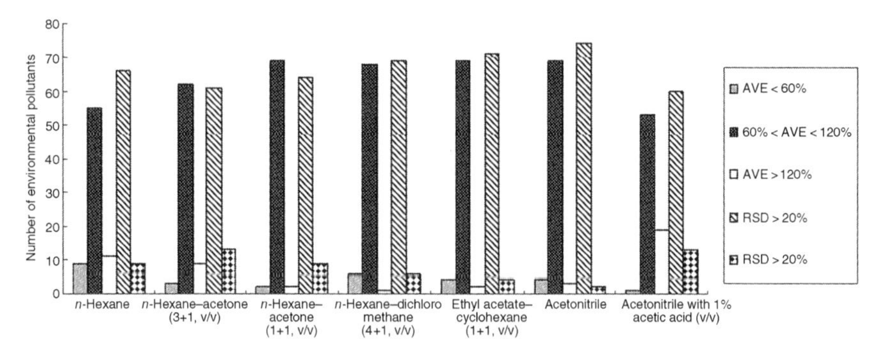

Figure 5.3. The comparison of the extraction efficiencies of seven solvents for 77 environmental pollutants in 5 g fat samples.

5.3.3 Comparison of Sample Extraction Methods

After the selection of acetonitrile as the extraction solvent, three extraction methods were compared:

1. *Accelerated Solvent Extraction:* Accurately weigh 5 g sample into a mortar containing 17 g diatomite, grind evenly and transfer to the sample cell, heat at 10.34 MPa, 80 °C for 5 min, extract with acetonitrile for 3 min, repeat twice, and purge with nitrogen for 100 s (to collect all acetonitrile from the sample residues).
2. *Homogeneous Extraction:* Accurately weigh 5 g sample, add 35 ml acetonitrile, homogenize at 12,000 rpm for 1 min, and repeat twice.
3. *Oscillation Extraction:* Accurately weigh 5 g sample into a mortar containing 10 g anhydrous sodium sulfate, grind evenly and transfer to an 80 ml centrifuge tube, add 35 ml acetonitrile, oscillation extract for 30 min, and repeat twice.

For each solvent, three standard added samples and one matrix standard are prepared for calculating recoveries of 77 environmental pollutants. The results of three standard fortified samples are averaged before the calculation of recoveries. Different extraction methods are applied, and the extracts are collected. After concentration and GPC cleanup, GC–MS/MS is used to detect the analytes, and internal standard is used for quantification.

In homogeneous extraction, the recoveries of 70 analytes are between 60 and 120%, accounting for 90.9% of the total; in oscillation extraction, the recoveries of 65 analytes are between 60 and 120%, accounting for 84.4% of the total; in ASE extraction, the recoveries of 64 analytes are between 60 and 120%, accounting for 83.1% of the total. In homogeneous extraction, the RSDs of 74 analytes are less than 20%, accounting for 96.1% of the total; in oscillation extraction, the RSDs of 66 analytes are less than 20%, accounting for 85.7% of the total; in ASE extraction, the RSDs of 69 analytes are less than 20%, accounting for 89.6% of the total. Overall, homogeneous extraction is better than the other methods. For octachlorobiphenyl, nonachlorobiphenyl, and decachlorobiphenyl, ASE is better than homogeneous extraction and oscillation extraction, but considering other environmental pollutants, homogeneous extraction is better than ASE and oscillation extraction as it is fast, easy, and can avoid the influence of the high temperature of ASE extraction. Thus, homogeneous extraction was selected.

5.3.4 Comparison of Sample Cleanup

Under the above-mentioned conditions, less fat can be dissolved by homogeneous extraction with acetonitrile. As SPE is simple, highly efficient, and uses less solvent, GPC is compared with SPE. Three SPE methods are selected.

Accurately weigh 5 g sample, add standard solution of 2 LOQ, add 35 ml acetonitrile, and homogeneously extract twice.

1. Transfer the concentrate with 2 ml acetonitrile–toluene (3 + 1, v/v) to NH_2-Carb cartridge cleanup, repeat three times, and then elute with 25 ml acetonitrile–toluene (3 + 1, v/v). Evaporate the eluate to about 1 ml using a rotary evaporator with a water bath at 45 °C. Add 5 ml *n*-hexane for solvent exchange, repeat twice, and then evaporate the eluate to about 1 ml using a rotary evaporator for detection.
2. Transfer the concentrate with 5 ml acetonitrile to neutral Al_2O_3 cartridge cleanup, repeat two times, and then elute with 10 ml acetonitrile. Concentrate the eluate and transfer the concentrate with 2 ml acetonitrile–toluene (3 + 1, v/v) to NH_2-Carb cartridge cleanup, repeat three times, and then elute with 25 ml acetonitrile–toluene (3 + 1, v/v). Evaporate the eluate to about 1 ml using a rotary evaporator with a water bath at 45 °C. Add 5 ml *n*-hexane for solvent exchange, repeat twice, and then evaporate the eluate to about 1 ml using a rotary evaporator for detection.
3. Transfer the concentrate with 5 ml acetonitrile to C18 cartridge cleanup, repeat two times, and then elute with 10 ml acetonitrile. Concentrate the eluate and transfer it with 2 ml acetonitrile–toluene (3 + 1, v/v) to NH_2-Carb cartridge cleanup, repeat three times, and then elute with 25 ml acetonitrile–toluene (3 + 1, v/v). Evaporatc thc cluatc to about 1 ml using a rotary cvaporator with a water bath at 45 °C. Add 5 ml *n*-hexane for solvent exchange, repeat twice, and then evaporate the eluate to about 1 ml using a rotary evaporator for detection.

Seven pesticides and seven PAHs cannot be detected by C18 + NH_2-Carb and Al_2O_3 + NH_2-Carb cleanup. It may be that these chemicals are adsorbed on the cartridge and thus are not completely eluted and have low recovery or no recovery. The results are shown in Figure 5.4.

Figure 5.4 shows that the recoveries of GPC cleanup and NH_2-Carb cartridge cleanup are better. The TIC plots of GPC cleanup and NH_2-Carb column cleanup show that GPC cleanup has low noise, good cleanup effect, and causes low pollution

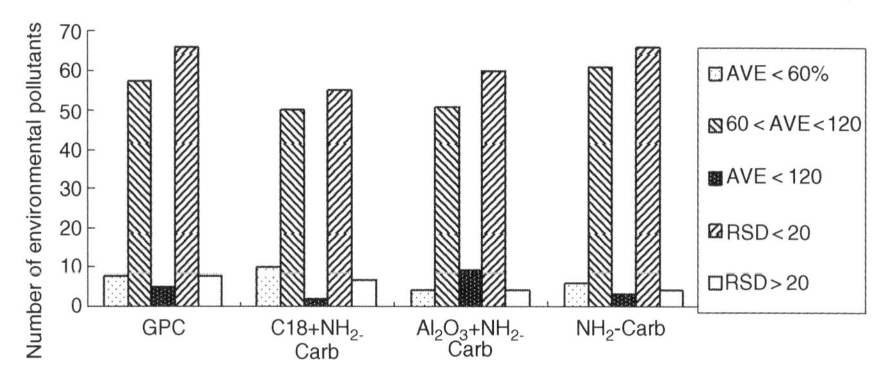

Figure 5.4. The comparison of the cleanup efficiencies of four cleanup methods for 77 environmental pollutants in 5 g fat samples.

on the instrument; thus, GPC cleanup was finally selected. The GC–MS/MS TIC plots of four cleanup methods are shown in Figures 5.5–5.8 for comparison.

5.3.5 Linear Range, LOD, and LOQ

Two hundred and ninety-five environmental pollutants are detected using the above method. The results show that the concentrations show a good linear relationship with the responses in the linear ranges, and the correlation coefficients are between 0.9684 and 1.0000. The correlation coefficients of 270 environmental pollutants are higher than 0.9900, accounting for 91.5% of the total. The linear ranges, linear equations, and correlation coefficients are given in Tables 5.1–5.3. The LODs (a S/N ratio of >5) of 295 pesticides are 0.1–233.0 μg/kg, and LODs of 275 pesticides are lower than 10.0 μg/kg, accounting for 93.2% of the total. The LOQ (S/N > 10) of the method is 0.2–466.0 μg/kg. The LOD and LOQ data are given in Table 5.4. For each concentration (LOQ, 2 LOQ, and 4 LOQ), five standard added samples and one matrix standard are prepared for calculating recoveries of 295 environmental pollutants. The results of five standard added samples are averaged before the calculation of recoveries.

5.3.6 Recoveries and Precisions

Pork samples are obtained from the markets in China. The fat of the pork is separated from other tissues manually and homogenized by grinding in a meat blender. Two hundred and ninety-five standard solutions are added to the homogenized samples to measure the recoveries of the samples spiked with LOQ, 2 LOQ, and 4 LOQ and to measure the precisions ($n = 6$); the results are given in Table 5.4. The recoveries of 259 standards with LOQ concentrations are between 60.0 and 120.0%, accounting for 87.8% of the total. The RSDs of 278 standards with LOQ concentrations are 20%, accounting for 94.2% of the total. The recoveries of 248 standards with 2 LOQ concentrations are between 60.0 and 120.0%, accounting for 84.1% of the total. The

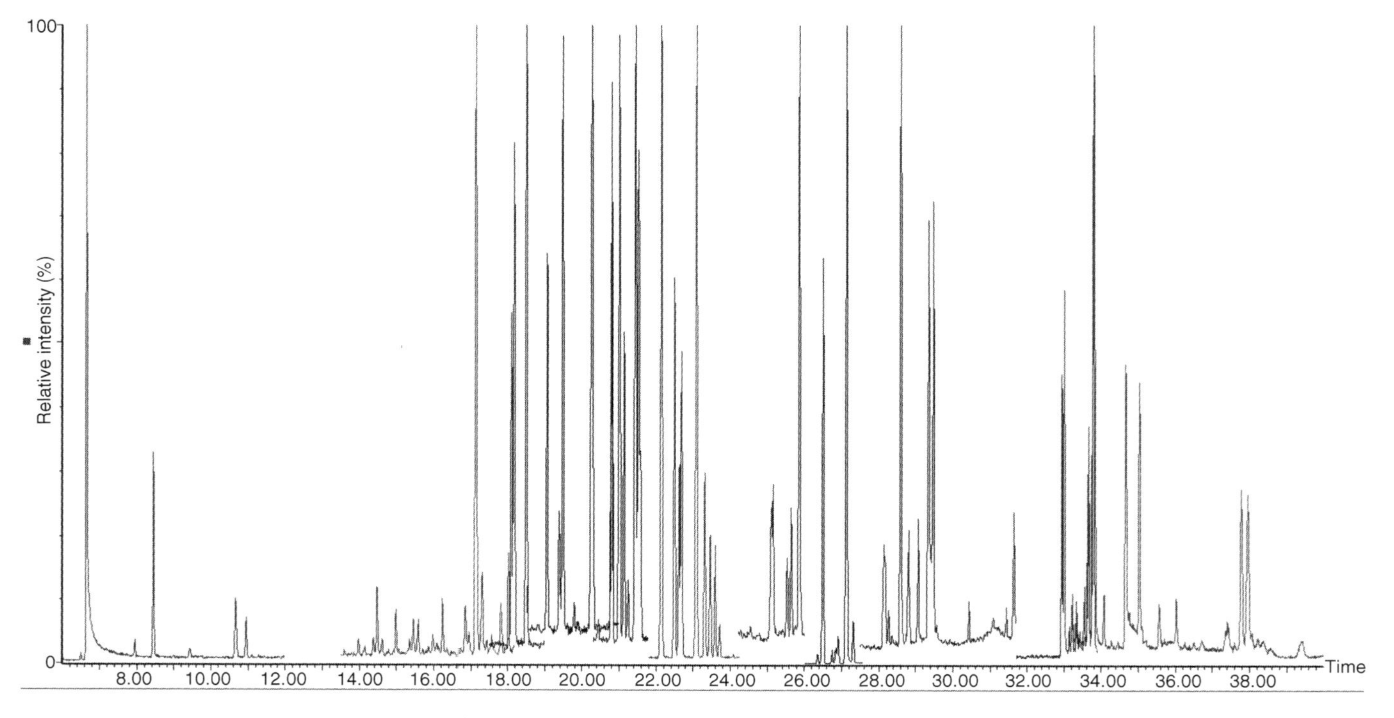

Figure 5.5. The GC–MS/MS TIC plot of GPC cleanup.

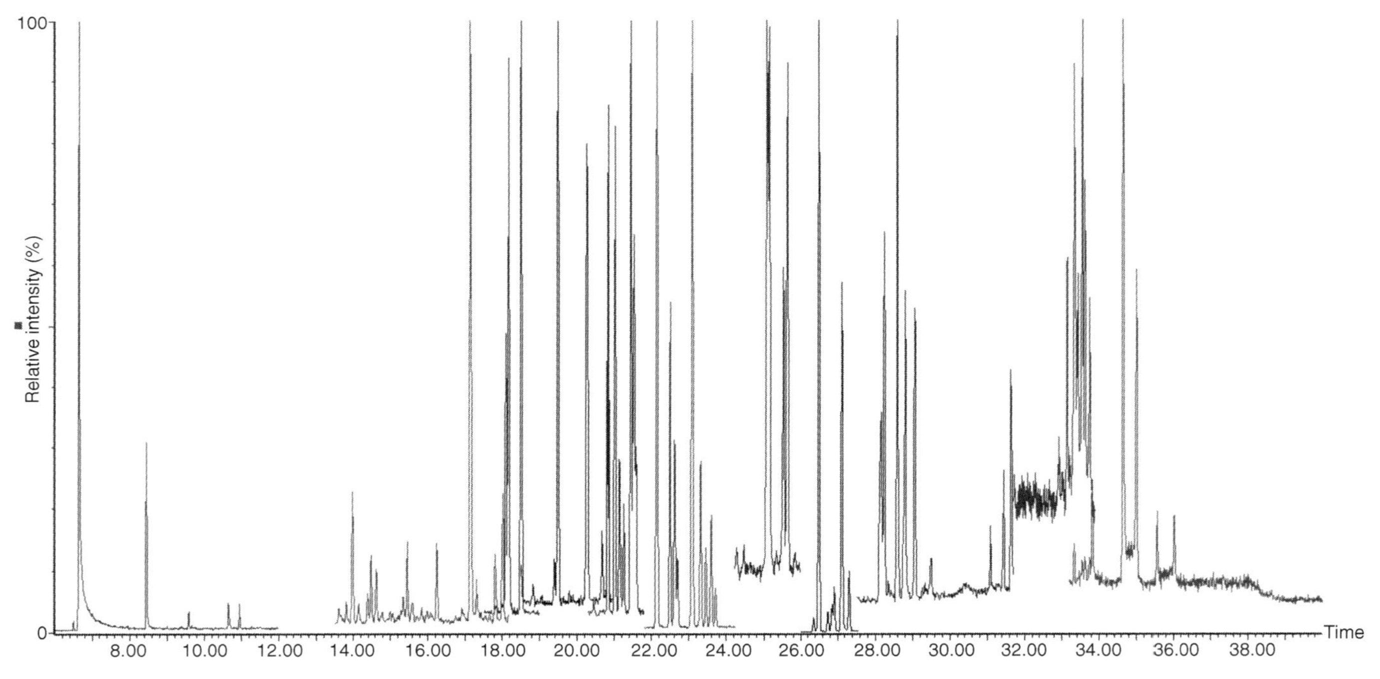

Figure 5.6. The GC–MS/MS TIC plot of C18 + NH_2-Carb cleanup.

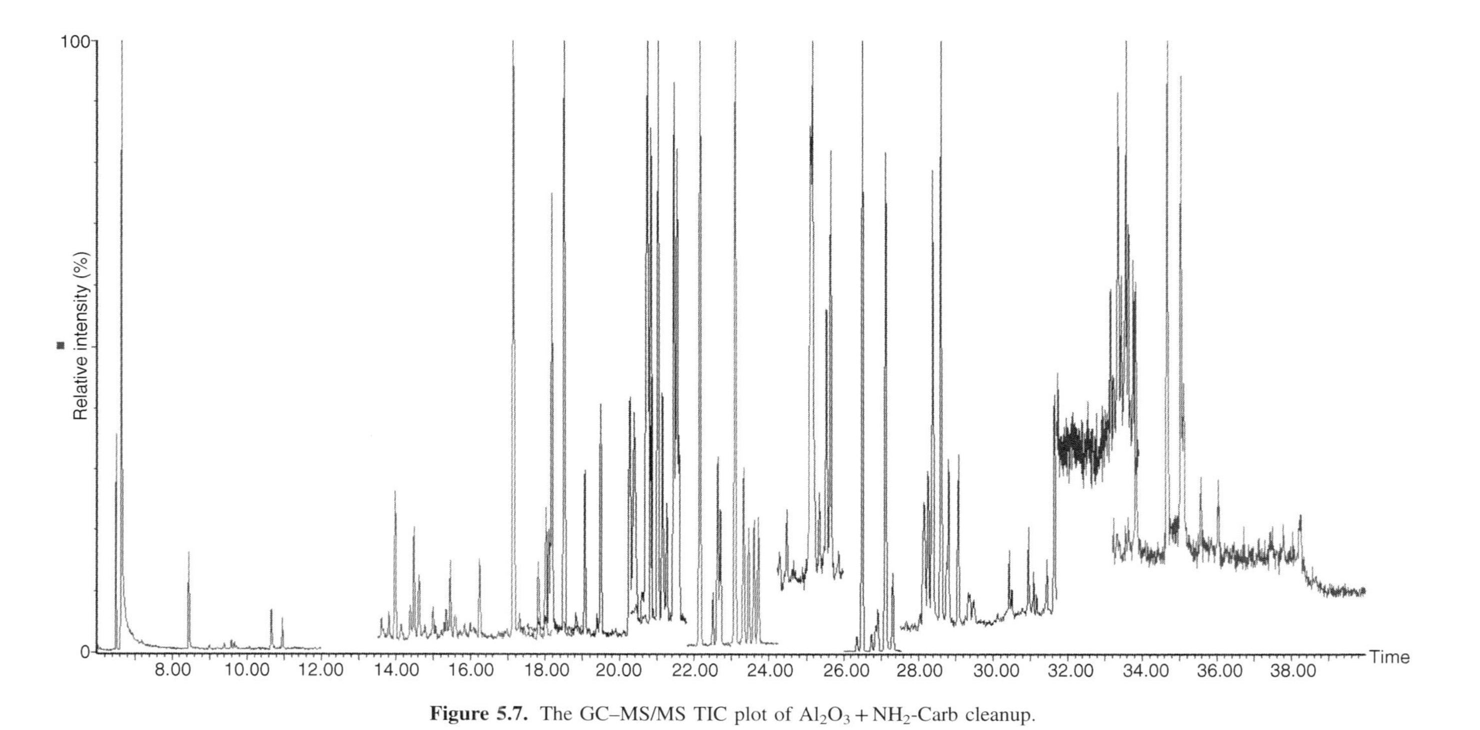

Figure 5.7. The GC–MS/MS TIC plot of $Al_2O_3 + NH_2$-Carb cleanup.

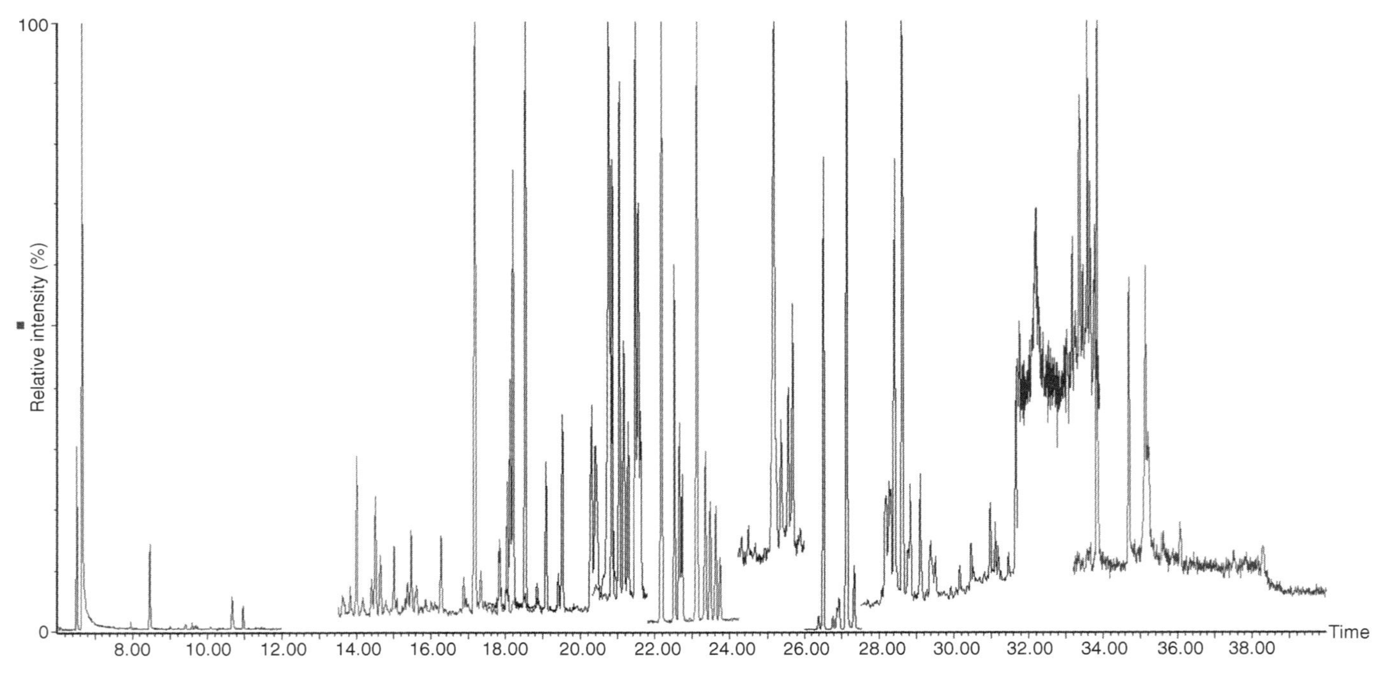

Figure 5.8. The GC–MS/MS TIC plot of NH_2-Carb cleanup.

RSDs of 277 standards with 2 LOQ concentrations are below 20%, accounting for 93.9% of the total. The recoveries of 252 standards with 4 LOQ concentrations are between 60.0 and 120.0%, accounting for 85.4% of the total. The RSDs of 287 standards with 4 LOQ concentrations are below 20%, accounting for 97.3% of the total. The above data show that the method has good reproducibility and is suitable for the analysis of actual body fat samples.

5.3.7 Actual Sample Analysis

Six hundred and thirty-three human body adipose samples from different regions of Jiangsu province of China (provided by Nanjing Institute of Environmental Science) were analyzed with the above method. Complete analysis of one sample, including sample preparation, instrument analysis, and data analysis, takes about 15 h. The results are given in Table 5.5.

Table 5.5 shows that 70 environmental pollutants are detected in 655 samples (including isomers, a total of 81), containing the following:

1. *POPs:* A total of seven kinds: DDT, HCH, aldrin, endrin, HCB, mirex, and chlordane; among them, the positive rate of 4,4′-DDT was the highest (99.1%), the positive rates of aldrin, *cis*-chlordane, *trans*-chlordane, and endrin were the lowest (0.2%), and the difference between them was 496 times; the average concentration of 4,4′-DDE was the highest (2220 μg/kg), the average concentration of aldrin was the lowest (4.6 μg/kg), and the difference between them was 483 times.
2. *PAHs:* A total of 15 kinds: phenanthrene, benzo(*k*)fluoranthene, fluoranthene, pyrene, benzo(*a*)anthracene, benzo(*g*,*h*,*i*)perylene, naphthalene, benzo(*a*)perylene, indeno(1,2,3-*cd*)pyrene, fluorine, acenaphthene, dibenzo(*a*,*h*)anthracene, acenaphthylene, benzo(*b*)fluoranthene, and anthracene; among them, the positive rate of fluorine was the highest (86.4%), the positive rate of dibenzo [*a*,*h*]anthracene was the lowest (21.8%), and the difference between them was 4.0 times; the average concentration of phenanthrene was the highest (163.0 μg/kg), the average concentration of benzo(*b*)fluoranthene was the lowest (5.1 μg/kg), and the difference between them was 32.2 times.
3. *PAEs:* A total of three kinds: phthalic acid bis-2-ethylhexyl ester, phthalic acid bis-butyl ester, and phthalic acid benzyl butyl ester; among them, the positive rate and average concentration of phthalic acid bis-2-ethylhexyl ester were the highest (46.9% and 970.6 μg/kg, respectively), the positive rate and average concentration of phthalic acid benzyl butyl ester were the lowest (7.6% and 10.3 μg/kg, respectively), and the difference of positive rate between them was 6.2 times and the difference of average concentration between them was 94.6 times.
4. *PCBs:* A total of 18 kinds: PCB 004, PCB 021, PCB 028, PCB 074, PCB 084, PCB 099, PCB 101, PCB 102, PCB 118, PCB 127, PCB 138, PCB 153, PCB 163, PCB 168, PCB 180, PCB 190, PCB 192, and PCB 193; among them, the

Table 5.5. Minimum, Maximum, Average, Positive Number, and Positive Rate of 70 Environmental Pollutants

No.	Species	Name	Minimum (μg/kg)	Maximum (μg/kg)	Average (μg/kg)	Positive Number	Positive Rate (%)
1	POPs	2,4′-DDD	2.7	55.4	5.2	40	6.3
2	POPs	2,4′-DDE	6.7	127.2	13.4	153	24.2
3	POPs	2,4′-DDT	2.1	363.4	7.1	568	89.7
4	POPs	4,4′-DDD	6.6	238.9	16.6	469	74.1
5	POPs	4,4′-DDE	22.8	28436	2220	621	98.1
6	POPs	4,4′-DDT	3.2	879.6	54.7	627	99.1
7	POPs	α-HCH	4.4	39.4	6.9	165	26.1
8	POPs	β-HCH	4.9	3636.0	305.4	590	93.2
9	POPs	γ-HCH	4.3	31.1	12.6	13	2.1
10	POPs	δ-HCH	7.6	401.1	204.3	2	0.3
11	POPs	Aldrin	4.6	4.6	4.6	1	0.2
12	POPs	Endrin	2206	2206	2206	1	0.2
13	POPs	Hexachlorobenzene	2.9	335.8	27.3	607	95.9
14	POPs	Mirex	2.8	34.4	4.8	181	28.6
15	POPs	*cis*-Chlordane	131.2	131.2	131.2	1	0.2
16	POPs	*trans*-Chlordane	337.6	337.6	337.6	1	0.2
17	PAHs	Acenaphthene	4.6	97.1	12.9	211	33.3
18	PAHs	Acenaphthylene	2.2	65.2	7.9	346	54.7
19	PAHs	Anthracene	0.1	73.6	5.2	513	81.0
20	PAHs	Benzo(*a*)anthracene	2.0	944.8	68.3	352	55.6
21	PAHs	Benzo(*a*)pyrene	1.8	695.4	48.6	330	52.1
22	PAHs	Benzo(*b*)fluoranthene	0.1	78.6	5.1	359	56.7
23	PAHs	Benzo(*g,h,i*)perylene	2.2	1033	66.7	338	53.4
24	PAHs	Benzo(*k*)fluoranthene	0.2	2360	132.5	385	60.8
25	PAHs	Dibenzo(*a,h*)anthracene	2.2	135.9	13.7	138	21.8
26	PAHs	Fluoranthene	3.0	1965	136.4	434	68.6
27	PAHs	Fluorene	6.6	179.2	20.5	547	86.4
28	PAHs	Indeno(1,2,3-*cd*)pyrene	2.2	386.6	26.8	242	38.2
29	PAHs	Naphthalene	1.1	348.3	15.9	534	84.4
30	PAHs	Phenanthrene	5.3	2600	163.0	503	79.5
31	PAHs	Pyrene	5.2	1884	112.4	379	59.9
32	PAEs	Phthalic acid benzyl butyl ester	6.6	31.1	10.3	48	7.6

33	PAEs	Phthalic acid bis-2-ethylhexyl ester	2.3	47274	970.6	297	46.9
34	PAEs	Phthalic acid bis-butyl ester	6.6	4555	74.6	239	37.8
35	Other pesticides	Chlorothalonil	4441	31392	7122	30	4.7
36	Other pesticides	Ethoprophos	1.9	711.0	239.6	3	0.5
37	Other pesticides	Nithophen	14.4	997.2	274.2	4	0.6
38	Other pesticides	Pyridaben	7.3	7.3	7.3	1	0.2
39	Other pesticides	Butachlor	1.0	430.6	30.3	18	2.8
40	Other pesticides	Chlorpyrifos	4.5	32.8	9.4	33	5.2
41	Other pesticides	Diazinon	2.7	3.4	2.9	3	0.5
42	Other pesticides	Trifluralin	4.5	1552	93.4	18	2.8
43	Other pesticides	Cyfluthrin	8.7	16.2	10.7	4	0.6
44	Other pesticides	Methamidophos	129.0	1033	245.4	53	8.4
45	Other pesticides	Alachlor	839.6	839.6	839.6	1	0.2
46	Other pesticides	Parathion-methyl	13.0	315.2	164.1	2	0.3
47	Other pesticides	Pirimicarb	3.3	4.6	3.9	2	0.3
48	Other pesticides	Carbofuran	4.3	5.4	4.9	2	0.3
49	Other pesticides	Endosulfan I	0.1	272.9	45.6	6	0.9
50	Other pesticides	Endosulfan II	0.1	240.6	27.6	16	2.5
51	Other pesticides	Cypermethrin	19.3	45.3	29.9	12	1.9
52	Other pesticides	Prometryne	3.4	3.4	3.4	1	0.2
53	Other pesticides	Fenvalerate 1	29.0	52.5	39.3	7	1.1
54	Other pesticides	Fenvalerate 2	29.3	74.8	45.7	9	1.4
55	Other pesticides	Buprofezin	14.5	406.9	82.1	9	1.4
56	Other pesticides	Lambda-cyhalothrin	38.5	38.5	38.5	1	0.2
57	Other pesticides	Tricyclazole	0.7	C	0.9	6	0.9
58	Other pesticides	Dicofol	3.4	558.8	11.9	268	42.3
59	Other pesticides	Triazophos	29.6	2667.5	526.7	8	1.3
60	Other pesticides	Chlordimeform	5.8	57.3	11.7	200	31.6
61	Other pesticides	Deltamethrin	45.3	71.8	57.6	4	0.6
62	Other pesticides	Oxyfluorfen	35.4	692.2	256.8	3	0.5
63	Other pesticides	Metolachlor	1.1	300.3	38.7	8	1.3
64	PCB	PCB 004	4.0	53.4	8.4	127	38.7
65	PCB	PCB 021	5.4	11.9	9.1	3	0.9
66	PCB	PCB 028	4.0	12.8	6.8	15	4.6
67	PCB	PCB 074	4.0	53.5	22.3	5	1.5

(continued)

Table 5.5 (*Continued*)

No.	Species	Name	Minimum (μg/kg)	Maximum (μg/kg)	Average (μg/kg)	Positive Number	Positive Rate (%)
68	PCB	PCB 084	6.2	6.2	6.2	1	0.2
69	PCB	PCB 099	5.0	6.2	5.6	2	0.6
70	PCB	PCB 101	4.0	12.9	5.8	10	3.0
71	PCB	PCB 102	4.0	4.0	4.0	1	0.2
72	PCB	PCB 118	4.0	43.5	6.4	55	16.8
73	PCB	PCB 127	4.5	8.4	6.4	4	1.2
74	PCB	PCB 138	4.0	50.5	6.6	87	26.5
75	PCB	PCB 153	6.7	31.3	10.4	44	13.4
76	PCB	PCB 163	4.0	43.8	6.3	86	26.2
77	PCB	PCB 168	4.0	42.3	6.9	113	34.5
78	PCB	PCB 180	4.0	10.8	5.1	13	4.0
79	PCB	PCB 190	5.4	5.4	5.4	1	0.2
80	PCB	PCB 192	4.0	15.8	6.1	21	6.4
81	PCB	PCB 193	4.0	6.3	4.5	5	1.5

positive rate of PCB 004 was the highest (38.7%), the positive rates of PCB 084, PCB 102, and PCB 190 were the lowest (0.3%), and the difference between them was 129 times; the average concentration of PCB 074 was the highest (22.3 μg/kg), the average concentration of PCB 102 was the lowest (4.0 μg/kg), and the difference between them was 5.6 times.

5. *Other Pesticides:* A total of 27 kinds: chlorothalonil, ethoprophos, nithophen, pyridaben, butachlor, chlorpyrifos, diazinon, trifluralin, cyfluthrin, methamidophos, alachlor, parathion-methyl, pirimicarb, carbofuran, endosulfan I, endosulfan II, cypermethrin, prometryne, fenvalerate 1, fenvalerate 2, buprofezin, lambda-cyhalothrin, tricyclazole, dicofol, triazophos, chlordimeform, deltamethrin, oxyfluorfen, and metolachlor; among them, the positive rate of dicofol was the highest (42.3%), the positive rates of pyridaben, alachlor, prometryne, and lambda-cyhalothrin were the lowest (0.2%), and the difference between them was 212 times; the average concentration of chlorothalonil was the highest (7122 μg/kg), the average concentration of tricyclazole was the lowest (0.9 μg/kg), and the difference between them was 8187 times.

Among these chemical pollutants, DDT, HCH, HCB, acenaphthylene, anthracene, benzo(*a*)anthracene, benzo(*a*)pyrene, benzo(*b*)fluoranthene, benzo(*g*,*h*,*i*)perylene, benzo(*k*)fluoranthene, fluoranthene, fluorine, naphthalene, phenanthrene, and pyrene had higher positive rates (>50%); chlorothalonil, 4,4′-DDE, endrin, phthalic acid bis-2-ethylhexyl ester, alachlor, triazophos, *trans*-chlordane, β-HCH, nithophen, oxyfluorfen, methamidophos, ethoprophos, δ-HCH, parathion-methyl, phenanthrene, fluoranthene, benzo(*k*)fluoranthene, *cis*-chlordane, and pyrene had higher average concentrations (>100 μg/kg).

From the statistical results above, the average positive rates of PAHs were the highest, those of POPs were the second, and those of other pesticides and PCBs were the lowest; the average concentrations of POPs, PAEs, and other pesticides were higher than those of others and the average concentrations of PCBs were the lowest. Therefore, comparing five types of chemical pollutants, POPs had higher average positive rates and average concentrations. Although PAHs had the highest average positive rates, their average concentrations were not high. In PAEs, the average positive rate and average concentration of phthalic acid bis-2-ethylhexyl ester were higher. The average positive rate and average concentration of PCBs were very low. The average positive rates of other pesticides were also very low, but their average concentrations were higher than those of others, for example, the average concentration of chlorothalonil was the highest in all detected chemical pollutants (7122 μg/kg). 4,4′-DDE, β-HCH, phenanthrene, fluoranthene, benzo(*k*)fluoranthene, and pyrene were the most noteworthy pollutants, as their positive rates and average concentrations were very high.

5.4 CONCLUSIONS

This chapter compares different extraction solvents, sample extraction methods, and cleanup methods and analyzes environmental pollutants by homogeneous extraction with acetonitrile, GPC cleanup, online concentration, GC–MS/MS, GC–MS (NCI source), and LC–MS/MS. This method is applied to the analysis of 633 actual body fat samples. The detected environmental pollutants basically accord with their usages in China. Thus, it is proven that this method is suitable for the screening analysis of different types of environmental pollutants in biological samples and can provide strong technical support for China's risk assessment of environmental pollution and environmental health management.

REFERENCES

1. Shi, Y.J.; Lv, L.Y.; Ren, C.H.; Liang, D. International developments on the study of persistent organic pollutants. *World Sci. Technol. Res. Dev.* **2003**, 25(2), 73–78.
2. World Health Organization. The biological detection of persistent organic pollutants (POPs). INFOSAN.
3. LeDoux, M. Analytical methods applied to the determination of pesticide residues in foods of animal origin: a review of the past two decades. *J. Chromatogr. A* **2011**, 1218(8), 1021–1036.
4. Gutiérrez Valencia, T.M.; García de Llasera, M.P. Determination of organophosphorus pesticides in bovine tissue by an on-line coupled matrix solid-phase dispersion–solid phase extraction–high performance liquid chromatography with diode array detection method. *J. Chromatogr. A* **2011**, 1218(39), 6869–6877.
5. Botella, B.; Crespo, J.; Rivas, A.; Cerrillo, I.; Olea-Serrano, M.F.; Olea, N. Exposure of women to organochlorine pesticides in Southern Spain. *Environ. Res.* **2004**, 96(1), 34–40.
6. Matsuoka, T.; Akiyama, Y.; Mitsuhashi, T. Screening method of pesticides in meat using cleanup with GPC and mini-column. *Shokuhin Eiseigaku Zasshi* **2009**, 50(2), 97–107.
7. Garrido Frenich, A.; Martínez Vidal, J.L.; Cruz Sicilia, A.D.; González Rodríguez, M.J.; Plaza Bolaños, P. Multiresidue analysis of organochlorine and organophosphorus pesticides in muscle of chicken, pork and lamb by gas chromatography–triple quadrupole mass spectrometry. *Anal. Chim. Acta* **2006**, 558(1–2), 42–52.
8. Ahmad, R.; Salem, N.M.; Estaitieh, H. Occurrence of organochlorine pesticide residues in eggs, chicken and meat in Jordan. *Chemosphere* **2010**, 78(6), 667–671.
9. Wu, G.; Bao, X.X.; Zhao, S.H.; Wu, J.J.; Han, A.L.; Ye, Q.F. Analysis of multi-pesticide residues in the foods of animal origin by GC–MS coupled with accelerated solvent extraction and gel permeation chromatography cleanup. *Food Chem.* **2011**, 126(2), 646–654.
10. Cheng, J.H.; Liu, M.; Zhang, X.Y.; Ding, L.; Yu, Y.; Wang, X.P.; Jin, H.Y.; Zhang, H.Q. Determination of triazine herbicides in sheep liver by microwave-assisted extraction and high performance liquid chromatography. *Anal. Chim. Acta* **2007**, 590(1), 34–39.
11. Przybylski, C.; Segard, C. Method for routine screening of pesticides and metabolites in meat based baby-food using extraction and gas chromatography–mass spectrometry. *J. Sep. Sci.* **2009**, 32(11), 1858–1867.
12. Pérez, A.; González, G.; González, J.; Heinzen, H. Multiresidue determination of pesticides in lanolin using matrix solid-phase dispersion. *J. AOAC Int.* **2010**, 93(2), 712–719.
13. Makabe, Y.; Miyamoto, F.; Hashimoto, H.; Nakanishi, K.; Hasegawa, Y. Determination of residual pesticides in processed foods manufactured from livestock foods and seafoods using ion trap GC/MS. *Shokuhin Eiseigaku Zasshi* **2010**, 51(4), 182–195.

14. Jacob, K.W. Polycyclic aromatic hydrocarbons of environmental and occupational importance. *Fresenius J. Anal. Chem.* **1986**, 323(1), 1–10.

15. Grote, M.; Schüürmann, G.; Altenburger, R. Modeling photoinduced algal toxicity of PAHs. *Environ. Sci. Technol.* **2005**, 39(11), 4141–4149.

16. Song, G.Q.; Lin, J.M. Sample pretreatment techniques for polycyclic aromatic hydrocarbons in environmental matrix. *Acta Sci. Circumst.* **2005**, 25(10), 1287–1296.

17. Jian, X.C. The protection of polycyclic aromatic hydrocarbons. *J. Environ. Prot.* **1995**, 10, 31–33.

18. Chen, B.H.; Wang, C.Y.; Chiu, C.P. Evaluation of analysis of polycyclic aromatic hydrocarbons in meat products by liquid chromatography. *J. Agric. Food Chem.* **1996**, 44(8), 2244–2251.

19. Jira, W. A GC/MS method for the determination of carcinogenic polycyclic aromatic hydrocarbons (PAH) in smoked meat products and liquid smokes. *Eur. Food Res. Technol.* **2004**, 218(2), 208–212.

20. Jira, W.; Ziegenhals, K.; Speer, K. Gas chromatography–mass spectrometry (GC–MS) method for the determination of 16 European priority polycyclic aromatic hydrocarbons in smoked meat products and edible oils. *Food Addit. Contam. Part A* **2008**, 25(6), 704–713.

21. Wang, L.X.; Li, H.; Zhang, J.X.; Zhou, Z. Simultaneous determination of 16 polycyclic aromatic hydrocarbons in roast by GPC–SPE–HPLC. *J. Hebei Acad. Sci.* **2009**, 26(4), 43–45.

22. Purcaroa, G.; Moret, S.; Contea, L.S. Optimisation of microwave assisted extraction (MAE) for polycyclic aromatic hydrocarbon (PAH) determination in smoked meat. *Meat Sci.* **2009**, 81(1), 275–280.

23. Smith, C.J.; Walcott, C.J.; Huang, W.; Maggio, V.; Grainger, J.; Patterson, D.G., Jr., Determination of selected monohydroxy metabolites of 2-, 3- and 4-ring polycyclic aromatic hydrocarbons in urine by solid-phase microextraction and isotope dilution gas chromatography–mass spectrometry. *J. Chromatogr. B* **2002**, 778(1–2), 157–164.

24. Djinovica, J.; Popovicb, A.; Jiraa, W. Polycyclic aromatic hydrocarbons (PAHs) in different types of smoked meat products from Serbia. *Meat Sci.* **2008**, 80(2), 449–456.

25. ISO 22959-2009. Animal and vegetable fats and oils: determination of polycyclic aromatic hydrocarbons by on-line donor–acceptor complex chromatography and HPLC with fluorescence detection.

26. Liu, Z.J.; Bai, X. Determination of PCBs in contaminated soil. *Environ. Sci. Manag.* **2005**, 30(4), 103–104.

27. Ren, R.; Zhang, D.X. *Chemistry and Environment*. Beijing, China: Chemical Industry Press; **2001**, p. 206.

28. Suiyuan, H. *Environmental Toxicology*. Beijing, China: Chemical Industry Press; **2002**, p. 92.

29. Liu, Z.R.; Chen, Z.M.; Zhao, G.Y.; Chen, D.H. *Environmental Chemistry Course*. Beijing, China: Chemical Industry Press; **2005**, p. 15.

30. Wang, L.S. *Environmental Organic Chemistry*, 2nd edition. Beijing, China: Chemical Industry Press; **2003**, p. 252.

31. Djordjevic, M.V.; Hoffmann, D.; Fan, J.; Prokopczyk, B.; Citron, M.L.; Stellman, S.D. Assessment of chlorinated pesticides and polychlorinated biphenyls in adipose breast tissue using a supercritical fluid extraction method. *Carcinogenesis* **1994**, 15 (11), 2581–2585.

32. González-Barros, S.T.; Simal Lozano, J.; Lage Yusty, M.A.; Alvarez Piñeiro, M.E. A simple ultraviolet method for discriminating between polychlorobiphenyls and organochloride pesticides coeluting under gas chromatography with electron-capture detection. *J. Chromatogr. Sci.* **1998**, 36(5), 263–268.
33. Kodba, Z.C.; Voncina, D.B. A rapid method for the determination of organochlorine, pyrethroid pesticides and polychlorobiphenyls in fatty foods using GC with electron capture detection. *Chromatographia* **2007**, 66(7–8), 619–624.
34. Kim, H.; Jeffrey, W.F. Determination of polychlorinated biphenyl 126 in liver and adipose tissues by GC–μECD with liquid extraction and SPE clean-up. *Chromatographia* **2008**, 68 (3–4),307–309.
35. Sannino, A.; Mambriani, P.; Bandini, M.; Bolzoni, L. Multiresidue method for determination of organochlorine insecticides and polychlorinated biphenyl congeners in fatty processed foods. *J. AOAC Int.* **1996**, 79(6), 1434–1446.
36. Valsamaki, V.I.; Boti, V.I.; Sakkas, V.A.; Albanis, T.A. Determination of organochlorine pesticides and polychlorinated biphenyls in chicken eggs by matrix solid phase dispersion. *Anal. Chim. Acta* **2006**, 573–574, 195–201.
37. Wang, Y.F.; Chen, X.H.; Fu, X.H. Determination of polychlorinated biphenyls and organochlorines in biological samples by gas chromatography–tandem mass spectrometry. *Chin. J. Chromatogr.* **2007**, 25(1), 112–114.
38. Jaborek-Hugo, S.; von Holst, C.; Allen, R.; Stewart, T.; Willey, J.; Anklam, E. Use of an immunoassay as a means to detect polychlorinated biphenyls in animal fat. *Food Addit. Contam.* **2001**, 18(2), 121–127.
39. Bordet, F.; Inthavong, D.; Fremy, J.M. Interlaboratory study of a multiresidue gas chromatographic method for determination of organochlorine and pyrethroid pesticides and polychlorobiphenyls in milk, fish, eggs, and beef fat. *J. AOAC Int.* **2002**, 85(6), 1398–1409.
40. Zhang, Z.; Ohiozebau, E.; Rhind, S.M. Simultaneous extraction and clean-up of polybrominated diphenyl ethers and polychlorinated biphenyls from sheep liver tissue by selective pressurized liquid extraction and analysis by gas chromatography–mass spectrometry. *J. Chromatogr. A* **2011**, 1218(8), 1203–1239.
41. Djordjevic, M.V.; Hoffmann, D.; Fan, J.; Prokopczyk, B.; Citron, M.L.; Stellman, S.D. Assessment of chlorinated pesticides and polychlorinated biphenyls in adipose breast tissue using a supercritical fluid extraction method. *Carcinogenesis* **1994**, 15(11), 2581–2585.
42. Chen, H.M.; Wang, C.; Wang, X. Analysis of the phthalates in cosmetics by capillary gas chromatography. *Chin. J. Chromatogr.* **2004**, 22(3), 224–227.
43. Casajuana, N.; Lacorte, S. New methodology for the determination of phthalate esters, bisphenol A, bisphenol A diglycidyl ether, and nonylphenol in commercial whole milk samples. *J. Agric. Food Chem.* **2004**, 52(12), 3702–3707.
44. Bartolomé, L.; Cortazar, E.; Raposo, J.C.; Usobiaga, A.; Zuloaga, O.; Etxebarria, N.; Fernández, L.A. Simultaneous microwave-assisted extraction of polycyclic aromatic hydrocarbons, polychlorinated biphenyls, phthalate esters and nonylphenols in sediments. *J. Chromatogr. A* **2005**, 1068(2), 229–236.
45. Ballesteros, O.; Zafra, A.; Navalón, A.; Vílchez, J.L. Sensitive gas chromatographic–mass spectrometric method for the determination of phthalate esters, alkylphenols, bisphenol A and their chlorinated derivatives in wastewater samples. *J. Chromatogr. A* **2006**, 1121(2), 154–162.

46. Kondo, F.; Ikai, Y.; Hayashi, R.; Okumura, M.; Takatori, S.; Nakazawa, H.; Izumi, S.; Makino, T. Determination of five phthalate monoesters in human urine using gas chromatography–mass spectrometry. *Bull. Environ. Contam. Toxicol.* **2010**, 85(1), 92–96.
47. Lin, Z.G.; Sun, R.N.; Zhang, L.L.; Zou, X.M.; Chen, M.Y.; Tu, F.Z.; Ma, Y.; Jiang, W.J. Simultaneous determination of 14 phthalate ester residues in animal innards by gas chromatography–mass spectrometry with electron impact ionization. *Chin. J. Chromatogr.* **2008**, 26(3), 280–284.
48. Colón, I.; Dimandja, J.M. High-throughput analysis of phthalate esters in human serum by direct immersion SPME followed by isotope dilution-fast GC/MS. *Anal. Bioanal. Chem.* **2004**, 380(2), 275–283.
49. Prieto, A.; Zuloaga, O.; Usobiaga, A.; Etxebarria, N.; Fernández, L.A. Development of a stir bar sorptive extraction and thermal desorption–gas chromatography–mass spectrometry method for the simultaneous determination of several persistent organic pollutants in water samples. *J. Chromatogr. A* **2007**, 1174(1–2), 40–49.
50. Rios, J.J.; Morales, A.; Márquez-Ruiz, G. Headspace solid-phase microextraction of oil matrices heated at high temperature and phthalate esters determination by gas chromatography multistage mass spectrometry. *Talanta* **2010**, 80(5), 2076–2082.
51. Pang, G.F.; Zhao, T.S.; Cao, Y.Z.; Fan, C.L. Cleanup with two Florisil column for gas chromatographic determination of multiple pyrethroid insecticides in products of animal origin. *J. AOAC Int.* **1994**, 77(6), 1634–1638.
52. Pang, G.F.; Cao, Y.Z.; Zhang, J.J.; Fan, C.L.; Liu, Y.M.; Li, X.M.; Jia, G.Q.; Li, Z.Y.; Shi, Y.Q.; Wu, Y.P.; Guo, T.T. Validation study on 660 pesticide residues in animal tissues by gel permeation chromatography cleanup/gas chromatography–mass spectrometry and liquid chromatography–tandem mass spectrometry. *J. Chromatogr. A* **2006**, 1125, 1–30.
53. Pang, G.F.; Cao, Y.Z.; Fan, C.L.; Jia, G.Q.; Zhang, J.J.; Li, X.M.; Liu, Y.M.; Shi, Y.Q.; Li, Z.Y.; Zheng, F.; Lian, Y.J. Analysis method study on 839 pesticide and chemical contaminant multiresidues in animal muscles by gel permeation chromatography cleanup, GC/MS, and LC/MS/MS. *J. AOAC Int.* **2009**, 92(3), S1–S72.

CHAPTER

6

ULTRAHIGH-PERFORMANCE LIQUID CHROMATOGRAPHY COUPLED WITH HIGH-RESOLUTION MASS SPECTROMETRY: A RELIABLE TOOL FOR ANALYSIS OF VETERINARY DRUGS IN FOOD

MARÍA DEL MAR AGUILERA-LUIZ, ROBERTO ROMERO-GONZÁLEZ, PATRICIA PLAZA-BOLAÑOS, JOSÉ LUIS MARTÍNEZ VIDAL, and ANTONIA GARRIDO FRENICH

6.1 INTRODUCTION

A wide variety of veterinary drugs (VDs) or veterinary medicinal products (VMPs) as they are referred to—particularly in the EU—are commonly used in livestock production for prevention (prophylaxis) and treatment of several types of pathologies of food-producing animals. Since the United States Food and Drug Administration (USFDA) approved their use as growth promoters in the 1950s, they have been administered to animals through feed at subtherapeutic doses or via drinking water, increasing food production and avoiding illnesses and infections. In the 1990s and at the beginning of 2000s, it was estimated that at least 70% [1] of food-producing animals were exposed to antimicrobials and 11,800 tonnes of antibiotics were administered to animals every year. Only 908 tonnes were used to treat active infections, whereas the rest was used to prevent infections or to promote growth [2]. Today, this figure may be higher due to current intense animal husbandry practices [3]. The average consumption in nine European countries (period 2005–2009) is higher than 2400 tonnes of active ingredient [4].

However, the widespread administration of VDs, and their uncontrolled or incorrect use (i.e., attention to withdrawal time) in some regions, entailed a risk to human health due to the possibility of introducing harmful residues into the food chain. The main concern regarding the ingestion of food-containing residues with antimicrobial activity relies on the fact that they might provoke allergies and gastric intestinal disturbances and potentially contribute to the development of resistant bacterial strains [5], although there is no robust evidence to support this last possibility. Therefore, the presence of VD residues, as well as metabolites and/or conjugates in animal food products, may have direct or indirect toxic effects on

High-Throughput Analysis for Food Safety, First Edition.
Edited by Perry G. Wang, Mark F. Vitha, and Jack F. Kay.

consumers [6]. In consequence, the control of the concentrations of these substances in the final food product has become a cause of concern for governments.

In the food safety framework, the EU and other countries such as the U.S. or Canada have strictly regulated and controlled the use of VDs, including growth promoters, particularly in food-producing animal species, by publishing different regulations and directives [7,8]. In addition, some of these substances, such as diethylstilbestrol, nitrofurans, and chloramphenicol, have been banned in farming production in the EU because either they are toxic compounds or complete toxicological dossiers are not available, whereas others, such as tetracyclines or sulfonamides, have been authorized as long as their concentrations in food of animal origin are below certain established limits.

In consequence, it is necessary to develop sensitive analytical methods that comply with the current legislation, allowing the control and determination of authorized and banned VDs in food and thus ensuring the safety of the food products derived from treated animals. In this sense, a high number of analytical methods have been developed to allow the identification of VD residues in several kinds of foodstuff [9–11]. Traditionally, these methods have been based on microbiological assays or immunoassay techniques [12,13]. However, they have been replaced by more selective and sensitive techniques, such as liquid chromatography (LC) coupled with fluorescence or UV detection [14,15]. These conventional detection techniques have been gradually replaced by mass spectrometry (MS) detection, bearing in mind that public health agencies rely on detection by MS for unambiguous confirmation of VDs in foodstuffs [16]. Thus, LC coupled with single quadrupole (Q) or triple quadrupole (QqQ) has been used for this purpose, although Q–MS has been highly replaced by QqQ tandem MS (QqQ–MS/MS), which provides more confidence in analyte identification and quantification.

LC coupled with Q–MS and QqQ–MS/MS (low-resolution mass spectrometry analyzers, (LRMS)) has been widely used for the determination of a relatively lower number of compounds (i.e., ≤100–200 compounds in a single run) [9,11]. However, LRMS has limitations due to the fact that there are >250 different chemical substances, which can be employed in husbandry practices, apart from other "cocktails" (mixtures of low amounts of several substances that exert a synergistic effect) [17], which have also to be monitored. This may imply a sensitivity decrease in multiresidue methods when conventional LC–QqQ–MS/MS is used. Besides, the need for developing quick and low-cost methods that cover all the established requirements by the regulations for residue control requires that typical methods and instruments be replaced by new ones, including high-resolution mass spectrometry analyzers (HRMS) such as time of flight (TOF), hybrid quadrupole-time of flight (QqTOF), or Orbitrap.

6.2 VETERINARY DRUG LEGISLATION

Bearing in mind the problems concerning the widespread use of VDs in livestock production, several measures have been taken by national governments to control their use in feed and their presence in food products and the environment as residues.

VDs were approved to be used as growth promoters in food-producing animals in the 1950s. Since then, the European countries have been concerned about their use. Studies demonstrating the negative effect on human health of VD residues led the EU to establish strict regulated controls of the use of these substances in living animals. The first documents were published in 1981 [18,19] and their subsequent modifications were developed for the establishment of a common framework regulating the production and distribution of VDs as well as the establishment of protocols for the analysis, production control, marketing, and free circulation of VDs among the Member States of the European Community (EC) [20–23]. All these directives were merged in a single document, Directive 2001/82/CE [24], which has been subsequently modified [25–27].

Regarding the food safety issue, the EU has defined a series of legislative documents in order to ensure a high standard of protection for consumers. Council Regulation 2377/90/EC [28] was the first document related to the control of VD residues. This document regulated the use of VDs, describing the procedure for establishing maximum residue limits (MRLs) and fixing the MRLs for veterinary medicinal products in foodstuffs of animal origin. This regulation classified pharmacologically active substances in four annexes: (I) substances for which a MRL has been fixed; (II) substances not subject to MRL; (III) substances for which a provisional MRL was established; and (IV) substances for which no MRL can be established because residues of those substances, at any concentration, constitute a hazard to human health or data were incomplete and could not permit a MRL to be set [28].

This Directive has been modified several times and it has been replaced by Commission Regulation (EU) 37/2010 [7]. This has simplified the initial classification system established in Regulation 2377/90/EC, thus applying the established Regulation (EC) No. 470/2009. In this way, all pharmacologically active substances (formerly separated in four groups) are now organized in two separate tables: allowed and prohibited substances. Moreover, the prohibition on the use of growth-promoting agents (GPAs) (e.g., hormones and β agonists) was laid down in Council Directive 96/22/EC [29].

Additionally, Council Directive 96/23/EC was established to describe guidelines for residue control and it divided all pharmacologically active substances into two groups:

- Group A, which comprises prohibited substances (listed in the prohibited substances in Regulation 37/2010/EC or in Annex IV of Regulation 2377/90/EC).
- Group B, which comprises substances with final and provisional MRLs (listed in the allowed substances in Regulation 37/2010/EC or in Annexes I and III of Regulation 2377/90/EC).

At this point, it is important to mention two concepts that have been developed after the establishment of the MRL: minimum required performance limit (MRPL) and zero tolerance. MRPL is defined as the lowest concentration that official control laboratories need to be able to achieve for detection and identification of non-authorized substances (e.g., substances for which the legal tolerance in principle is zero). For instance, MRPL values have been established for chloramphenicol, medroxyprogesterone acetate, malachite green, and nitrofurans by Decision 2002/657/EC [30]. The first MRPL list was published in Annex II of Commission Decision 2003/181/EC [31], and the most recent modification was set in the Decision of the Commission 2004/25/CE [32].

On the other hand, the principle of zero tolerance was established by the EU for certain residues of veterinary medical products in foodstuffs. Zero tolerance is applied to all substances that are either not approved or whose use is explicitly prohibited, such as Group A substances.

Apart from the EU, legislation and regulations have been established regarding human health, food safety, and environmental protection in different countries. For instance, in the U.S., tolerances for VDs/GPAs in foodstuffs can be found in the Code of Federal Regulations, namely, Title 21 (Food and Drugs, 556) [33], and these values are different from those established by the EU. Therefore, for some VDs, MRLs are comparable (e.g., the tolerance for tylosin in chicken is set at 0.2 mg/kg in the U.S., whereas it has been set at 0.1 mg/kg in the EU), whereas for other substances, differences are considerably large (e.g., the tolerance for the sum of tetracycline residues is set at 2 mg/kg in calves, swine, sheep, chicken, and turkeys in the U.S., whereas the MRL is set at 0.1 mg/kg in the EU). The differences are due to the fundamentally different ways the limits are assigned in the U.S. compared with the international standard protocol. The U.S. assigns the entire acceptable daily intake (ADI) to the single edible tissue in the species for which the drug use is sought. All other countries assign the ADI across a basket of foods as set out by the Joint FAO/WHO Expert Committee on Food Additives (JECFA).

The Canadian Food and Drug Regulations indicate that all VDs must be authorized by Health Canada prior to their sale and administration to prevent and treat diseases in animals. This department consults on the development of these laws with the Canadian public industry, nongovernmental organizations (NGOs), and other interested organizations, and MRLs for VD residues in food have also been established (Table III in Division 15 of Part B of the Food and Drug Regulations [34]). A common policy, between Health Canada and the Canadian Food Inspection Agency (CFIA), regarding the use of administrative MRLs (AMRLs), was established in 2002 as a mechanism for applying limits to authorized drugs prior to their promulgation in the Regulations. In 2004, the VDD of Health Canada made a commitment to establish MRLs with every notice of compliance (NOC) for food-producing animal drugs. MRLs and AMRLs enhance health protection by identifying and measuring the risks of VD residues to the health of consumers and as a result, appropriate action can be taken to protect Canadians from those risks.

In addition, the Asian-Pacific area also shows concern about food safety by establishing MRLs for different substances in food. For instance, in Japan, the Ministry of Health, Labour and Welfare (MHLW), through the Food Safety Basic Law, establishes an extensive "positive list" with the MRLs for pesticides, veterinary drugs, and feed additives [35]. This positive list covers 758 substances having at least provisional MRLs and indicates a safe concentration of 0.01 mg/kg for those substances without established MRLs. In Australia, the Australian Pesticides and Veterinary Medicines Authority (APVMA), an Australian government statutory authority, is responsible for centralizing the registration of all agricultural and veterinary chemical products into the Australian marketplace. This organization sets MRLs for agricultural and veterinary chemicals in agricultural produce, particularly those entering the food chain [36].

Internationally, CODEX Alimentarius has also established MRLs for a veterinary drug or certain veterinary drugs, specifying the species and the target tissue that should be analyzed [37]. Furthermore, they provide information related to acceptable daily intake as well as estimated dietary exposure. Many of these CODEX MRLs are higher than or equal to the MRLs set by the EU, with some exceptions such as azaperone.

Due to the high interest concerning veterinary medicines, the European Medicines Agency (EMA) was established in 1995 [38]. The main task of this agency is to provide scientific advice to the Community institutions and the Member States of the EU in relation to authorization and supervision of medicinal products for human and veterinary use. It contributed to international activities of the EU through its work with the European Pharmacopoeia, the World Health Organization (WHO), the International Conference on Harmonization of Technical Requirements for Registration of Pharmaceuticals for Human Use (ICH), and the International Cooperation on Harmonization of Technical Requirements for Registration of Veterinary Medicinal Products (VICH). Furthermore, the European Food Safety Authority (EFSA) was created in 2002 after several food alerts that emerged in the late 1990s [39]. This institution is an essential tool for coordination and integration of the European food safety politics.

In the EU, there is no obligation to use standardized analytical methods in residue control of food-producing animals, whereas application of standardized methods is mandatory in the U.S. In the EU, analytical methods must meet minimum performance criteria (e.g., detection limits, selectivity, and specificity) as set out in Commission Decision 2002/657/EC [16]. This Decision sets up the procedures for validation and performance criteria of the analytical methods. Moreover, the concept of identification points (IPs) was introduced by defining criteria for ion intensities and ion ratios. In this sense, it is important to highlight that this Decision indicates that chromatographic methods without the use of mass spectrometric detection are not suitable as confirmatory methods. Other definitions introduced in the 2002/657 EC guideline are the decision limit CCα and the detection capability CCβ, which are intended to replace the limits of detection and quantification, respectively. In this context, CCα is defined as the limit at and above which a sample is considered to be noncompliant, with an error probability of α (5%), and it is a crucial limit for confirmatory methods. CCβ is the smallest amount of the substance that can be detected and/or quantified in a sample, with an error probability of β (1% for banned substances and 5% for group B substances). A significant advantage of the EU approach is the high degree of flexibility, which allows ready adoption of analytical methods to technical developments and a faster answer to newly emerging problems. Instead of using standardized methods, the EU is focused on describing performance characteristics, limits, and criteria that have to be fulfilled by the applied methods.

Finally, it is important to indicate that Codex Alimentarius has also established performance criteria for analytical methods for veterinary drugs (CAC/GL 71–2009) [40] and they will be established for marine toxins and pesticides in the near future.

6.3 ANALYTICAL TECHNIQUES FOR VD RESIDUE ANALYSIS

As aforementioned, the presence of VD residues and their associated harmful effects on humans make their control an important issue in ensuring consumer protection. Analytical methods used to monitor VDs in feed and food are essential to ensure human and animal health, monitor consumer exposure, reduce the impact of chemicals on the environment, support the enforcement of laws and regulations, and facilitate international trade of animal food products. However, the development of these methods is a difficult task because the generic term "veterinary drugs" is complex and covers several classes of chemical compounds that exhibit many different chemical properties. Among these different classes of compounds, two relevant groups are antimicrobial medicines (e.g., antibiotics or dyes) and drugs exhibiting growth-promoting properties (e.g., steroids, β-agonists, thyrostats, or growth hormones). In addition, these compounds can be present in a wide variety of complex foodstuff samples.

Basically, analytical methods can be classified into two main groups: screening methods and confirmatory methods. Screening methods are used first to determine the absence/presence of an analyte or a group of analytes at the concentration of interest. These types of methods provide a qualitative binary response and samples are classified as negative or nonnegative. Afterward, nonnegative samples must be analyzed by a confirmation/quantification method to determine the concentration of the target analyte(s) in those samples. This strategy is suitable to reduce the number of samples to be quantitatively analyzed, and it can be applied in routine laboratory analysis with high throughput.

Traditionally, VD residues in food samples have been detected by microbiological or immunochemical techniques, which provide a rapid detection of certain compounds [41,42]; for instance, in many areas, rapid milk testing for antimicrobials were used. However, several drawbacks related to the lack of selectivity and the inability of providing quantitative determination (requiring another technique for that purpose) have led them to be replaced by physicochemical techniques. In this way, the combination of MS with gas chromatography (GC) or LC is extensively used in the simultaneous identification and quantification of VD residues in feed and food [43,44]. Nowadays, analytical strategies to determine VDs are predominantly based on LC–MS due to its applicability for direct determination of polar compounds, such as most VDs; other less polar compounds are also LC amenable after appropriate derivatization [45]. However, GC–MS-based assays are favored and used for specific applications.

In this context, the combination of LC and MS appears to be a suitable approach that fulfills key requirements in terms of sensitivity, selectivity, and confirmation for rapid and reliable determination of analytes at low concentrations in complex matrices. Tandem MS (MS/MS) or multidimensional MS (MS^n) has presented remarkable advantages in the field of residue analysis at trace concentrations in food samples. QqQ analyzers operating in selected reaction monitoring (SRM) mode [9,46] or ion trap (IT) analyzers [47,48] operating in product ion scan (PIS) mode have evolved from single analyte to multianalyte monitoring. However, the reported methods based on LC–QqQ–MS/MS or LC–IT–MS/MS have been limited to one class or similar classes of VD compounds [49–51]. In addition, the number of analytes that could be monitored during one chromatographic run is necessarily limited (e.g., 100 in 10–20 min analysis time) because of sensitivity and/or limited number of MS/MS data acquisition time windows that fit in one chromatogram. This limitation is also more remarkable in IT instruments due to their lower scan speed in comparison with QqQ systems and the type of acquisition mode. On the other hand, by definition, SRM in QqQ and PIS in IT methods are focused on target compounds, but in certain circumstances, retrospective analyses are demanded. Therefore, there is an inherent limitation when these analyzers are used: their inability to detect new/additional residues for which no MS/MS conditions are programmed in the method. For this aim, it is always necessary to have standards to optimize MS/MS conditions. Furthermore, the lack of commercial availability of some reference substances (e.g., active compounds, degradation products, and metabolites) requires instrumentation that does not need previous individual compound-specific instrument tuning.

Due to these new needs, other LC systems coupled with full scan technologies are being successfully implemented in the area of VD residue analysis, such as linear ion traps (LITs) [52], TOF, or more recent technologies such as Orbital trap (Orbitrap™) [53] and a new generation of hybrid instruments such as QqTOF [54], quadrupole-linear ion trap (QqLIT), or linear-orbital trap (LTQ-Orbitrap). These HRMS technologies offer several advantages, which can be summarized as follows:

i. Mass accuracy, which is one of the most important characteristics of HRMS coupled with LC because it is necessary for an adequate confirmation and peak assignment.
ii. Resolving power, which allows the reconstruction of highly selective accurate mass chromatograms.
iii. Ability to analyze a theoretically unlimited number of compounds in a sample because these instruments operate basically in the full scan mode.
iv. Possibility of performing retrospective analyses, which means that they let a careful examination of old raw data sets to search for additional residues without reinjecting the samples.

v. High-throughput capabilities owing to these instruments requiring neither optimizing the MS conditions to detect each analyte nor readjusting the retention time windows such as in SRM-based methods.
vi. High-throughput capabilities considering that screening and identification can be carried out in one single injection.

Consequently, they provide high specificity due to both high mass accuracy and resolving power, and they permit the study of target, posttarget, and unknown residues. These advantages, together with improved quantification characteristics, make the recent HRMS technologies adequate for both screening and confirmation methods.

In the following sections, a brief discussion of the chromatographic conditions as well as HRMS analyzers used for the determination of VDs is given.

6.3.1 Chromatographic Separation

Chromatographic separation of VDs has currently relied on the use of LC, due to low volatility and thermolabile characteristics of many of these compounds, although some of them such as chloramphenicol, florfenicol, and thiamphenicol have been traditionally determined by GC [55]. The conventional LC systems coupled with MS analyzers, in particular HPLC–QqQ–MS/MS, have become the techniques of choice in the field of analysis of VD residues in foodstuffs. However, new high-resolution analyzers, based on full scan MS, require an adequate chromatographic separation, since a faster elution with superior resolution and improved sensitivity is needed to increase peak intensity and minimize interferences from coeluting peaks. Besides, the high number of compounds to be separated may increase run times, which will become relatively long by employing conventional LC systems. Therefore, advances in conventional LC by utilization of ultrahigh performance liquid chromatography (UHPLC) allowed low-dead-volume and high-pressure (1000 bar) LC equipment, providing new strategies to improve resolution, maintaining or even shortening run times. An essential aspect of the UHPLC concept is the use of sub-2 μm particle for the stationary phase, while maintaining other aspects of the column geometry. This allows faster separation and/or increased peak capacity (i.e., the number of peaks that can be separated in a given time window). Today, new high-resolution analytical techniques for unambiguous identification of VDs in foodstuffs of animal origin are mainly based on UHPLC–HRMS [56,57].

In this context, chromatographic separation of VDs with UHPLC analysis usually involves reversed-phase LC (RPLC) using alkyl-bonded silica columns (C18 and C8), although other columns such as phenyl [58,59] or C12-based stationary phases [60] have also been used.

In relation to the mobile phase, it is usually selected as a compromise between optimal chromatographic separation, adequate ionization efficiency, and overall MS performance. The most suitable solvents for LC–MS and specifically for LC–HRMS are water, methanol, and acetonitrile. Consequently, the mobile phases employed for the separation of a single family or multiple families of VDs have been mixtures of water–methanol or water–acetonitrile. In addition, some additives or modifiers of the mobile phase have been employed to enhance ion abundance, diminish the formation of sodium adducts, and improve chromatographic peak shape, such as formic acid [61–63], ammonium formate [64], and ammonium acetate [65,66]. Mobile phases containing nonvolatile compounds such as phosphate buffers should be avoided because they can clog the interface and produce buildup of deposits in the ion source. Other additives have been required for the chromatographic separation of some families, such as tetracyclines, aminoglycosides, or ivermectins. For instance, for the elution of tetracyclines [67], oxalic acid is usually used to minimize the effect of residual silanols on the stationary phase and to avoid the formation of complexes with traces of metals. Furthermore, triethylamine (TEA) has also been employed to block silanol groups of the chromatographic column [67]. Aminoglycosides comprise a family of compounds that are not easily separated under generic chromatographic conditions due to their high polarity. Therefore, poor chromatographic retention on classical RP-C18 columns is observed, hindering their inclusion in multiresidue methods developed for VDs [61]. However, the use of ion pairing agents, such as heptafluorobutyric acid (HFBA) added to the LC mobile phase, improves the chromatographic performance by increasing the retention of these compounds on RPLC columns [68]. On the other hand, traces of sodium hydroxide (NaOH) have been used in the mobile phase to convert ivermectin to the sodium adduct in order to enhance the sensitivity of this compound [59]. In other analyses, the addition of ammonium formate in the mobile phase has been used to favor the formation of ammonium adduct [64].

6.3.2 High-Resolution Mass Spectrometers

6.3.2.1 TOF–MS Analyzer

The TOF–MS analyzer is an attractive instrument to carry out multiresidue analyses due to its potentially unlimited *m/z* range and high-speed acquisition capabilities with high sensitivity and mass accuracy. It has been widely used during the past years to carry out multiresidue analysis of pesticides [69], pharmaceuticals [70], or toxins [71]. Nevertheless, its application in the field of VDs has been very recent [57].

This analyzer is the simplest mass spectrometer, where ions that have same kinetic energy but different *m/z* values are separated in a field-free flight tube and reach the detector at different times [72]. In this way, a complete mass spectrum is obtained simply by allowing sufficient time for all of the ions of interest to reach the detector. This allows fast full spectral acquisition rates and full spectral sensitivity at high mass resolution (around 10,000 expressed at full peak width at one-half maximum, FWHM, and defined as $m/\Delta m$, where m is the mass of the ion and Δm is the width of the peak at the half height of the peak) with high mass accuracy. According to these characteristics, TOF is an interesting choice for posttarget and nontarget analyses because it is possible to monitor every potential contaminant ionized in the source (in the defined *m/z* range) in a sample without reinjecting it [73].

Since the release of the first TOF–MS analyzer, which showed limited resolving power (≈300 FWHM) and mass accuracy, this technology has undergone numerous modifications. Some of these were reflectron and orthogonal acceleration. The first of these solutions, reflectron, allowed reaching mass resolving power approaching 10,000 FWHM and mass accuracy <10 ppm [74,75]. With the development of off-axis or orthogonal acceleration TOF–MS (oa-TOF–MS) of ions, it has become the catalyst for the current range of resolution of TOF–MS instrument with greatly improved resolving power and mass accuracy. Nowadays, instruments offer a mass resolving power of >18,000 FWHM [76].

Despite these improvements, early TOF–MS instrumentation was hindered by the narrow dynamic range of the detector. This resulted in the notion that TOF–MS could not be used for quantitative purposes. Consequently, in the past years, significant enhancements related to the dynamic range have been made. There have been improvements in other parameters as well, such as mass accuracy, resolution, sensitivity and scan speed, and type of detector [77].

The linearity of TOF–MS measurements is limited because of the way the ions are detected. Regarding the type of detector, two types of detector systems are used in TOF–MS: the analog-to-digital converter (ADC) and the time-to-digital converter (TDC) [77]. ADC detectors suffer from inherent background noise, whereas one of the major problems associated with TDC detectors is saturation, affecting linearity and mass accuracy [77]. In order to overcome these problems, a traveling wave-based radio frequency-only stacked ring ion guide (TWIG) has been used. TWIG has been used to extend the dynamic range of TDC-equipped TOF–MS [77]. This technique is called dynamic range enhancement (DRE). DRE has improved the linear dynamic range up to four orders of magnitude.

In order to maintain the stability of the mass axis, TOF–MS instruments require the use of a continuous internal lock mass or the periodic recalibration by switching to a discontinuous lock spray [70]. The introduction of the reference compound(s) together with the mobile phase, however, can lead to matrix ionization suppression, thus decreasing the reference compound sensitivity and increasing mass errors. Consequently, an additional electrospray source that orthogonally generates lock mass ions or a baffle, which periodically switches between two positions [78], is used to eliminate the potential risk of interferences by isobaric sample compounds, as well as the probability of signal suppression.

Therefore, the modern benchtop TOF–MS instruments characterized by technological advances in reflectron technology, orthogonal injection, and DRE data acquisition have resulted in a new generation of MS analyzers for LC that have improved mass resolution (6,000–20,000 FWHM), mass accuracy (2–10 ppm), dynamic range (four to five orders of magnitude), and sensitivity (fmol). These developments in TOF–MS instrumentation makes this analyzer a powerful analytical tool, enabling target, posttarget, and nontarget analyses, as well as identification, confirmation, and quantification of comprehensive lists of analytes in a single injection.

6.3.2.2 QqTOF–MS Analyzer

The development of the hybrid QqTOF–MS analyzer was closely related to the progress in TOF–MS. For this analyzer, Q refers to a mass-resolving quadrupole; q refers to a radio frequency-only quadrupole or hexapole collision cell, and TOF refers to a TOF–MS analyzer [75]. Hybrid QqTOF–MS instruments combine the advantages of quadrupole and TOF–MS analyzers. In this way, this configuration shows the benefits of high sensitivity, mass resolution, and mass accuracy. Therefore, QqTOF offers a great potential for screening analysis and confirmation of positive samples on the basis of acquisition of product ion full scan spectra at high resolution. Moreover, it is possible to perform MS/MS experiments with accurate mass measurements, facilitating the structural elucidation of nontarget/unknown compounds by valuable fragmentation information. Thus, Q permits precursor ion selection, but at low resolution, which will be fragmented in q and their fragments will be monitored in the TOF. The measurement of accurate mass in TOF allows the assignment of the elemental composition of a compound, whereas QqTOF also allows the establishment of the elemental composition of all product ions obtained [79].

QqTOF–MS instruments allow the possibility of performing MS/MS acquisitions in several ways. As aforementioned, the simplest method is the selection of the precursor ion in Q and its subsequent fragmentation in q; then, all product ions can be scanned in the TOF. For this mode, it is necessary to have a prior knowledge of the ions, usually by previous full scan MS. However, it cannot be used for nontarget and unknown compounds and only one precursor ion can be monitored in each acquisition function. Therefore, alternative acquisition modes have been developed to improve this. One of them was the data-dependent acquisition (DDA) mode, in which the MS instrument switches from full scan MS mode to MS/MS mode when an eluting peak rises above a predefined threshold [80]. Another alternative is the acquisition mode named MS^E, which involves the simultaneous acquisition of exact mass data at high and low collision energy [81]. These two types of monitoring modes permit the acquisition of full scan and MS/MS spectra during the analysis for all the ions generated in the ion source, increasing the qualitative data obtained.

6.3.2.3 Orbitrap–MS Analyzer

Since its introduction in 2005, the Orbitrap–MS analyzer has proven to be a valuable analytical tool with a range of applications in different fields of chemistry such as proteomics [82], metabolomics [83], environmental analysis [84], and food safety [85,86]. This analyzer uses the principle of orbital trapping in electrostatic fields [87] and it is the latest MS analyzer developed so far; in other words, it is not a hybrid system but a totally new mass analyzer.

Orbitrap–MS consists of an inner and an outer electrode, which are shaped to create a quadro-logarithmic electrostatic potential. Ions oscillate harmonically along its axis (z-direction), with a frequency dependent on their m/z values, generating an image current transient that is converted to a frequency spectrum using a Fourier transform [88]. In this way, Fourier transformation of the acquired transient together with the use of a C-shaped storage trap (called C-trap), employed to store and cool ions before injection into the Orbitrap, allows wide mass range detection with high resolving power, mass accuracy, and dynamic range [89]. Thus, this device performs as a high-resolution mass analyzer, providing high mass accuracy (2–5 ppm) and a mass resolving power up to 100,000 FWHM (*m/z* 200) that allows a proper discrimination between isobaric interferences and ions of interest in most cases [88] and provides a large dynamic range over which accurate masses can be determined [89].

The first commercial form of this instrument was a hybrid analyzer, which coupled a LIT mass spectrometer with an Orbitrap–MS via a RF-only trapping quadrupole with a curved axis (LTQ–Orbitrap by Thermo Fisher). The combination of LIT and Orbitrap–MS allowed high-quality accurate mass MS^n spectra. In this way, the results generated by MS/MS provided important information on structural characteristics of analytes and allowed utmost confidence in target analyte identification. However, the main pitfall of this instrument is the scan speed, which is sometimes insufficient for chromatographic analyses due to the LIT analyzer.

A nonhybrid benchtop version of the Orbitrap mass analyzer was next commercialized: the Exactive™ Orbitrap (single-stage Orbitrap). This instrument was reduced in both size and cost in comparison with the first hybrid system, facilitating its use in routine applications. The new system is capable of generating fragmentation information by a nonselective MS/MS mode using higher energy collision dissociation (HCD) in a collision cell without precursor ion selection. In this way, structural information can be obtained for compounds of interest and fragment ions can be used for confirmation in targeted analyses, as well as for identification and confirmation in nontarget or unknown analyses. The Exactive technology currently provides resolutions ranging from 10,000 to 100,000 FWHM (*m/z* 200), depending on the goal of the analysis (Figure 6.1). The resolution is an important advantage in the area of VD residue analysis owing to the fact that higher resolution and selectivity than TOF instruments can be achieved. Therefore, the analyte and the matrix-related isobaric interferences can be differentiated, allowing for more sensitive detection of analytes in complex matrices (Figure 6.2). Furthermore, the dynamic range with accurate mass measurements is wider than those provided by other HRMS analyzers such as TOF–MS or QqTOF–MS.

6.3.2.4 Other Analyzers

In this section, the application of other analyzers such as IT (also known as quadrupole IT, QIT), LIT, and QqLIT is briefly discussed. These instruments belong to the group of LRMS detectors because they generally produce unit resolution. Despite the fact that they are not HRMS analyzers, they can be mentioned here because of their ability to carry out the identification and structural elucidation of target analytes, impurities, or metabolites.

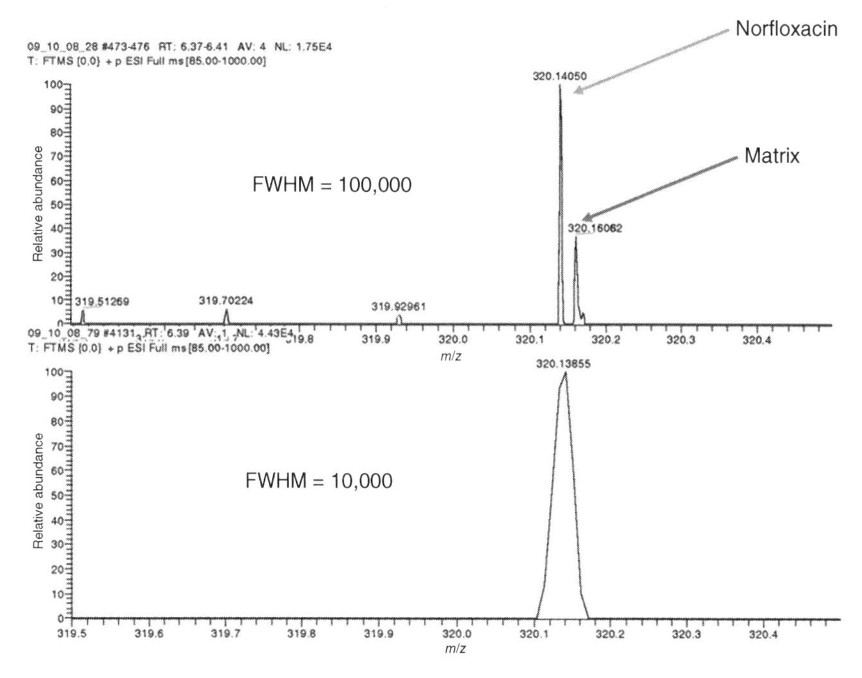

Figure 6.1. Example of use of the high resolving power of single-stage Orbitrap for the determination of norfloxacin and its isobaric interference. *Source*: Ref. [90], Figure 3, p. 1239. Reproduced with permission of Springer.

IT analyzers consist of a metal ring electrode between two hyperbolic end-cap electrodes to form a three-dimensional ion trap. The oscillating potential difference established between the ring and end-cap electrodes creates a field to store or pass ions in and out of the traps. The trapped ions precess in the trapping field with a frequency that is dependent on their *m/z*. The IT mass analyzer allows working in multiple stage mass fragmentation (MS^n) mode and product ion scan MS/MS [88] mode, obtaining good quality full scan spectra with relatively low amounts of analyte. It is the so-called tandem-in-time MS. Only a limited number of ions can be monitored simultaneously due to the occurrence of space charge effects and ion–molecule reactions, which negatively affect sensitivity [91].

These drawbacks were overcome by LITs, which are also known as 2D ion traps. They consist of a mass analyzer based on a four-rod quadrupole and end electrodes that confine ions radially by a two-dimensional quadrupole ion trap [92]. The advantages of a LIT versus a 3D IT include enhanced ion trapping capacity and reduced space charge effects due to increased ion storage volume. More ions can be introduced into the LIT, resulting in increased sensitivity and a larger dynamic range compared with a 3D ion trap [79].

Significant improvements in LIT technology have been achieved by the implementation of the hybrid QqQLIT platform where two quadrupoles precede a LIT mass

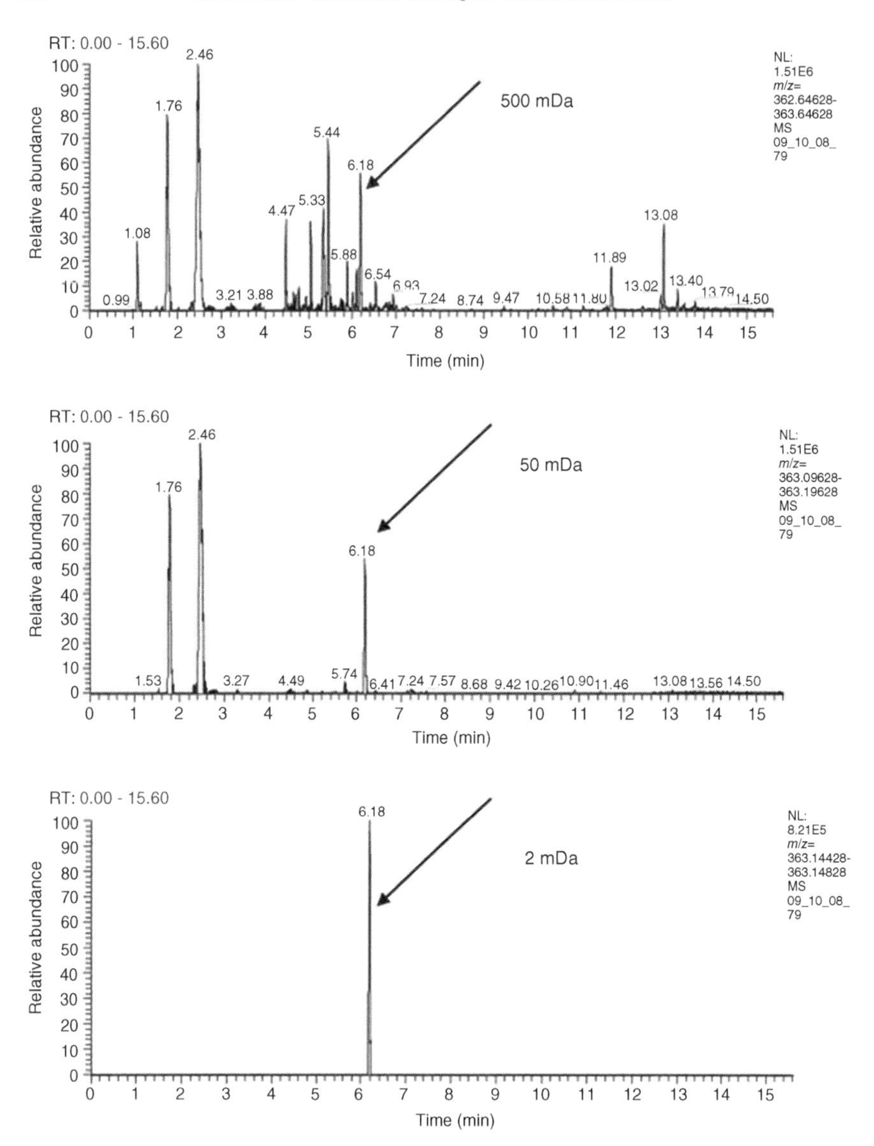

Figure 6.2. Influence of mass width on the elimination of matrix components. *Source*: Ref. [56], Figure 4, p. 64. Reproduced with permission of Elsevier.

analyzer. These hybrid instruments combine the attributes of QqQ modes for quantitative analysis and the sensitivity of LIT scan modes such as enhanced product ion scan mode, time-delayed fragmentation, and MS^3 for the confirmation of analytes or characterization of unknowns [93]. The use of third quadrupole (Q3) as a LIT with axial ion ejection significantly improved ion trap performance by enhancing full scan sensitivity in both precursor and product ion scan modes, while maintaining complete QqQ operational modes, that is, MRM mode and precursor and constant neutral loss scanning.

6.4 FOOD CONTROL APPLICATIONS

This section focuses on the description of relevant analytical methods reported in the literature for the detection of VDs in a variety of food samples employing HRMS analyzers. These analytical methods have been divided according to their objectives into two groups: screening and quantitative/confirmation methods. Additionally, three tables summarizing these methods are also included.

6.4.1 Screening Applications

In screening methods, several modalities can be applied, basically targeted and nontargeted screening, including in this last group the posttargeted and unknown analyses, where there is no previous information or restriction on the compounds to be sought in the sample.

In the development of screening methods, at least two relevant issues must be considered for their successful implementation: first, the preparation of the sample (which is not the aim of this chapter and can be found elsewhere) [94], and, second, the detection technique. The first challenge can be overcome by developing/applying generic extraction procedures that are able to cover a wide range of compounds showing different chemical properties, such as dilute-and-shoot [95] and QuEChERS approaches [44]. In the first example, although extraction time is longer than 1 h, several samples can be extracted in parallel, allowing simultaneous determination of several families of compounds and increasing sample throughput. An obvious advantage is the use of a single extraction procedure for different groups of compounds, which are usually named as "multiclass" methods, a step forward in multiresidue methods. The reduction in the number of methods to be applied in a single sample clearly benefits sample throughput. On the other hand, the QuEChERS approach allows the extraction of >20 VDs in <10 min. Subsequently, efficient sample preparation approaches must be combined with determination techniques, which can provide a response for all compounds at their required target limits with high specificity and selectivity.

Multiresidue screening methods are generally developed for rapidly assessing the presence/absence of contaminants in a complex sample, and currently they have been developed most in the area of pesticide and VD residue analyses. For instance, >200 compounds can be analyzed in <20 min [96], >250 compounds in 31 min [95], and 150 compounds in 9 min [57] in one single run. Methods developed for multicomponent screening should be able to detect as many pollutants as possible in a single run. In recent years, there have been important improvements in this field. First, the possibility of replacing conventional HPLC with UHPLC was evaluated to improve throughput. Second, the availability of HRMS instruments represented a promising alternative to LRMS screening (i.e., by QqQ) for two main characteristics: (i) accurate mass measurements and (ii) full scan acquisition of all ions generated in the ionization source. Therefore, the selectivity needed for screening methods is obtained from extracted ion chromatograms (XICs) of the accurate mass of the ions of interest using filters based on narrow mass windows (e.g., 20 mDa, 10 ppm, etc.). This characteristic also allows the performance of retrospective analysis, which means that there is always the possibility of looking for the presence of any compound at any time from stored full acquisition data without reinjecting the sample. Thus, it can improve long-term high throughput because it eliminates the need for new runs (e.g., reinjection of the sample) and, therefore, saves time.

Taking into account all these advantages, multiresidue screening of VD residues in food samples by UHPLC–HRMS has been developed in the past years using several analyzers such as TOF and Orbitrap. Tables 6.1 and 6.2 show some screening methods developed for VDs in food samples by UHPLC–HRMS.

One of the first multiscreening methods based on UHPLC–TOF–MS was developed by Stolker et al. [97]. This approach used full scan accurate mass screening, enabling the analysis of 101 VDs and metabolites in raw milk samples in <10 min. The satisfactory quantitative results obtained during validation showed the feasibility of the system for quantification (Table 6.1). Moreover, mass accuracy was a valuable tool to differentiate between positive and negative samples. Following this approach, two years later, the same group reported another UHPLC–TOF–MS multiresidue and multimatrix screening and quantification method for the same analytes in eggs, fish, and meat (Table 6.1) [73]. This last study was developed using a Bruker micrOTOF system, whereas in the previous study, they used a Waters LCT Premier TOF–MS. When LCT Premier TOF–MS was used, the authors found that for >80% of the compounds, the mass accuracy was within the 10 ppm acceptability limit and most of the compounds that did not comply eluted in the region where most of the matrix compounds eluted [97]. In this study they found an average mass measurement error of 3 ppm (median 2.5 ppm) with little difference between the three matrices. While for >98% of the studied compounds the mass accuracy was below the 10 ppm limit, individual analyte measurement exceptions up to 20 ppm were encountered. These latest results were comparable to or better than the linearity determined for the same compounds in a milk matrix using a Waters–Micromass LCT Premier TOF–MS and drug-specific extraction windows. The results obtained in the validation of the proposed method showed that this method provided satisfactory performance characteristics for >90% of the compounds in meat, for >80% of the compounds in fish, and for >70% of the compounds in eggs, clearly showing the influence of the matrix on method performance. Considering that some differences between matrices and concentrations were observed, the authors concluded that the TOF–MS itself was not able to distinguish between unlimited numbers of compounds in any matrix. In the case of simple matrices, such as milk or meat, the instrument appeared to be less limited in terms of matrix effects, but for more complicated matrices such as fish and eggs, more complicated sample preparation techniques seemed to be required. Additionally, in this study an alternative validation procedure requiring a lower number of samples to be analyzed was suggested and tested for the screening method. Repeatability, reproducibility, and CCβ results were similar for meat and fish and slightly different for eggs. The results showed that the high number of samples required in the 2002/657/EC [16] for the determination of specificity ($n = 20$), detection capability ($n = 20$), can be reduced by about 50% using a different strategy.

Table 6.1. Methodologies Used for the Detection of Veterinary Drug Residues in Food Samples Employing TOF–MS and QqTOF–MS Analyzers[a]

Number of Compounds	Matrix	Analyzer	Resolution	Mode	Mass Tolerance	Ionization Mode	Lower Limits	Reference
Seven VDs (multiclass)	Salmon	TOF–MS	9,500 ± 500 FWHM	Determination and confirmation	0.01 Da	ESI (+)	LOD: 1–3 μg/kg LOQ: 3–9 μg/kg (<MRLs) CCα: 103–218 μg/kg CCβ: 107–234 μg/kg	[62]
101 VDs (multiclass)	Raw milk	TOF–MS	10,000 FWHM	Screening and quantification	10 ppm	ESI (+)	CCα: 1–50 μg/l CCβ: 1.2–141.8 μg/l LOQs: <7 μg/l for >90% of the compounds	[97]
>100 VDs (multiclass)	Meat, liver, and kidney	TOF–MS	12,000 FWHM	Quantification	60 ppm	ESI (+)	(Muscles) CCα: 30.9–369.8 μg/kg (with MRL) and 1.1–6.0 μg/kg (Liver) CCα: 29.2–1675.4 μg/kg (with MRL) and 0.4–20.0 μg/kg (Kidney) CCα: 39.0–1175.1 μg/kg (with MRL) and 0.9–20.0 μg/kg	[61]
Eight Tetracyclines	Honey	TOF–MS (HPLC coupled with two detectors online, DAD and TOF)	5,000–10,000 FWHM	Determination and confirmation	5 ppm	ESI (+/−)	LOD: 0.05–0.76 μg/kg	[67]
Eight Quinolones	Pig liver	TOF–MS	10,000 ± 500 FWHM	Determination and characterization	5 ppm	ESI (TIS) (+)	LOD: 0.5–2.0 μg/kg LOQ:1.5–6.0 μg/kg CCα: 164–818 μg/kg CCβ: 175–828 μg/kg (with MRL 100 μg/kg)	[98]
100 VDs	Eggs, fish, and meat	TOF–MS	10,000 FWHM	Screening	20 ppm	ESI (+)	(Meat) CCβ: 2–2.5 VL (Fish) CCβ: 2–4 VL (Eggs) CCβ 2–4 VL	[73]

(*continued*)

Table 6.1 (*Continued*)

Number of Compounds	Matrix	Analyzer	Resolution	Mode	Mass Tolerance	Ionization Mode	Lower Limits	Reference
150 VDs (multiclass)	Raw milk	TOF–MS	≈7,000 FWHM	Screening and quantification	0.02 Da	ESI (+)	LOD: 0.5–25.0 μg/kg CCα: 0.3–222.0 μg/kg (with MRL) and 0.1–43 μg/kg CCβ: 0.5–245.0 μg/kg (with MRL) and 0.2–73.0 μg/kg	[57]
13 VDs (multiclass)	Shrimp	TOF–MS	<10,000 and 18,000 FWHM	Detection and quantification	0.01 Da	ESI (+)	LOD: 0.06–7.00 μg/kg	[99]
25 VDs (multiclass)	Milk	QqTOF–MS	9,000 and 15,000 FWHM	Screening, confirmation, and quantification	10 ppm	ESI (+)	LOD: 1–10 μg/l LOI: 2–50 μg/l	[100]
Six anabolic steroids	Raw meat and baby food	QqTOF–MS	8,033 FWHM	Confirmation and quantification	<20 ppm	APCI (+) and TIS (−)	(Raw meat) CCα: 0.05–0.75 μg/kg and CCβ 0.47–5.20 μg/kg (Baby food) CCα: 0.03–0.33 μg/kg and CCβ: 0.13–2.13 μg/kg	[101]
Six macrolides	Eggs, raw milk, and honey	QqTOF–MS	15,000 FWHM	Quantification and confirmation, identification of degradation products	0.08 Da	ESI (+)	LODs: 0.2–1.0 mg/kg	[65]

[a] *Abbreviations:* APCI: atmospheric pressure chemical ionization; CCα: decision limit; CCβ: detection capability; ESI: electrospray ionization; FWHM: full peak width at one-half maximum; LOI: limit of identification (the lowest level at which predominant product ions are observed above the noise in the MS/MS spectra with accurate mass within 10 ppm of theoretical value); LOD: limit of detection; LOQ: limit of quantification; QqTOF: quadrupole time-of-flight mass spectrometry analyzer; TIS: turbo ion spray; TOF–MS: time-of-flight mass spectrometry analyzer; VL: validation level. It is equal to the MRL level, the minimum required performance limit (MRPL) level, or a "specific level of interest" if no MRL or MRPL level was available, as is the case for many of these drugs in these matrices; VDs: veterinary drugs.

A similar application of the first screening method developed by Stolker [97] was subsequently published by Ortelli et al. [57]. This work describes the use of UHPLC–TOF–MS for the screening of 150 VDs and metabolites in raw milk, estimating a routine application of >50 samples per day. According to the high sensitivity and selectivity of TOF–MS, the limits of detection ranged from 0.5 to 25 µg/l, and they were far below the corresponding MRL for the majority of the compounds (Table 6.1). Apart from some problems with avermectins, the method allowed suitable screening and quantification for the rest of VDs.

These reports demonstrated the suitability of LC–TOF–MS for the screening and quantification of VD residues in different food matrices. However, in terms of unequivocal confirmation, the main drawback relies on the fact that only MS data of the protonated molecules are used in these reports. Even with very low mass error (e.g., <10 ppm) the confirmation of VD identities can be insufficient. In this way, Turnipseed et al. [100] took advantage of the ability of QqTOF–MS to carry out screening and unambiguous confirmation of 25 VDs in milk samples with precursor ion selection. Screening of residues was accomplished by collecting TOF data, while MS/MS data generated for the $[M+H]^+$ ions (SRM mode) were employed to confirm the presence of VD residues in the samples by monitoring product ions. Nevertheless, screening and confirmation were carried out using different methods and, in consequence, nonnegative samples were reinjected for confirmation. Although the method was intended to be qualitative, an evaluation of the MS data indicated a linear response and acceptable recoveries for the majority of target compounds. Moreover, several metabolites were identified evaluating MS and MS/MS data (Figure 6.3). For example, several plausible metabolites of enrofloxacin, some of them not previously observed in milk, were found in the samples such as ciprofloxacin (another fluoroquinolone that differs from enrofloxacin by an ethyl group) or des-enrofloxacin (*m/z*

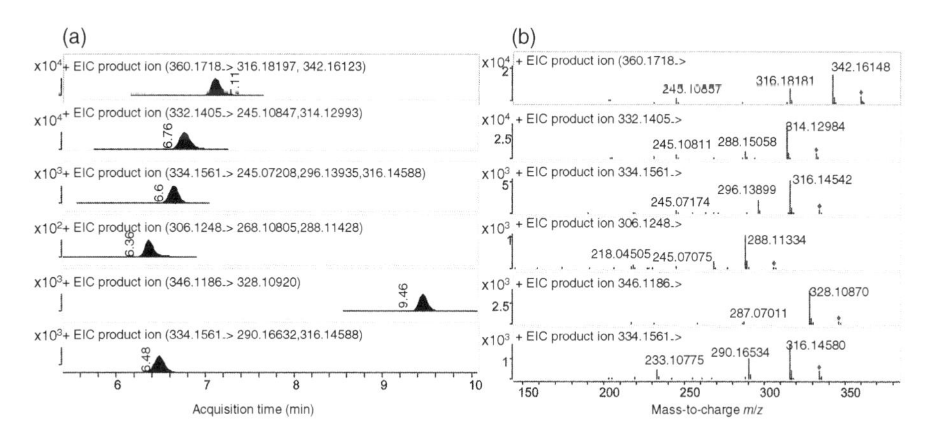

Figure 6.3. (a) Extracted MS/MS ion chromatograms for compounds in an incurred enrofloxacin milk sample. From the top, traces for enrofloxacin, ciprofloxacin, and the proposed metabolites desethylene enrofloxacin, desciprofloxacin, and oxociprofloxacin are shown. For comparison, the bottom MS/MS chromatogram is for a standard of pefloxacin. (b) Product ion spectra for these compounds. *Source*: Ref. [100], Figure 4, p. 7577. Reproduced with permission of American Chemical Society.

334.1562), a known metabolite that results from the loss of ethylene from enrofloxacin.

Even though the application of TOF–MS and QqTOF–MS analyzers was successful for the screening of VDs in food samples, the requirement of high resolution for the correct assignment of analyte masses in complex samples becomes a key factor when these methods are applied. In consequence, after the successful applications of other HRMS, such as single-stage Orbitrap–MS or LTQ–Orbitrap–MS in biological matrices [106,107], they were applied in food safety applications.

One of these first studies evaluated the mass resolution required for trace analysis in different food commodities. Kellmann et al. [103] evaluated the resolution power for the screening and quantification of 151 pesticides, VDs, mycotoxins, and plant toxins using UHPLC coupled with single-stage Orbitrap–MS in <25 min. In this work, the authors evaluated the application of the full scan mass spectrometry for generic screening assays and showed the importance of the resolving power requirements for accurate mass measurements. Analyses were performed with resolving power settings varying from 10,000 to 100,000 FWHM in two different matrices: honey as a representative matrix of intermediate complexity, and horse feed as a realistic worst-case sample of high complexity. The results showed that 50,000 FWHM or higher were suggested for low concentrations of analytes in complex matrices such as animal feed, whereas a resolving power of 25,000 FWHM was sufficient for less complex matrices such as honey. Additionally, the authors highlighted that although a maximum resolving power can provide the best qualitative and quantitative performance (Table 6.2), the use of the minimum-required resolving power allows the use of faster scan rates (better compatibility with fast LC separation).

More recently, new approaches for Orbitrap–MS have been reported for the screening of target VD residues, such as the development of a screening method for the detection of 63 VDs in muscle tissues using UHPLC–LTQ–Orbitrap–MS [68]. Target compounds were identified by their accurate masses and LC retention time, employing a narrow mass window of 5 ppm and a resolving power of 60,000 FWHM (Table 6.2). This screening method was also applied to the identification of nontarget compounds. For instance, apart from sulfadimethoxine, *N*-4-acetyl-sulfadimethoxine is another of its metabolites, which was identified in beef muscle. It was further confirmed by acquisition of product ions with CID fragmentation experiments, and by comparison with a chemical standard. Despite the fact that MRLs are only set for the parent compound and metabolites are not considered, this type of finding can be interesting for the improvement of MRLs and tolerance definitions. It has been reported that in many cases, metabolites or transformation products of the parent compound can be more toxic than the original drug [108].

Another report detailed the development of a screening method for the detection of 29 VDs in milk- and powdered milk-based formula samples by UHPLC coupled with single-stage Orbitrap–MS [64]. A rapid screening method (<4 min) was used and the samples could be tested regarding the presence or absence of the compounds below the established cutoff values of <5 μg/kg for the majority of the studied compounds (Table 6.2). The cutoff values were evaluated as the concentration at which the sensitivity rate is 95%, when the β-type error has been set at 5%. For nonnegative

Table 6.2. Developed Methodologies for the Detection of Veterinary Drug Residues in Food Samples Employing LTQ–Orbitrap–MS and Single-Stage Orbitrap MS Analyzers[a]

Number of Compounds	Matrix	Analyzer	Resolution	Mode	Mass Tolerance	Ionization Mode	Lower Limits	Reference
Nine coccidiostats drugs	Animal feed	LTQ–Orbitrap–MS	7,500 and 60,000 FWHM	Screening	±5 ppm	ESI (+)	LOD: 15 μg/l	[102]
63 VDs (multiclass)	Muscle tissues	LTQ–Orbitrap–MS	60,000 FWHM	Screening	±5 ppm	ESI (+)	LOD: 1–308 μg/kg	[68]
Chloramphenicol	Meat products	LTQ–Orbitrap	60,000 FWHM	Confirmatory and quantitative	±5 ppm	ESI (−)	LOQ: 0.1 μg/kg	[85]
68 VDs (multiclass) (in addition to pesticides, mycotoxins, and plan toxins)	Honey and animal feed	Single-stage Orbitrap–MS	25,000, 50,000, and 100,000 FWHM	Screening and quantification	<5 ppm and <2 ppm (resolution 100,000)	ESI (+)	LOD: ≤25 μg/kg	[103]
100 VDs (multiclass)	Liver, muscles, kidneys, fish, and honey	Single-stage Orbitrap– MS	50,000 FWHM	Detection and quantification	10 ppm	ESI (+)	(Kidney) CCα: 1.0–1181.2 μg/kg and CCβ: 1.1–2337.0 μg/kg (Honey) CCα: 0.9–327.5 μg/kg and CCβ: 1.1–391.0 μg/kg	[53]
112 Target VDs and semitarget 116 VDs (multiclass)	Fish	Single-stage Orbitrap–MS	50,000 and 100,000 FWHM	Semitargeted screening	10 ppm	ESI (+)	Not indicated	[104]
13 Anthelmintic drugs	Milk and meat products	Single-stage Orbitrap–MS	50,000 FWHM	Quantification	10 ppm	ESI (+)	(Milk) CCα: 1.01–1.18 μg/l and 10.4–112.1 μg/l (with MRL) CCβ: 1.16–1.81 μg/l and 11.5–147.9 μg/l (with MRL) LOD: 0.5 μg/l (Muscles) CCα: 1.1–1.2 μg/kg and 23.1–245.3 μg/kg (with MRL) CCβ: 1.3–1.7 μg/kg and 31–292 μg/kg (with MRL) LODs: 0.5–2.0 μg/kg	[105]

(continued)

Table 6.2 (*Continued*)

Number of Compounds	Matrix	Analyzer	Resolution	Mode	Mass Tolerance	Ionization Mode	Lower Limits	Reference
29 VDs (multiclass)	Milk- and powdered milk-based infant formulas	Single-stage Orbitrap–MS	50,000 FWHM at *m/z* 200	Screening, confirmation and quantification	<5 ppm	ESI (+)	Cutoff: 1.1–50.0 μg/kg (milk) and 1.3–50.0 μg/kg (Milk) CCα: 6.1–117.9 μg/kg CCβ: 9.3–226.6 μg/kg LOQ: 0.2–5.0 μg/kg (Powdered milk) (without MRLs): CCα: 4.1–26.0 μg/kg CCβ: 8.1–40.5 μg/kg LOQs: 0.5–5.0 μg/kg (except for spiramycin and ivermectin 25 μg/kg)	[64]

[a]Other abbreviations not indicated in Table 6.1: LTQ–Orbitrap: hybrid linear ion trap (LIT)–Orbitrap mass spectrometry analyzer.

samples, the analytes were confirmed using the data obtained by the fragmentation of the native ions in the HCD collision cell. The authors showed the ability of HRMS detectors to acquire with and without fragmentation in the HCD collision cell, and therefore the possibility of performing the screening and confirmation with a single injection.

Finally, the ability of the proposed method was tested to quantify the confirmed analytes showing good quantitative results for all the studied analytes. The linearity for all VDs was acceptable in a range of concentrations from 5 to 100 μg/kg, showing determination coefficients (R^2) higher than 0.98. The limit of quantification was ≤5 μg/kg for milk- and powdered milk-based formula samples, except for spiramycin in powdered milk-based formula (Table 6.2), being in all cases lower than the MRLs established by the EU.

Once the applicability of UHPLC-coupled single-stage Orbitrap–MS technique was demonstrated for the analysis of target and nontarget compounds, Kaufmann et al. [104] tried to go beyond these applications by developing a semitargeted screening of VDs (112 target VDs and 116 posttarget VDs) in fish samples by UHPLC single-stage Orbitrap–MS with a chromatographic run time of 14 min. In this work, qualitative and quantitative analyses of the 112 targeted analytes were based on the traditional approach of target analysis by external standards. However, different procedures were developed for the detection of 116 additional compounds, which were monitored without having access to reference materials. This additional data evaluation can be classified as a posttargeted analysis of VDs. First, the detection procedure was based on theoretical exact masses and narrow mass windows (10 ppm) because one can include posttarget analytes considering that the elemental composition is known and ionization is adequate. Although the XICs were extracted using narrow mass windows, some matrix components interfered in the determination of the selected compounds. Therefore, any signal observed in the chromatogram could potentially be the compound. The measurement of accurate masses and relative isotopic abundance (RIA) of suspected peaks was then applied. None of the 116 monitored compounds could be detected in the investigated samples using this strategy. In consequence, other alternatives were tested, such as the search for generic product ions characteristic of certain VD families; specific RIA by specific software, which searches for a particular RIA in the entire chromatographic time and mass scan range; and specific neutral losses. Any of these searching strategies were clearly feasible for the posttarget study proposed, and, consequently, the authors emphasized that the evaluated procedures still showed some limitations for their application in posttarget analysis.

The IT technology has been used for identification and confirmation of VD residues in food samples, but only a few works have focused on the development of screening methods. Among these studies, the method proposed by Baiocchi et al. [109], in which they describe a screening and confirmation method to monitor eight synthetic corticosteroids in bovine liver tissues employing LC–IT–MS, can be highlighted. In this work, selected ion monitoring (SIM) and SRM detection modes were checked and compared, concluding that the best sensitivity was obtained in MS/MS mode (Table 6.3). In a different study, Heller et al. [110] demonstrated the

Table 6.3. Developed Methodologies for the Detection of Veterinary Drug Residues in Food Samples Employing IT and QIT Analyzers[a]

Number of Compounds	Matrix	Analyzer	Acquisition Mode	Mode	Ionization Mode	Lower Limits	Reference
Eight synthetic corticosteroids	Liver	IT–MS	MS/MS and MS^3	Screening and confirmation	APCI (+)	LOQ: 0.85–10.03 μg/kg	[109]
11 β-lactams	Kidney	IT–MS	SRM–MS^n and full scan MS^n ($n = 2$ or 3)	Confirmatory analysis	ESI (+)	LOC: 10–500 μg/kg	[49]
7 VDs (multiclass)	Eggs	IT– MS	SRM mode (MS/MS mode–MS^2)	Screening and confirmation	ESI (+)	LOD (for screening method): <10 μg/kg	[110]
Five moenomycins	Medicated chicken feed	IT–MS	Product ion MS^2 and MS^3 scans	Structural characterization	ESI (+/−)	Not indicated	[112]
29 VDs (multiclass)	Eggs	IT–MS	Product ion MS^2 scans	Screening	ESI (+)	LCQ Classic IT: LOD: 10–50 μg/kg Deca XP Plus IT: LOD:10–20 μg/kg (not LOD for β-lactams)	[111]
38 VDs (multiclass)	Fish	IT–MS	MS^2 and MS^3	Screening and confirmation	ESI (+/−)	LOC: 0.01–1.0 μg/kg	[59]
Moenomycin A	Feed	IT–MS	SRM mode	Quantitative and identification analysis	ESI (−)	LOD: 30 μg/kg LOQ: 100 μg/kg	[66]
42 VDs (multiclass)	Honey	QTRAP–MS	SRM mode	Screening, confirmation, and quantification	ESI (TIS) (+)	LOD: 27–80 μg/kg	[113]

[a]Other abbreviations not indicated in Table 6.1: IT–MS: ion trap mass spectrometry analyzer; LOC: limit of confirmation; QTRAP: quadrupole coupled linear ion trap analyzer; SRM: selected reaction monitoring mode.

application of IT acquisition modes for screening (a single product ion) or confirmation (multiple product ions). They developed a screening method for nonpolar VDs (five ionophores and two macrolides) in eggs. The proposed screening method was validated by analysis of control, fortified, and incurred eggs (when the hens were dosed with the parent drugs). In this way, it was able to screen and confirm these residues below 10 μg/kg (Table 6.3). Two years later, the same authors [111] developed a method for the screening and confirmation of 29 polar VD residues (sulfonamides, tetracyclines, and fluoroquinolones) in eggs by HPLC–IT–MS (employing two generations of ion trap instruments). This study was based on the method described previously, employing a single product ion for screening and multiple product ions for confirmation purposes. LODs ranged from 10 to 50 μg/kg (with LCQ Classic IT–MS) and from 10 to 20 μg/kg (with Deca XP Plus–IT–MS) (Table 6.3). The method was applied to the analysis of eggs from dosed hens, and the ability to detect incurred residues was demonstrated. Smith et al. [59] also proposed the use of LC–IT–MS to screen and confirm 38 compounds from a variety of VD classes in four species of fish. MS^2 or MS^3 spectra were monitored for each compound, allowing screening and confirmation simultaneously. The method was able to confirm all quinolones and fluoroquinolones, macrolides, malachite green, and most of the imidazoles at 0.01 μg/kg and florfenicol amine, metronidazole, sulfonamides, tetracyclines, and most of the β-lactams at 0.1 μg/kg, while ivermectin and penicillin G were detectable only at 1 μg/kg in fortified samples (Table 6.3). Additionally, the authors demonstrated that an easy modification of the method could be used to analyze metabolites, detecting the presence of metabolite hydroxymetronidazole and the metabolites of albendazole (sulfone, sulfoxide, and aminosulfone).

The hybrid QLIT instrument has also been used for screening purposes using the MS/MS mode. Hammel et al. [113] employed HPLC–QLIT–MS/MS using SRM mode to carry out a multiscreening of 42 VDs in honey (Table 6.3). The two most intense fragment ions were selected for each compound, being the most intense SRM used for quantitative analysis and the second SRM for analyte confirmation. This proposed screening method utilized an internal limit at 20 μg/kg, without specifying if such limit can be considered as the cutoff concentration of the method. Finally, it was validated and good performance data were obtained for 37 analytes out of the 42 studied.

6.4.2 Confirmation and Quantification Methods

HRMS analyzers have also been applied for the quantitative analysis of VDs, although they have not been easily accepted as quantification tools due to the predominance of QqQ or IT as the only analyzers capable of obtaining accurate trace-level quantifications, as evidenced by the following quote: "Obviously, a Q–TOF mass spectrometer instrument will most probably never take the place of our trusted QqQ or IT instruments" [63]. However, numerous improvements in HRMS techniques have increased selectivity, sensitivity, and dynamic range to be adequate for quantitative purposes and some published work has demonstrated their applicability and achievements for VDs analysis. This fact allows that all the information needed is acquired in the same run, and consequently, screening and confirmation/quantification can be performed in one single analysis, increasing sample throughput.

In this section, the confirmation and quantification capabilities of HRMS systems for the analyses of VDs are addressed. To this aim, a number of relevant studies are discussed.

The first application of TOF–MS for quantitative analysis was developed by Hernando et al. [62] for the simultaneous analysis of seven VDs in fish. For that purpose, HPLC–TOF–MS with a resolution power of 9,500 FWHM was used. This method achieved CCα and CCβ values in the range of 103–218 and 107–234 μg/kg, respectively, for substances with MRL values in the range of 100–200 μg/kg, demonstrating the feasibility of HRMS for quantitative approaches (Table 6.1). However, sensitivity needed to be improved because for some analytes (e.g., malachite green and leucomalachite green), CCα and CCβ were above the established MRLs.

Two subsequent works described methods for the quantification of tetracyclines in honey samples [67] and quinolones in pig liver samples [98], using TOF–MS. The first [67] was based on the use of HPLC and two different online detectors, diode array (DAD) and TOF–MS. In this work, the HPLC–TOF–MS method was validated in terms of specificity, linearity, sensitivity, precision, accuracy, recovery, and ion suppression. Eventually, the proposed method allowed the detection of eight tetracyclines at concentrations between 0.05 and 0.76 μg/kg (Table 6.1), allowing their quantification with very low mass accuracy (less than or equal to ± 5.3 ppm) even at a low concentration. Additionally, the authors highlighted the ability of the TOF–MS to unambiguously identify these compounds because of the sensitivity, mass accuracy, and true isotopic pattern provided by the TOF analyzer. In the second work [98], a LC–TOF–MS method was developed for the determination and characterization of quinolones regulated by the EU in pig liver samples below the MRLs. Satisfactory quality parameters were established for the developed method according to the FDA and European Community guidelines. The authors emphasized the improved selectivity reached with this analyzer, especially for two quinolones, oxolinic acid and flumequine, showing identical nominal mass (*m/z* 262) and a common fragment (*m/z* 244). However, they could be easily discriminated by the corresponding XICs of accurate mass, *m/z* 262.0710 and 262.0874 for oxolinic acid and flumequine, respectively, setting a suitable extraction mass window. Apart from this, they showed the ability of this analyzer to carry out the reliable identification of fragments of flumequine, generating characteristic fragments at *m/z* 244, 220, and 202. The *m/z* 244 was clearly due to the loss of H_2O, whereas *m/z* 220 may have been due to the fragmentation of the $[M+H]^+$ by loss of C_2H_4N or C_3H_6. In this way, the use of accurate mass of each fragment and their errors allowed proposing structures for each ion, and concluding that *m/z* 220 was due to $[M+H–C_3H_6]^+$ and *m/z* 202 corresponded to $[M+H–H_2O–C_3H_6]^+$.

After these single-class multiresidue methods, other multiclass methods have been reported. Villar-Pulido et al. [99] showed a quantitative determination method of 13 different VDs in shrimp using UHPLC–TOF–MS. The unambiguous identification was carried out by measuring accurate mass and retention time, but additionally, the confirmation was based on accurate mass measurements of their fragment ions, obtaining mass errors <2 ppm in most cases. The optimized UHPLC–TOF–MS method showed excellent sensitivity for the studied analytes, with LODs ranging from 0.06 to 7.00 μg/kg, demonstrating enough sensitivity to be applied for quantitative trace analysis (Table 6.1). Finally, it was applied to real samples and it allowed the detection and quantification of benzalkonium chloride-C_{12} in one sample. Another interesting publication [61] described a quantitative method for 100 VDs in meat matrices by UHPLC–TOF–MS. This was the first fully validated quantitative method covering a high number of analytes. It allowed the identification and quantification of a wide range of VDs with different polarity and p*K* values (benzimidazoles, quinolones, lincomycin, macrolides, nitroimidazoles, penicillins, sulfonamides, tetracyclines, tranquilizers, and others). However, it was not able to detect other compounds such as chloramphenicol, nitrofurans, and aminoglycosides, and quantify other apolar drugs such as benzimidazoles, avermectins, and ionophores; this fact was then supported by the validation procedure, and consequently, the authors concluded that UHPLC–TOF–MS was not ideal for the measurement at very low concentrations of certain banned drugs. Taking into account the results obtained after trying some extraction and cleanup protocols, the authors questioned the initial euphoria that UHPLC–TOF–MS would be able to detect and quantify an unlimited number of drug residues. However, this method could detect and quantify other analytes that had MRL values at 100 or even 1000 μg/kg (Table 6.1). Additionally, bearing in mind the high number of analyses that were needed to carry out the validation, they showed the need for redefining validation guidelines for multiresidue methods, which cover hundreds of compounds.

An additional approach developed by the same authors was based on the use of single-stage Orbitrap–MS analyzer operating at 50,000 FWHM [53]. The resolution power (<15,000 FWHM) of TOF–MS technology was not selective enough for monitoring low residue concentrations for some compounds in the studied matrices; this was the main reason adduced to increase mass resolution. When liver and kidney extracts obtained according to the previous validated multiresidue method were analyzed by a single-stage Orbitrap–MS analyzer, extensive signal suppression was observed. The phenomenon was termed postinterface signal suppression, because the suppression of signals did not occur in the electrospray interface but in the C-trap device. Thus, a low analyte concentration can be easily detected in a pure standard, but it can no longer be reliably detected in a dirty matrix sample. It is important to note that the reported postinterface signal suppression affects only the intensity of low-mass ions, but does not cause mass shifts of the affected ions. This problem was partially fixed by a more extensive protein removal step. As a consequence, the proposed sample cleanup had to be intensified by more extensive deproteination steps, and instrumental settings had to be reoptimized to eliminate these suppression effects. Finally, the resulting method proved to be capable of detecting all analytes included in the original TOF–MS-based method, and significantly better performance (e.g., linearity, reproducibility, and detection limits) was obtained (Table 6.2). Although the average recovery was lower than that obtained for the previous TOF–MS-based method, the average signal suppression for all compounds improved. In conclusion, the authors attributed all these improvements to the higher resolution (50,000 versus 15,000 FWHM) and the superior mass stability of the Orbitrap over the previously used TOF instrument.

Single-stage Orbitrap–MS was also used for the determination of some anthelmintic drugs and phenylbutazone residues in milk and muscles [105], obtaining good performance characteristics compared with the QqQ–MS/MS method (Table 6.2).

Another example of the successful application of Orbitrap–MS, using HPLC–LTQ–Orbitrap was the determination of sub-μg/kg concentrations of chloramphenicol in meat products [114]. Because of the higher mass accuracy of the extracted ion obtained with LTQ–Orbitrap–MS, which minimizes matrix interferences, this method consisted of a simpler sample preparation compared to that required by a traditional analysis of chloramphenicol. Selectivity is achieved here by the MS instrument and not by the application of tedious extraction methods. The high resolution and high accurate mass used to detect chloramphenicol greatly reduced matrix interferences and increased the signal to noise ratio, achieving LOQ of 0.1 μg/kg when isotope internal standard calibration was used (Table 6.2). The application of HRMS detected the presence of false positives of chloramphenicol in some of the samples analyzed by QqQ. In spite of the fact that the diagnostic ion found in the suspected samples by LC–QqQ–MS/MS was outside the range of maximum permitted tolerances, the inexistence of chloramphenicol was unambiguously demonstrated by HRMS.

The application of HRMS technologies as confirmation tools was shown by Marchesini et al. [115,116], who applied TOF–MS technology for unambiguous identification of fluoroquinolones in chicken. These works showed the feasibility of coupling the simultaneous screening of fluoroquinolones using a dual surface plasmon resonance biosensor immunoassay in parallel with LC–TOF–MS for their confirmation. Six fluoroquinolones were simultaneously screened at or below their MRLs in chicken muscles [115] and the noncompliant samples were further concentrated and fractionated with gradient LC. The effluent was split toward two 96-well fraction collectors resulting in two identical 96-well plates. One fraction was rescreened with the dual biosensor to identify the immunoactive fractions and the second one was analyzed with high-resolution LC–TOF–MS [115] and with nano-LC–TOF–MS [116]. Both studies demonstrated the possibility to screen and identify known fluoroquinolones and the potential for discovering and identifying unknown compounds.

Even though published screening methods employing IT technology to analyze VDs in food samples can be easily found, in general, this analyzer has been more frequently used for quantification and confirmation. LC–IT–MS has been demonstrated to be an adequate confirmatory tool within the group of LRMS analyzers, due to the different scanning modes. Two of these modes have been compared by Fagerquist et al. [49]. A confirmatory method of 11 β-lactam antibiotics in kidney using HPLC–IT–MS/MS with SRM and MS^n scanning modes was developed. They compared the advantages of SRM–MS^n ($n = 2$ or 3) and full scan MS^n ($n = 2$ or 3) for analysis of unknown incurred tissue. They found that the SRM–MS^n mode provided rapid and unambiguous identification of analytes in the unknown incurred tissues, whereas full scan MS^n mode required manual mass 'filtering' for each analyte to identify the presence of a compound. In addition, they did not observe any significant differences in the absolute intensity of fragment ions when using SRM–MS^n compared with full scan MS^n. However, they highlighted that both SRM–MS^n and full scan MS^n required previous knowledge of possible analytes present in the sample in order to set instrument parameters for the detection of those analytes. Gallo et al. [66] proposed an HPLC–IT–MS/MS method for confirmatory analysis of moenomycin A in feed. The analysis was performed using SRM mode and proved to be highly selective and reliable for unambiguous identification of moenomycin A (Table 6.3).

The capability of the MS^n acquisition mode to characterize and study fragmentation patterns of the major components of moenomycin was shown by Eichhorn et al. [112]. In this work, five moenomycins (A, A_{12}, C_1, C_3, and C_4) isolated from a commercial chicken feed were chromatographically separated and identified, and their fragmentation patterns were explored using IT–MS employing full scan MS, MS^2, and MS^3.

6.4.3 Comparison Studies

The relatively recent use of HRMS techniques and their successful applications reported in the field of food safety has provoked their unavoidable comparison with traditional LRMS techniques. Some comparative studies on the performance of LRMS and HRMS analyzers, such as QqQ–MS and TOF–MS, QqTOF–MS, and Orbitrap–MS, have been developed. One of the first studies was developed by Gentili et al. [101], which compared HPLC–QqQ–MS/MS and HPLC–QqTOF–MS/MS systems in terms of sensitivity and specificity for the analysis of six hormones in meat and baby food. The results showed that the QqQ–MS/MS achieved at least 20-fold higher sensitivity compared with the QqTOF–MS/MS instrument for almost all of the analytes that were studied. Nevertheless, in terms of selectivity, QqTOF–MS/MS offered the highest performance. Additionally, the low values of CCα and CCβ obtained with HPLC–QqQ–MS/MS system for all the analytes in meat demonstrated its applicability to satisfying the MRPLs of anabolic agents (1 μg/kg for the analyzed anabolic agents, except for zeranol proposed at 2 μg/kg), while HPLC–QqTOF–MS/MS did not satisfy these conditions, obtaining CCβ > 1 μg/kg. Subsequently, Wang et al. [65] compared UHPLC–QqTOF–MS and HPLC–QqQ–MS/MS for quantification and identification of six macrolides and degradation products in eggs, milk, and honey. Both techniques demonstrated suitable quantitative performance in terms of trueness (the closeness of agreement between the average value obtained from a large series of test results and an accepted reference value) and repeatability. However, LODs obtained with the UHPLC–QqTOF–MS method (0.2–1.0 μg/kg) were higher than those obtained with the HPLC–QqQ–MS/MS (0.01–0.5 μg/kg), demonstrating the higher sensitivity of this technique in comparison with the HRMS instrument used. On the other hand, UHPLC–QqTOF–MS provided unambiguous confirmation of positive findings and the identification of degradation products based on accurate mass measurements. This was tested by the identification of tylosin B, a degradation product of tylosin A in honey samples, which cannot be purchased from commercial sources. As a result of this work, the authors considered both techniques as complementary and they showed their combination as a powerful tool for analysis of macrolide residues and their degradation products in food matrices.

Hermo et al. [98] developed an HPLC–TOF–MS method for the determination and characterization of eight quinolones in liver, which was subsequently compared with both HPLC–Q–MS and HPLC–QqQ–MS/MS methods. The obtained LOQs were 1–4 times higher using Q–MS than TOF–MS, whereas the LOQs were 1.5–6 times higher using TOF–MS than QqQ–MS/MS, and thus, sensitivity followed this order: QqQ > TOF > Q. It must be noted that the LOQs obtained with TOF–MS were sufficient to detect quinolones below the EU MRLs. On the other hand, CCα and CCβ values were comparable to those obtained using Q–MS or QqQ–MS/MS. Despite these results, the authors highlighted the advantage of the LC–TOF–MS system to provide high-quality data by exact mass measurement. These results indicated that TOF–MS can be considered a very compelling instrument for use as a confirmatory method in the EU legislative context. These three analyzers were employed to analyze different pig liver samples and only one sample contained enrofloxacin, detected with the three approaches (Figure 6.4). When this sample was quantified, the concentration

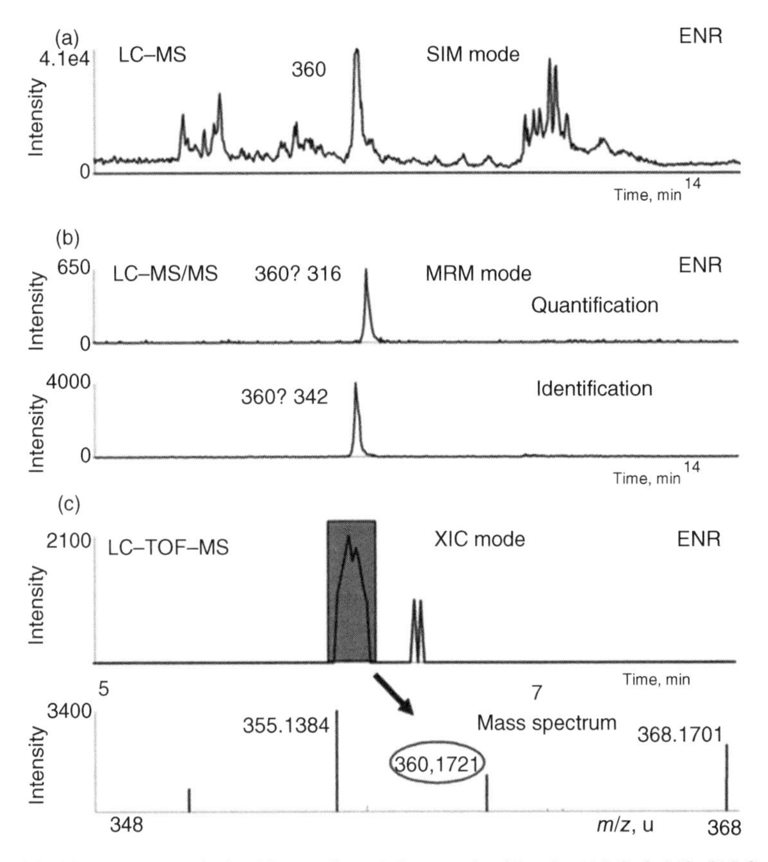

Figure 6.4. Chromatograms obtained for enrofloxacin in a sample of liver by (a) LC–Q–MS, (b) LC–QqQ–MS/MS, and (c) TOF–MS. *Source*: Ref. [98], Figure 6, p. 13. Reproduced with permission of Elsevier.

of enrofloxacin (2 μg/kg) was higher than the LOQ of TOF–MS and QqQ–MS/MS and around the LOQ of Q–MS. Therefore, the results indicated that the use of TOF–MS was a feasible alternative to Q and QqQ in multiresidue analysis in food samples.

A recent study attempted to determine the mass resolution and corresponding mass window width required with HRMS techniques to obtain selectivity comparable to that of MS/MS [56]. They investigated the "vacant *m/z* space" available in MS/MS and HRMS spectra of an analyzed blank sample. The idea behind this concept was the assumption that monitoring a large number of such randomly selected masses shows a number of traces containing one or more chromatographic peaks caused by endogenous matrix compounds. The measurement of the number and the intensity of the detected peaks provides information regarding the degree of selectivity obtained by the analyzer. To carry out this comparison, 100 dummy transitions and exact masses (traces) were created by a random generator and monitored in blank extracts of fish, kidney, liver, and honey by UHPLC–QqQ–MS/MS and UHPLC-single-stage Orbitrap–MS. With this last analyzer, different resolution power and mass window were applied (10,000 FWHM:20 mDa, 25,000 FWHM:8 mDa, 50,000 FWHM:4 mDa, and 100,000 FWHM:2 mDa). Although these dummy transitions did not correspond to any particular analyte, they corresponded to the typical precursor and product ion mass range, as commonly observed for VDs. As expected, most extracted dummy traces were free of chromatographic peaks, but the dummy traces that contained chromatographic peaks were integrated. In the next step, the SRM and HRMS peak areas obtained were standardized. This was done by determining the response of seven typical VDs (peak area/concentration), and an average response for all these analytes was calculated. Then, each dummy peak area was divided by this average response to produce a standardized concentration for each matrix-related dummy peak area. This was done for MS/MS and HRMS. As a result of these experiments, the authors concluded that a HRMS resolution of 50,000 FWHM and a corresponding mass window of 10 ppm provide selectivity as good as or slightly higher than MS/MS. Moreover, the false positive found in a sample of honey, when it was analyzed by MS/MS technique, supported their conclusion.

The same authors evaluated the general quantitative and confirmative performance of MS/MS against the latest generation of Orbitrap–MS and TOF–MS technology [105,117]. First, they compared UHPLC-single-stage Orbitrap–MS operated at 50,000 FWHM with UHPLC–QqQ–MS/MS for quantification of anthelmintic drugs and phenylbutazone residue in milk and muscles [105]. The results showed that repeatability of both technologies were equal, but significantly higher sensitivity was obtained for critical compounds (avermectins) by single-stage Orbitrap detection. The LODs obtained with single-stage Orbitrap–MS were 0.5 μg/kg in milk and from 0.5 to 2 μg/kg in muscles, whereas LODs for QqQ–MS/MS ranged from 0.5 to 10 μg/kg in milk and from 1 to 10 μg/kg in muscle samples. Therefore, it was shown that analytes with poor fragmentation properties (e.g., sodium-cationized molecules), such as avermectin and ivermectin, can be more easily quantified by single-stage HRMS than by MS/MS.

In the second of these studies [117], the quantitative performance of HRMS-based detection (a single-stage Orbitrap–MS operated at 50,000 FWHM and a TOF–MS operated at 12,000 FWHM) versus a unit mass resolution-based MS/MS (QqQ technology) detection was compared. The comparison covered a limited set of 36 analyte residues present at trace concentrations in honey. Complete validation was performed for the honey matrix on single-stage Orbitrap–MS operated at 50,000 FWHM, TOF–MS, and QqQ–MS/MS. Low recoveries at high-spiked concentrations could be observed for several compounds when they were monitored by TOF–MS. In addition, the low recovery rate was indicated by a poor determination coefficient covering the two orders of magnitude dynamic range (corresponding to 10–1000 μg/kg). This phenomenon was attributed to the saturation of the TDC detector used in the TOF–MS instrument, which limits the dynamic range, as was previously explained (Section 6.3.2.1). Furthermore, in many cases, recovery apparently increased at the lowest concentration with TOF–MS and to a lesser degree with MS/MS. Most of these cases could be explained by the increasing relative importance of coeluting endogenous and exogenous compounds appearing at the same accurate mass MS/MS trace. However, single-stage Orbitrap–MS analyzer did not show this behavior at 50,000 FWHM. On the other hand, the results showed that determination coefficients and relative standard deviation (RSD) values were poorest for the utilized 12,000 FWHM TOF–MS instrument, whereas performance was slightly better using single-stage Orbitrap–MS instead of MS/MS instruments. As a result, the authors concluded that an equal or even a slightly better quantitative performance was observed for the single-stage Orbitrap–MS-based approach referring to precision, trueness, and dynamic range. A direct comparison of the sensitivity was not possible because the sensitivity of MS/MS strongly depends on the number of transitions to be monitored and in this work only 36 analytes were studied. Hence, although the sensitivity was higher for unit mass resolution MS/MS, it is not true when a large number of analytes have to be detected and quantified.

Additionally, the confirmatory capabilities of HRMS versus MS/MS were also compared [117]. This included the critical evaluation of precision and accuracy of ion ratios obtained by the use of MS/MS collision chambers, which includes precursor selection versus nonprecursor-selected fragmentation as obtained by HRMS technology. Two different nonprecursor-selected fragmentation techniques, as provided by the single-stage Orbitrap–MS instrument, were evaluated: fragmentation in the electrospray ionization (ESI) interface and fragmentation in the higher collision-induced dissociation cell (HCD). It was observed that many ESI-fragmented compounds produced ion ratios where the second ion was hardly visible and such ratios provided poor diagnostic information, while the ratios obtained by ions produced in HCD were higher, and most compounds showed acceptable ion ratios. Hence, the precision of fragmentations in single-stage Orbitrap–MS with HCD was significantly better than those in the ESI interface. HCD even appeared to produce better ion ratio precision than a classical collision chamber of an MS/MS instrument. However, poorer accuracy (fortified matrix extracts versus pure standard solution) of ion ratios was observed when comparing data obtained by Orbitrap–MS versus MS/MS. Additionally, the observed higher absolute ion ratio deviations demonstrated that fragmentation ratios based on nonprecursor-selected experiments (Orbitrap utilizing ESI or HCD fragmentations) seemed to be affected to a certain degree by matrix effects.

Nielen et al. [118] discussed the mass resolution and accuracy for LC–MS screening and confirmation of targeted analytes and for the identification of unknown compounds employing HRMS detectors. The experiments were based on the screening of the anabolic steroid stanozolol and the designer β-agonist "Clenbuterol-R" using screening resolutions of 10,000 FWHM or higher with mass windows of ±50 mDa, employing different HRMS analyzers. It was observed that accurate mass determination without proper mass resolving power criteria led to false negative results in MS screening as well as in MS/MS confirmation. Finally, the authors concluded that only resolutions of 70,000 FWHM or higher allowed reliable accurate masses of elemental compositions differing in one CO, C_2H_4, or N_2 substructure to distinguish between the analytes and coeluting substances. A lack of mass resolving power was demonstrated for anabolic steroid stanozolol analyzed using the LC–QqTOF–MS.

Recently, two HRMS techniques, QqTOF–MS and single-stage Orbitrap–MS, were compared with the traditional LRMS detector, QqQ–MS/MS [64], to evaluate their ability to serve as screening tools. In this work, two screening methods based on the use of UHPLC–QqTOF–MS (8,000 FWHM) and UHPLC coupled single-stage Orbitrap–MS (50,000 FWHM) were developed and compared for the determination of several classes of VDs in milk- and powdered milk-based formula samples. The performance characteristics of these screening methods were compared in terms of the uncertainty region and cutoff values. Better results were obtained using the Orbitrap-based screening method, obtaining narrower uncertainty regions in all cases and lower cutoff values (Figure 6.5). For the Orbitrap–MS screening method, cutoff values were ≤5.0 μg/kg, while for QqTOF–MS, the values ranged from 5.0 to 7.5 μg/kg or higher, although in all cases and using both analyzers, the obtained cutoff values were lower than the MRLs established by the EU in the selected matrices. Therefore, the results clearly suggested that the higher resolving power of the Orbitrap–MS analyzer ensured adequate high full scan selectivity, which enabled the detection of the analytes at lower concentrations in these matrices. Additionally, the results obtained with HRMS techniques were compared with two screening methods developed with a LRMS analyzer, QqQ [42], one of them based on the selection of neutral loss or product ions (method A), whereas the other one was based on the use of a SRM transition for each compound (method B). HRMS analyzers provided better results than LRMS analyzers. In terms of screening validation parameters, such as cutoff and uncertainty region, it could be indicated that the single-stage Orbitrap screening method provided better results than the QqTOF and QqQ screening methods.

The applicability of UHPLC combined with full scan accurate mass TOF–MS and LTQ–Orbitrap–MS was also evaluated for the analysis of hormone and coccidiostats [102]. UHPLC–LTQ–Orbitrap–MS analysis was performed at a resolving power of 60,000 FWHM and it enabled the detection at accurate mass measurement (<3 ppm error) of all 14 steroid esters at low ng/kg concentrations, despite the complex matrix background. A 5 ppm mass tolerance window proved to be essential to generate highly selective reconstructed ion chromatograms, having reduced background from the hair matrix. UHPLC–LTQ–Orbitrap–MS at a lower resolving power of 7,500

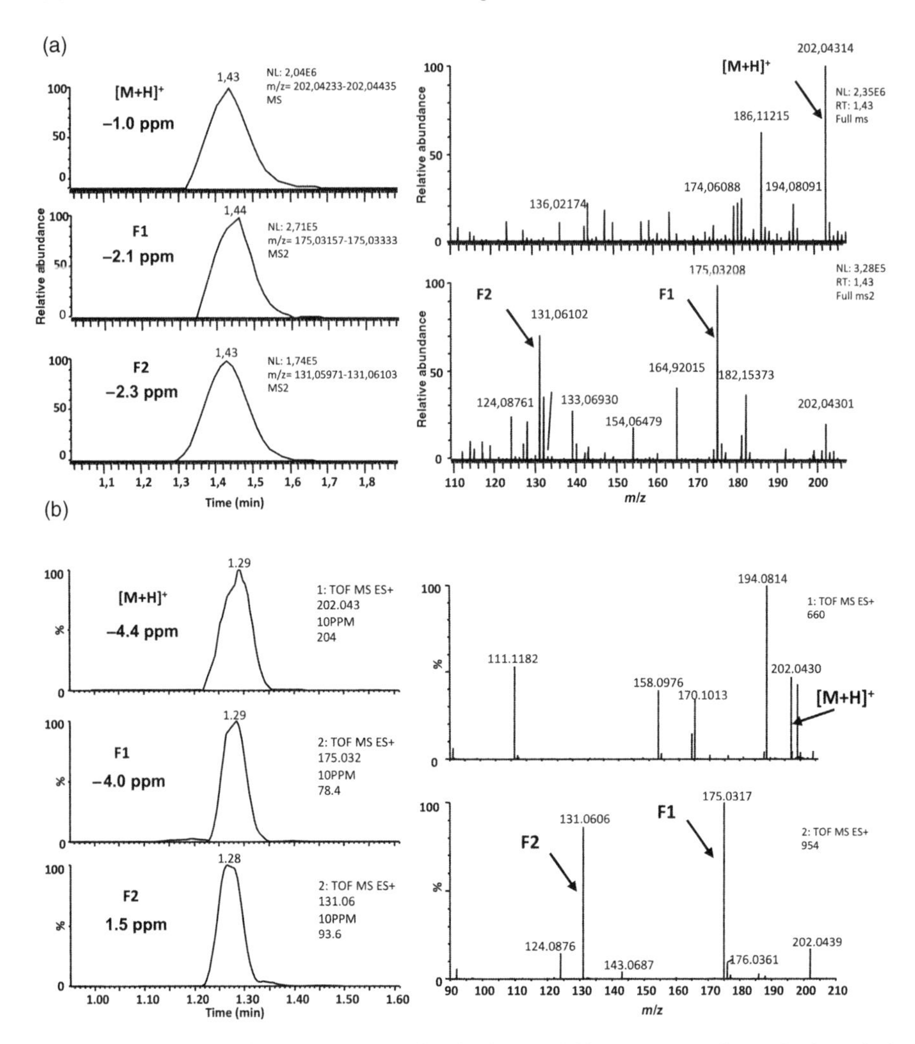

Figure 6.5. (a) Extracted ion chromatogram using single-stage Orbitrap, corresponding to the theoretical *m/z* of the protonated molecule of thiabendazole (*m/z* 202.04334) and its fragments (fragment 1 (F1): *m/z* 175.0325 and fragment 2 (F2): *m/z* 131.0604, mass tolerance 5 ppm), and the spectrum acquired at the elution time of the compound in a milk matrix-matched standard at 25 μg/kg. (b) Extracted ion chromatogram using QqTOF, corresponding to the theoretical *m/z* of the protonated molecule of thiabendazole (*m/z* 202.0433) and its fragments (fragment 1 (F1): *m/z* 175.0324 and fragment 2 (F2): *m/z* 131.0604, mass tolerance 10 ppm), and the spectrum acquired at the elution time of the compound in a milk matrix-matched standard at 25 μg/kg. *Source*: Ref. [64], Figure 2, p. 9358. Reproduced with permission of Elsevier.

FWHM and UHPLC–TOF–MS at a mass resolving power of 10,000 FWHM failed to detect all of the steroid esters in hair extracts owing to the inability to mass resolve analyte ions from the coeluting isobaric matrix compounds. However, when the comparison was made with coccidiostats in feed samples, the resolution power was not as critical as for steroids in hair samples. UHPLC–TOF–MS allowed the detection of all coccidiostats, whereas UHPLC–LTQ–Orbitrap–MS at a resolving power of 7,500 only missed the detection of amprolium at the lowest concentration (15 μg/kg). The successful detection of coccidiostats in feed extracts in contrast to the failure to detect steroid esters in hair extracts at a mass resolving power of 7,500 was most likely explained by the difference in the relative concentrations of compounds versus the sample matrix. This indicated that a medium resolving power of 7,500–10,000 was sufficient to prevent inaccurate mass assignment owing to matrix interferences, although for complex matrices, higher resolution should be needed [90].

6.5 CONCLUSIONS AND FUTURE TRENDS

It has been shown throughout this chapter that in the past several years there has been an increase in the number of studies reporting HRMS-based approaches for VDs determination in food. The reported works derived from HRMS measurements are still scarce, which reflects the fact that HRMS has not yet been widely used in routine residue analysis. This could be explained by the relatively recent availability of modern HRMS instrumentations; the resistance to leave/change a successful, well-established, and time-proven technology such as MS/MS (QqQ, IT, etc.); the lack of speed; the difficulty of the data processing software; and some hardware limitations of HRMS instruments. However, taking into account the successful applications discussed in this chapter, it is expected that HRMS based on resolutions higher than 10,000 FWHM will be implemented and used together with LRMS techniques in the field of VDs analysis in food in the near future. The latest generation of HRMS instrumentation is capable of providing higher resolution, superior sensitivity, larger dynamic ranges, and improved speed, and therefore a shift toward the application of HRMS and away from LRMS has been observed. Nevertheless, their implementation will be effectively accelerated as long as the available software improves significantly. Current limitations concerning processing speed, ease of use, confirmation tools (e.g., utilization of exact masses and isotopic ratio), and flexible report generation are hindering their application in routine analysis.

Additionally, it is important to highlight the high-throughput abilities of HRMS techniques. The use of full scan-based MS techniques simplified method setup because of the application and selection of generic acquisition parameters. The use of this acquisition mode also implies monitoring of every analyte ionized in the source, and therefore, there is no theoretical limit on the number of compounds per method. This allows the development of multiclass methods with a high number of monitored compounds, which also reduces the number of analyses per sample, increasing sample throughput. Furthermore, it is possible to perform screening and confirmation processes in one single injection, increasing again sample throughput.

The first multicompound analyses employing HRMS, which allowed determination of over 100 compounds, consisted of a preliminary screening using the LC–HRMS platform for initial qualitative identification followed by a quantitative

SRM confirmation using the LC–QqQ–MS platforms. The multiscreening step consisted of full scan accurate MS techniques (e.g., TOF–MS or Orbitrap MS) and required an additional MS/MS-based quantification and confirmation step. However, HRMS has been demonstrated to be an attractive tool for trace analyte detection by screening analysis, as well as for confirmation and quantification in challenging matrices.

Currently, available resolution achieved with HRMS means that its selectivity exceeds that provided by currently used LRMS instruments for confirmation. HRMS has been shown to be capable of detecting compounds at trace concentrations and identifying coeluted compounds, and can potentially suppress matrix compounds. Additionally, because HRMS instruments acquire full scan accurate data, they allow retrospective data analysis based on an *a posteriori* hypothesis. This permits the detection of target compounds as well as the monitoring of metabolites or degradation products, which are not commercially available for the analytes investigated. Thus, it has been demonstrated that LC–HRMS opens the possibility of identifying target and nontarget compounds and improves reliability and robustness. In consequence, laboratory throughput is enhanced because the samples do not need to be injected again if the monitoring of additional analytes is requested.

Additionally, HRMS permits the elucidation of the elemental composition of analytes based on exact masses and isotopic patterns (RIA). For reliable structural assignment, high resolution (>10,000 FWHM) is recommended and the accuracy of the isotope intensities must be excellent to allow elemental formula fits for substances without highly characteristic isotope patterns (compounds without Cl or Br atoms) as is the case of some families of VDs (such as macrolides and quinolones).

On the other hand, it can be concluded that HRMS analyzers such as the Orbitrap, and to a lesser extent, TOF–MS or QqTOF–MS, are also suitable for quantification, obtaining performance characteristics similar to conventional LRMS analyzers, where a large number of analytes have to be detected and quantified. Therefore, UHPLC coupled with TOF–MS, QqTOF–MS, single-stage Orbitrap–MS, and LQT–Orbitrap–MS offers unsurpassed performance for screening purposes. They can also effectively provide quantitative information and are accurate and sensitive enough to differentiate between positive and negative samples in VDs.

ACKNOWLEDGMENTS

The authors gratefully acknowledge the Spanish Ministry of Economy and Competitiveness (MINECO) and FEDER for financial support (Project Ref. AGL2010-21370). M.M.A.L. acknowledges her grant (F.P.U.) (MINECO, Ref. AP2008-02811). P.P.B gratefully acknowledges personal funding through the Agrifood Campus of International Excellence, ceiA3 (Spanish Ministry of Education, Culture and Sport). R.R.G. is also grateful for personal funding through the Ramón y Cajal Program (MINECO-ESF).

REFERENCES

1. Mellon, M.; Benbrook, C.; Benbrook, K.L. Hogging It: Estimates of antimicrobial abuse in livestock. Cambridge: UCS Publications; **2001**.
2. http://foodawareness.org/antibiotic_resistance_64.html (accessed September, 2012).
3. Bogialli, S.; Di Corcia, A. Recent applications of liquid chromatography–mass spectrometry to residue analysis of antimicrobials in food of animal origin. *Anal. Bioanal. Chem.* **2009**, 395, 947–966.
4. European Medicines Agency. European Surveillance of Veterinary Antimicrobial Consumption (ESVAC). Available at http://www.ema.europa.eu/ema/index.jsp?curl=pages/regulation/document_listing/document_listing_000302.jsp (accessed July 2013).
5. Di Corcia, A.; Nazzari, M. Liquid chromatographic–mass spectrometric methods for analyzing antibiotic and antibacterial agents in animal food products. *J. Chromatogr. A* **2002**, 974, 53–89.
6. Moreno-Bondi, M.C.; Marazuela, M.D.; Herranz, S.; Rodriguez, E. Antibiotics in food and environmental simples. *Anal. Bioanal. Chem.* **2009**, 395, 921–946.
7. Commission Regulation (EU) No. 37/2010 of 22 December 2009 on pharmacologically active substances and their classification regarding maximum residue limits in foodstuffs of animal origin. *Off. J. Eur. Commun.* **2010**, L15, 1–72. Available at http://eur-lex.europa.eu/LexUriServ/LexUriServ.do?uri=CONSLEG:2010R0037:20120409:ES:PDF (accessed September 2012).
8. The Code of Federal Regulations. Title 21, Food and Drugs. Available at http://www.gpo.gov/fdsys/browse/collectionCfr.action?collectionCode=CFR (accessed September 2012).
9. Aguilera-Luiz, M.M.; Martínez Vidal, J. L.; Romero-González, R.; Garrido Frenich, A. Multi-residue determination of veterinary drugs in milk by ultra-high-pressure liquid chromatography–tandem mass spectrometry. *J. Chromatogr. A* **2008**, 1205, 10–16.
10. Garrido Frenich, A.; Aguilera-Luiz, M.M.; Martínez Vidal, J.L.; Romero-González, R. Comparison of several extraction techniques for multiclass analysis of veterinary drugs in eggs using ultra-high pressure liquid chromatography–tandem mass spectrometry. *Anal. Chim. Acta* **2010**, 661, 150–160.
11. Nebot, C., Iglesias, A.; Regal, P.; Miranda, J.; Cepeda, A.; Fente, C. Development of a multi-class method for the identification and quantification of residues of antibiotics, coccidiostats and corticosteroids in milk by liquid chromatography–tandem mass spectrometry. *Int. Dairy J.* **2012**, 22, 78–85.
12. Lamar, J.; Petz, M. Development of a receptor-based microplate assay for the detection of beta-lactam antibiotics in different food matrices. *Anal. Chim. Acta* **2007**, 586, 296–303.
13. Adrian, J.; Pinacho, D.G.; Granier, B.; Diserens, J.M.; Sánchez-Baeza, F.; Marco, M.P. A multianalyte ELISA for immunochemical screening of sulfonamide, fluoroquinolone and β-lactam antibiotics in milk samples using class-selective bioreceptors. *Anal. Bioanal. Chem.* **2008**, 391, 1703–1712.
14. Borrás, S.; Companyó, R.; Guiteras, J. Analysis of sulfonamides in animal feeds by liquid chromatography with fluorescence detection. *J. Agric. Food Chem.* **2011**, 59, 5240–5247.
15. Christodoulou, E.A.; Samanidou, V.F.; Papadoyannis, I.N. Validation of an HPLC–UV method according to the EU Decision 2002/657/EC for the simultaneous determination of 10 quinolones in chicken muscle and egg yolk. *J. Chromatogr. B* **2007**, 859, 246–255.

16. Commission Decision 2002/657/EC of 12 August 2002 implementing Council Directive 96/23/EC concerning the performance of analytical methods and the interpretation of results. *Off. J. Eur. Union* **2002**, L221, 8–36. Available at http://eur-lex.europa.eu/LexUriServ/LexUriServ.do?uri=CONSLEG:2010R0037:20120801:ES:PDF (accessed September 2012).
17. Reig, M.; Toldrá, F. Veterinary drug residues in meat: concerns and rapid methods for detection. *Meat Sci.* **2008**, 78, 60–67.
18. Directive of Council 81/851/EEC of 28 September 1981 on the approximation of the laws of the Member States relating to veterinary medicinal products. *Off. J. Eur. Commun.* **1981**, L317, 1–15. Available at http://eur-lex.europa.eu/LexUriServ/LexUriServ.do?uri=CELEX:31981L0851:en:NOT (accessed September 2012).
19. Directive of Council 81/852/EEC of 28 September 1981 on the approximation of the laws of the Member States relating to analytical, pharmaco-toxicological and clinical standards and protocols in respect of the testing of veterinary medicinal products. *Off. J. Eur. Commun.* **1981**, L317, 16–28. Available at http://eur-lex.europa.eu/LexUriServ/LexUriServ.do?uri=OJ:L:1981:317:0016:0028:EN:PDF (accessed September 2012).
20. Council Directive 90/676/EEC Of 13 December 1990 amending Directive 81/851/EEC on the approximation of the laws of the Member States relating to veterinary medicinal products. *Off. J. Eur. Commun.* **1991**, L373, 15–25. Available at http://eur-lex.europa.eu/LexUriServ/LexUriServ.do?uri=CELEX:31990L0676:en:NOT (accessed September 2012).
21. Council Directive 90/677/EEC of 13 December 1990 extending the scope of Directive 81/851/EEC on the approximation of the laws of the Member States relating to veterinary medicinal products and laying down additional provisions for immunological veterinary medicinal products. *Off. J. Eur. Commun.* **1990**, L373, 26–28. Available at http://eur-lex.europa.eu/LexUriServ/LexUriServ.do?uri=CELEX:31990L0677:en:NOT (accessed September 2012).
22. Council Directive 92/74/EEC of 22 September 1992 widening the scope of Directive 81/851/EEC on the approximation of provisions lay down by law, regulation or administrative action relating to veterinary medicinal products and laying down additional provisions on homeopathic veterinary medicinal products. *Off. J. Eur. Commun.* **1992**, L297, 12–15. Available at http://eur-lex.europa.eu/LexUriServ/LexUriServ.do?uri=CELEX:31992L0074:en:NOT (accessed September 2012).
23. Council Directive 93/40/EEC of 14 June 1993 amending Directives 81/851/EEC and 81/852/EEC on the approximation of the laws of the Member States relating to veterinary medicinal products. *Off. J. Eur. Commun.* **1993**, L214, 31–39. Available at http://eur-lex.europa.eu/LexUriServ/LexUriServ.do?uri=CELEX:31993L0040:en:NOT (accessed September 2012).
24. Directive 2001/82/EC of the European Parliament and of the Council of 6 November 2001 on the Community code relating to veterinary medicinal products. *Off. J. Eur. Commun.* **2011**, L311, 1–66. Available at http://eur-lex.europa.eu/LexUriServ/LexUriServ.do?uri=OJ:L:2001:311:0001:0066:EN:PDF (accessed September 2012).
25. Common Position (EC) No. 62/2003 adopted by the Council on 29 September 2003 with a view to the adoption of a Directive 2003 of the European Parliament and of the Council of amending Directive 2001/82/EC on the Community code relating to veterinary medicinal products. *Off. J. Eur. Union* **2003**, C 297 E/72-100. Available at http://eur-lex.europa.eu/LexUriServ/LexUriServ.do?uri=OJ:C:2003:297E:0072:0100:EN:PDF (accessed September 2012).

26. Directive 2004/28/EC of the European Parliament and of the Council of 31 March 2004 amending Directive 2001/82/EC on the Community code relating to veterinary medicinal products. *Off. J. Eur. Union* **2004** L136, 58–84. Available at http://eur-lex.europa.eu/smartapi/cgi/sga_doc?smartapi!celexplus!prod!DocNumber&lg=en&type_doc=Directive&an_doc=2004&nu_doc=28 (accessed September 2012).
27. Directive 2009/53/EC of the European Parliament and of the Council of 18 June 2009, amending Directive 2001/82/EC and Directive 2001/83/EC, as regards variations to the terms of marketing authorisation for medicinal products. *Off. J. Eur. Union* **2009** L168, 33–34. Available at http://eur-lex.europa.eu/Notice.do?val=497321:cs&lang=es&list=497321:cs,491368:cs,487055:cs, &pos=1&page=1&nbl=3&pgs=10&hwords=&checktexte=checkbox&visu=#texte (accessed September 2012).
28. Council Regulation (EEC) No. 2377/90 of 26 June 1990 laying down a Community procedure for the establishment of maximum residue limits of veterinary medicinal products in foodstuffs of animal origin. *Off. J. Eur. Commun.* **1990** L224, 1–139. Available at http://eur-lex.europa.eu/LexUriServ/LexUriServ.do?uri=CONSLEG:1990R2377:20090902:ES:PDF (accessed September 2012).
29. Council Directive (EEC) No. 96/22/CE of 29 April 1996 concerning the prohibition on the use in stockfarming of certain substances having a hormonal or thyrostatic action and of beta-agonists, and repealing Directives 81/602/EEC, 88/146/EEC and 88/299/EEC. *Off. J. Eur. Commun.* L125, 10–32. Available at http://eur-lex.europa.eu/LexUriServ/LexUriServ.do?uri=CONSLEG:1996L0022:20081218:ES:PDF (accessed September 2012).
30. Commission Decision of 12 August 2002 implementing Council Directive 96/23/EC concerning the performance of analytical methods and the interpretation of results. *Off. J. Eur. Commun.* L221, 1–8. Available at http://eur-lex.europa.eu/LexUriServ/LexUriServ.do?uri=CONSLEG:2002D0657:20040110:ES:PDF (accessed September 2012).
31. Commission Decision of 13 March 2003 amending Decision 2002/657/EC as regards the setting of minimum required performance limits (MRPLs) for certain residues in food of animal origin. *Off. J. Eur. Union* L71, 17–18. Available at http://eur-lex.europa.eu/LexUriServ/LexUriServ.do?uri=OJ:L:2003:071:0017:0018:EN:PDF (accessed September 2012)
32. Commission Decision of 22 December 2003 amending Decision 2002/657/EC as regards the setting of minimum required performance limits (MRPLs) for certain residues in food of animal origin. *Off. J. Eur. Union* **2003**, L6, 38–39. Available at http://crl.fougeres.anses.fr/publicdoc/l_00620040110en00380039.pdf (accessed September 2012).
33. Tolerances for residues on new animal drugs in food. Code of Federal Regulations, Food and Drugs, Part 556, Title 21; Office of the Federal Register, National Archives and Records Administration, Washington, DC, April 1, 2011. Available at http://www.gpo.gov/fdsys/pkg/CFR-2011-title21-vol6/pdf/CFR-2011-title21-vol6.pdf (accessed September 2012).
34. Available at http://www.hc-sc.gc.ca/dhp-mps/vet/mrl-lmr/index-eng.php (accessed September 2012).
35. Australian Pesticides and Veterinary Authority. Maximum residue limits (MRLs) in food and animal feedstuff, Table 1, maximum residue limits of agricultural and veterinary chemicals and associated substances in food commodities. Available at http://www.apvma.gov.au/residues/docs/mrl_table1_september_2012.pdf (accessed September 2012).

36. Ministry of Health. Labour and Welfare, Positive List System for Agricultural Chemical Residues in Foods, MRLs Data Base. Available at http://www.mhlw.go.jp/english/topics/foodsafety/positivelist060228/index.html (accessed September 2012).
37. Codex Alimentarius. Veterinary Drug Residues in Food, 34th Session of the Codex Alimentarius Commission, **2011**. Available at http://www.codexalimentarius.net/vetdrugs/data/vetdrugs/index.html?lang=en (accessed September 2012).
38. Available at http://www.ema.europa.eu/ema/index.jsp?curl=/pages/home/Home_Page.jsp (accessed September 2012).
39. Available at http://www.efsa.europa.eu (accessed September 2012).
40. Available at http://www.codexalimentarius.org/standards/list-of-standards/ (accessed July 2013).
41. Lee, J.B.; Chung, H.H.; Chung, Y.H.; Lee, K.G. Development of an analytical protocol for detecting antibiotic residues in various foods. *Food Chem.* **2007**, 105, 1726–1731.
42. Virolainen, N.E.; Pikkemaat, M.G.; Alexander Elferink, J.W.; Karp, M.T. Rapid detection of tetracyclines and their 4-epimer derivatives from poultry meat with bioluminescent biosensor bacteria. *J. Agric. Food Chem.* **2008**, 56, 11065–11070.
43. Dasenaki, M.E.; Thomaidis, N.S. Multi-residue determination of seventeen sulfonamides and five tetracyclines in fish tissue using a multi-stage LC–ESI–MS/MS approach based on advanced mass spectrometric techniques. *Anal. Chim. Acta* **2010**, 672, 93–102.
44. Martínez Vidal, J.L.; Garrido Frenich, A.; Aguilera-Luiz, M.M.; Romero-González, R. Development of fast screening methods for the analysis of veterinary drug residues in milk by liquid chromatography–triple quadrupole mass spectrometry. *Anal. Bioanal. Chem.* **2010**, 397, 2777–2790.
45. Gentilli, A.; Perret, D.; Marchese, S. Liquid chromatography–tandem mass spectrometry for performing confirmatory analysis of veterinary drugs in animal-food products. *Trends Anal. Chem.* **2005**, 24, 622–634.
46. Carretero, V.; Blasco, C.; Picó, Y. Multi-class determination of antimicrobials in meat by pressurized liquid extraction and liquid chromatography–tandem mass spectrometry. *J. Chromatogr. A* **2008**, 1209, 162–173.
47. Msagati, T.A.M.; Nindi, M.M. Determination of β-lactam residues in foodstuffs of animal origin using supported liquid membrane extraction and liquid chromatography–mass spectrometry. *Food Chem.* **2007**, 100, 836–844.
48. Cherlet, M.; De Baere, S.; De Backer, P. Quantitative determination of dexamethasone in bovine milk by liquid chromatography–atmospheric pressure chemical ionization–tandem mass spectrometry. *J. Chromatogr. B* **2004**, 805, 57–65.
49. Fagerquist, C.K.; Lightfield, A.R. Confirmatory analysis of β-lactam antibiotics in kidney tissue by liquid chromatography/electrospray ionization selective reaction monitoring ion trap tandem mass spectrometry. *Rapid Commun. Mass Spectrom.* **2003**, 17, 660–671.
50. Chico, J.; Rúbies, A.; Centrich, F.; Companyó, R.; Prat, M.D.; Granados, M. High-throughput multiclass method for antibiotic residue analysis by liquid chromatography–tandem mass spectrometry. *J. Chromatogr. A* **2008**, 1213, 189–199.
51. Ben, W.; Qianga, Z.; Adams, C.; Zhang, H.; Chen, L. Simultaneous determination of sulfonamides, tetracyclines and tiamulin in swine wastewater by solid-phase extraction and liquid chromatography–mass spectrometry. *J. Chromatogr. A* **2008**, 1202, 173–180.
52. Wu, X.; Zhang, G.; Wu, Y.; Hou, X.; Yuan, Z. Simultaneous determination of malachite green, gentian violet and their leuco-metabolites in aquatic products by high-performance

liquid chromatography–linear ion trap mass spectrometry. *J. Chromatogr. A* **2007**, 1172, 121–126.

53. Kaufmann, A.; Butcher, P.; Maden, K.; Walker, S.; Widmer, M. Development of an improved high resolution mass spectrometry based multi-residue method for veterinary drugs in various food matrices. *Anal. Chim. Acta* **2011**, 700, 86–94.
54. Huang, J.F.; Zhang, H.J.; Lin, B.; Yu, Q.W.; Feng, Y.Q. Multiresidue analysis of β-agonists in pork by coupling polymer monolith microextraction to electrospray quadrupole time-of-flight mass spectrometry. *Rapid Commun. Mass Spectrom.* **2007**, 21, 2895–2904.
55. Le Bizec, B.; Pinel, G.; Antignac, J.P. Options for veterinary drug analysis using mass spectrometry. *J. Chromatogr. A* **2009**, 1216, 8016–8034.
56. Kaufmann, A.; Butcher, P.; Maden, K.; Walker, S.; Widmer, M. Comprehensive comparison of liquid chromatography selectivity as provided by two types of liquid chromatography detectors (high resolution mass spectrometry and tandem mass spectrometry): "Where is the crossover point?" *Anal. Chim. Acta* **2010**, 673, 60–72.
57. Ortelli, D.; Cognard, E.; Jan, P.; Edder, P. Comprehensive fast multiresidue screening of 150 veterinary drugs in milk by ultra-performance liquid chromatography coupled to time of flight mass spectrometry. *J. Chromatogr. B* **2009**, 877, 2363–2374.
58. Andersen, W.C.; Turnipseed, S.B.; Karbiwnyk, C.M.; Lee, R.H.; Clark, S.B.; Rowe, W. D.; Madson, M.R.; Miller, K.E. Multiresidue method for the triphenylmethane dyes in fish: malachite green, crystal (gentian) violet, and brilliant green. *Anal. Chim. Acta* **2009**, 637, 279–289.
59. Smith, S.; Gieseker, C.; Reimschuessel, R.; Decker, C.S.; Carson, M.C. Simultaneous screening and confirmation of multiple classes of drug residues in fish by liquid chromatography–ion trap mass spectrometry. *J. Chromatogr. A* **2009**, 1216, 8224–8232.
60. Kantiani, L.; Farré, M.; Grases i Freixiedas, J. M.; Barceló, D. Development and validation of a pressurised liquid extraction liquid chromatography–electrospray–tandem mass spectrometry method for β-lactams and sulfonamides in animal feed. *J. Chromatogr. A* **2010**, 1217, 4247–4254.
61. Kaufmann, A.; Butcher, P.; Maden, K.; Widmer, M. Quantitative multiresidue method for about 100 veterinary drugs in different meat matrices by sub 2-μm particulate high-performance liquid chromatography coupled to time of flight mass spectrometry. *J. Chromatogr. A* **2008**, 1194, 66–79.
62. Hernando, M.D.; Mezcua, M.; Suárez-Barcena, J.M.; Fernández-Alba, A.R. Liquid chromatography with time-of-flight mass spectrometry for simultaneous determination of chemotherapeutant residues in salmon. *Anal. Chim. Acta* **2006**, 562, 176–184.
63. Van Bocxlaer, J.F.; Vande Casteele, S. R.; Van Poucke, C.J.; Van Peteghem, C.H. Confirmation of the identity of residues using quadrupole time-of-flight mass spectrometry. *Anal. Chim. Acta* **2005**, 529, 65–73.
64. Romero-González, R.; Aguilera-Luiz, M.M.; Plaza-Bolaños, P.; Garrido Frenich, A.; Martínez Vidal, J.L. Food contaminant analysis at high resolution mass spectrometry: application for the determination of veterinary drugs in milk. *J. Chromatogr. A* **2011**, 1218, 9353–9365.
65. Wang, J.; Leung, D. Analyses of macrolide antibiotic residues in eggs, raw milk, and honey using both ultra-performance liquid chromatography/quadrupole time-of-flight

mass spectrometry and high-performance liquid chromatography/tandem mass spectrometry. *Rapid Commun. Mass Spectrom.* **2007**, 21, 3213–3222.

66. Gallo, P; Fabbrocino, S.; Serpe, L.; Fiori, M.; Civitareale, C.; Stacchini, P. Determination of the banned growth promoter moenomycin A in feed stuffs by liquid chromatography coupled to electrospray ion trap mass spectrometry. *Rapid Commun. Mass Spectrom.* **2010**, 24, 1017–1024.
67. Carrasco-Pancorbo, A.; Casado-Terrones, S.; Segura-Carretero, A.; Fernández-Gutiérrez, A. Reversed-phase high-performance liquid chromatography coupled to ultraviolet and electrospray time-of-flight mass spectrometry on-line detection for the separation of eight tetracyclines in honey samples. *J. Chromatogr. A* **2008**, 1195, 107–116.
68. Hurtaud-Pessel, D.; Jagadeshward-Reddy, T.; Verdon, E. Development of a new screening method for the detection of antibiotic residues in muscle tissues using liquid chromatography and high resolution mass spectrometry with a LC–LTQ–Orbitrap instrument. *Food Addit. Contam.* **2011**, 28, 1340–1351.
69. Ferrer, I.; Thurman, E.M. Multi-residue method for the analysis of 101 pesticides and their degradates in food and water samples by liquid chromatography/time-of-flight mass spectrometry. *J. Chromatogr. A* **2007**, 1175, 24–37.
70. Williamson, L.N.; Bartlett, M.G. Quantitative liquid chromatography/time-of-flight mass spectrometry. *Biomed. Chromatogr.* **2007**, 21, 567–576.
71. Ortelli, D.; Edder, P.; Cognard, E.; Jan, P. Fast screening and quantitation of microcystins in microalgae dietary supplement products and water by liquid chromatography coupled to time of flight mass spectrometry. *Anal. Chim. Acta* **2008**, 617, 230–237.
72. Dass, C. *Fundamentals of Contemporary Mass Spectrometry*, 1st edition. Hoboken, NJ: John Wiley & Sons, Inc.; **2007**.
73. Peters, R.J.B.; Bolck, Y.J.C.; Rutgers, P.; Stolker, A.A.M.; Nielen, M.W.F. Multi-residue screening of veterinary drugs in egg, fish and meat using high-resolution liquid chromatography accurate mass time-of-flight mass spectrometry. *J. Chromatogr. A* **2009**, 1216, 8206–8216.
74. Ardrey, R.E. *Liquid Chromatography–Mass Spectrometry: An Introduction*, 1st edition. Chichester, UK: John Wiley & Sons, Ltd.; **2003**.
75. Chernushevich, I.V.; Loboda, A.V.; Thomson, B.A. An introduction to quadrupole–time-of-flight mass spectrometry. *J. Mass Spectrom.* **2011**, 36, 849–865.
76. Hager, J.W. Recent trends in mass spectrometer development. *Anal. Bioanal. Chem.* **2004**, 378, 845–850.
77. Díaz, R.; Ibáñez, M.; Sancho, J.V.; Hernández, F. Building an empirical mass spectra library for screening of organic pollutants by ultra-high-pressure liquid chromatography/hybrid quadrupole time-of-flight mass spectrometry. *Rapid Commun. Mass Spectrom.* **2011**, 25, 355–369.
78. Wolff, J.C.; Eckers, C.; Sage, A.B.; Giles, K.; Bateman, R. Accurate mass liquid chromatography/mass spectrometry on quadrupole orthogonal acceleration time-of-flight mass analyzers using switching between separate sample and reference sprays. 2. Applications using the dual-electrospray ion source. *Anal. Chem.* **2001**, 73, 2605–2612.
79. Wang, J.; MacNeil, J.D.; Kay, J.F. *Chemical Analysis of Antibiotic Residues in Food*, 1st edition. Hoboken, NJ: John Wiley & Sons, Inc.; **2012**.
80. Plumb, R.S.; Johnson, K.A.; Rainville, P.; Smith, B.W.; Wilson, I.D.; Castro-Perez, J.M.; Nicholson, J.K. UPLC/MS; a new approach for generating molecular fragment

information for biomarker structure elucidation. *Rapid Commun. Mass Spectrom.* **2006**, 20, 1989–1994.

81. Weaver, P.J.; Laures, A.M.F.; Wolff, J.C. Investigation of the advanced functionalities of a hybrid quadrupole orthogonal acceleration time-of-flight mass spectrometer. *Rapid Commun. Mass Spectrom.* **2007**, 21, 2415–2421.
82. Zang, Y.; Ficarro, S.B.; Li, S.; Marto, J.A. Optimized Orbitrap HCD for quantitative analysis of phosphopeptides. *J. Am. Soc. Mass Spectrom.* **2009**, 20, 1425–1434.
83. Dervilly-Pinela, G.; Weigel, S.; Lommen, A.; Chereau, S.; Rambaud, L.; Essers, M.; Antignac, J. P.; Nielen, M. W.F.; Le Bizec, B. Assessment of two complementary liquid chromatography coupled to high resolution mass spectrometry metabolomics strategies for the screening of anabolic steroid treatment in calves. *Anal. Chim. Acta* **2011**, 700, 144–154.
84. Pinhancos, R.; Maass, S.; Ramanathan, D.M. High-resolution mass spectrometry method for the detection, characterization and quantitation of pharmaceuticals in water. *J. Mass Spectrom.* **2011**, 46, 1175–1181.
85. Xu, H.; Zhang, J.; He, J.; Mi, J.; Liu, L. Rapid detection of chloramphenicol in animal products without clean-up using LC–high resolution mass spectrometry. *Food Addit. Contam.* **2011**, 28, 1364–1371.
86. Mol, H.G.J.; Van Dam, R.C.J.; Zomer, P.; Mulder, P.P.J. Screening of plant toxins in food, feed and botanicals using full-scan high-resolution (Orbitrap) mass spectrometry. *Food Addit. Contam.* **2011**, 28, 1405–1423.
87. Hu, Q.; Noll, R.J.; Li, H.; Makarov, A.; Hardman, M.; Cooks, R.G. The Orbitrap: a new mass spectrometer. *J. Mass Spectrom.* **2005**, 40, 430–443.
88. Makarov, A.; Denisov, E.; Kholomeev, A.; Balschun, W.; Lange, O.; Strupat, K.; Horning, S. Performance evaluation of a hybrid linear ion trap/Orbitrap mass spectrometer. *Anal. Chem.* **2006**, 78, 2113–2120.
89. Makarov, A.; Denisov, E.; Lange, O.; Horning, St. Dynamic range of mass accuracy in LTQ Orbitrap hybrid mass spectrometer. *J. Am. Soc. Mass Spectrom.* **2006**, 17, 977–982.
90. Kaufmann, A. The current role of high-resolution mass spectrometry in food analysis. *Anal. Bioanal. Chem.* **2012**, 403, 1233–1249.
91. Wang, J. Analysis of macrolide antibiotics, using liquid chromatography–mass spectrometry, in food, biological and environmental matrices. *Mass Spectrom. Rev.* **2009**, 28, 50–92.
92. Schwartz, J.C.; Senko, M.W. A two-dimensional quadrupole ion trap mass spectrometer. *J. Am. Soc. Mass Spectrom.* **2002**, 13, 659–669.
93. Malik, A.K.; Blasco, C.; Picó, Y. Liquid chromatography–mass spectrometry in food safety. *J. Chromatogr. A* **2010**, 1217, 4018–4040.
94. Garrido Frenich, A. Plaza-Bolaños, P.; Aguilera-Luiz, M.M.; Martínez Vidal, J.L. Recent advances in the analysis of veterinary drugs and growth-promoting agents by chromatographic techniques. In: Quintín, T.J., editor. *Chromatography Types, Techniques and Methods*, 1st edition. New York: Nova Publishers; **2010**. pp. 1–102.
95. Mol, H.G.J.; Plaza-Bolaños, P.; Zomer, P.; de Rijk, T.C.; Stolker, A.A.M.; Mulder, P.P.J. Toward a generic extraction method for simultaneous determination of pesticides, mycotoxins, plant toxins, and veterinary drugs in feed and food matrixes. *Anal. Chem.* **2008**, 80, 9450–9459.

96. Gómez-Pérez M.L.; Plaza-Bolaños P.; Romero-González R.; Martínez Vidal J.L.; Garrido Frenich A. Comprehensive qualitative and quantitative determination of pesticides and veterinary drugs in honey using liquid chromatography–Orbitrap high resolution mass spectrometry. *J. Chromatogr. A* **2012**, 1248, 130–138.
97. Stolker, A.A.M.; Rutgers, P.; Oosterink, E.; Lasaroms, J.J.P.; Peters, R.J.B.; van Rhijn, J.A.; Nielen, M.W.F. Comprehensive screening and quantification of veterinary drugs in milk using UPLC–ToF–MS. *Anal. Bioanal. Chem.* **2008**, 391, 2309–2322.
98. Hermo, M.P.; Barrón, D.; Barbosa, J. Determination of multiresidue quinolones regulated by the EU in pig liver samples: high-resolution time-of-flight mass spectrometry versus tandem mass spectrometry detection. *J. Chromatogr. A* **2008**, 1201, 1–14.
99. Villar-Pulido, M.; Gilbert-López, B.; García-Reyes, J.F.; Ramos Martos, N.; Molina-Díaz, A. Multiclass detection and quantitation of antibiotics and veterinary drugs in shrimps by fast liquid chromatography time-of-flight mass spectrometry. *Talanta* **2011**, 85, 1419–1427.
100. Turnipseed, S.B.; Storey, J.M.; Clark, S.B.; Miller, K.E. Analysis of veterinary drugs and metabolites in milk using quadrupole time-of-flight liquid chromatography–mass spectrometry. *J. Agric. Food Chem.* **2011**, 59, 7569–7581.
101. Gentili, A.; Sergi, M.; Perret, D.; Marchese, S.; Curini, R.; Lisandrin, S. High- and low-resolution mass spectrometry coupled to liquid chromatography as confirmatory methods of anabolic residues in crude meat and infant foods. *Rapid Commun. Mass Spectrom.* **2006**, 20, 1845–1854.
102. van der Heeft, E.; Bolck, Y.J.C.; Beumer, B.; Nijrolder, A.W.J.M.; Stolker, A.A.M.; Nielen, M.W.F. Full-scan accurate mass selectivity of ultra-performance liquid chromatography combined with time-of-flight and Orbitrap mass spectrometry in hormone and veterinary drug residue analysis. *J. Am. Soc. Mass Spectrom.* **2009**, 20, 451–463.
103. Kellmann, M.; Muenster, H.; Zomer, P.; Mol, H. Full scan MS in comprehensive qualitative and quantitative residue analysis in food and feed matrices: how much resolving power is required? *J. Am. Soc. Mass Spectrom.* **2009**, 20, 1464–1476.
104. Kaufmann, A.; Butcher, P.; Maden, K.; Walker, S.; Widmer, M. Semi-targeted residue screening in complex matrices with liquid chromatography coupled to high resolution mass spectrometry: current possibilities and limitations. *Analyst* **2011**, 136, 1898–1909.
105. Kaufmann, A.; Butcher, P.; Maden, K.; Walker, S.; Widmer, M. Quantification of anthelmintic drug residues in milk and muscle tissues by liquid chromatography coupled to Orbitrap and liquid chromatography coupled to tandem mass spectrometry. *Talanta* **2011**, 85, 991–1000.
106. Thevis, M.; Makarov, A.A.; Horning, S.; Schänzer, W. Mass spectrometry of stanozolol and its analogues using electrospray ionization and collision-induced dissociation with quadrupole-linear ion trap and linear ion trap–Orbitrap hybrid mass analyzers. *Rapid Commun. Mass Spectrom.* **2005**, 19, 3369–3378.
107. Virus, E.D.; Sobolevsky, T.G.; Rodchenkov, G.M. Introduction of HPLC/Orbitrap mass spectrometry as screening method for doping control. *J. Mass Spectrom.* **2008**, 43, 949–957.
108. Martínez Vidal, J. L., Plaza-Bolaños, P., Romero-González, R., Garrido Frenich, A. Determination of pesticide transformation products: a review of extraction and detection methods. *J. Chromatogr. A* **2009**, 1216, 6767–6788.

109. Baiocchi, C.; Brussino M.; Pazzi, M.; Medana C.; Marini, C.; Genta, E. Separation and determination of synthetic corticosteroids in bovine liver by LC–ion-trap–MS–MS on porous graphite. *Chromatographia* **2003**, 58, 11–14.

110. Heller, D.N.; Nochetto, C.B. Development of multiclass methods for drug residues in eggs: silica SPE cleanup and LC–MS/MS analysis of ionophore and macrolide residues. *J. Agric. Food Chem.* **2004**, 52, 6848–6856.

111. Heller, D.N.; Nochetto, C.B.; Rummel, N.G.; Thomas, M.H. Development of multiclass methods for drug residues in eggs: hydrophilic solid-phase extraction cleanup and liquid chromatography/tandem mass spectrometry analysis of tetracycline, fluoroquinolone, sulfonamide, and β-lactam residues. *J. Agric. Food Chem.* **2006**, 54, 5267–5278.

112. Eichhorn, P.; Aga, D.S. Characterization of moenomycin antibiotics from medicated chicken feed by ion-trap mass spectrometry with electrospray ionization. *Rapid Commun. Mass Spectrom.* **2005**, 19, 2179–2186.

113. Hammel, Y.A.; Mohamed, R.; Gremaud, E.; LeBreton, M.H.; Guy, P.A. Multi-screening approach to monitor and quantify 42 antibiotic residues in honey by liquid chromatography–tandem mass spectrometry. *J. Chromatogr. A* **2008**, 1177, 58–76.

114. Xu, H.; Zhang, J.; He, J.; Mi, J.; Liu, L. Rapid detection of chloramphenicol in animal products without clean-up using LC–high resolution mass spectrometry. *Food Addit. Contam.* **2011**, 28, 1364–1371.

115. Marchesini, G. R.; Haasnoot, W.; Delahaut, P.; Gerçek, H.; Nielen, M.W.F. Dual biosensor immunoassay-directed identification of fluoroquinolones in chicken muscle by liquid chromatography electrospray time-of-flight mass spectrometry. *Anal. Chim. Acta* **2007**, 586, 259–268.

116. Marchesini, G.R.; Buijs, J.; Haasnoot, W.; Hooijerink, D.; Jansson, O.; Nielen, M.W.F. Nanoscale affinity chip interface for coupling inhibition SPR immunosensor screening with nano-LC TOF MS. *Anal. Chem.* **2008**, 80, 1159–1168.

117. Kaufmann, A.; Butcher, P.; Maden, K.; Walker, S.; Widmer, M. Quantitative and confirmative performance of liquid chromatography coupled to high resolution mass spectrometry compared to tandem mass spectrometry. *Rapid Commun. Mass Spectrom.* **2011**, 25, 979–992.

118. Nielen, M.W.F.; van Engelen, M.C.; Zuiderent, R.; Ramaker, R. Screening and confirmation criteria for hormone residue analysis using liquid chromatography accurate mass time-of-flight, Fourier transform ion cyclotron resonance and Orbitrap mass spectrometry techniques. *Anal. Chim. Acta* **2007**, 586, 122–129.

CHAPTER

7

A ROLE FOR HIGH-RESOLUTION MASS SPECTROMETRY IN THE HIGH-THROUGHPUT ANALYSIS AND IDENTIFICATION OF VETERINARY MEDICINAL PRODUCT RESIDUES AND OF THEIR METABOLITES IN FOODS OF ANIMAL ORIGIN

ERIC VERDON, DOMINIQUE HURTAUD-PESSEL, and JAGADESHWAR-REDDY THOTA

7.1 INTRODUCTION

The analysis of residues of veterinary medicinal products (VMPRs) is of growing interest due to their impact on human health [1,2]. For public health reasons, it is necessary to ensure that food products of animal origin are free from VMPR contamination or that they are safe in terms of public health before bringing them to the market. In today's global market, concerns related to food safety are becoming increasingly important. Across the world, many countries are rigorously following scientifically well-established, biological and physicochemical methods to determine VMPRs in biological sample matrices. The physicochemical analyses of drug residues (from medicinal products) or contaminants (toxins and organic pollutants) are currently performed using mainly molecular separation techniques (chromatography) involving the collection of spectral information. However, as indicated in Chapter 2, there has recently been a rapid advancement in the use of mass spectrometry, particularly coupled with liquid chromatography (LC) [3,4]. For example, liquid chromatography with triple quadrupole(QqQ) mass analyzers (LC–MS/MS or LC–QqQ) are widely used in the food testing laboratories across the world for qualitative and quantitative analyses of targeted veterinary drug residues at sub-ng/g level that are present in complex biological matrices. Preselected targeted analysis benefits from the high sensitivity and selectivity of tandem mass spectrometry systems operated in the multiple reaction monitoring (MRM) or selective reaction monitoring (SRM) modes. However, these modes have some limitations: (i) The screening of targeted analytes above the instrument capacity in a single run is not feasible. Additionally, it needs the availability of all targeted standards to record the MRM values and conditions. (ii) The detection of unexpected/untargeted compounds

High-Throughput Analysis for Food Safety, First Edition.
Edited by Perry G. Wang, Mark F. Vitha, and Jack F. Kay.

that might also be present in the biological system is not possible. Unlike MRM/SRM modes, the high-resolution (HR) full scan MS analysis detects ionized compounds in a single run. The selectivity is obtained after acquisition by searching for a specific molecular mass. Identification of biomarkers or untargeted compounds is possible after getting the mass spectra with high mass accuracy at sufficient mass resolution. In this case, the analyzer's resolution is the determining factor; it provides precise measurements of accurate mass for each analyte ion.

High mass resolution can be achieved by sophisticated mass analyzers such as time-of-flight (ToF), Fourier transform orbital trap (FTMS–Orbitrap), and Fourier transform ion cyclotron resonance (FTICR) mass spectrometers, as discussed in Chapter 2. These instruments play a key role in the detection of analytes with minute differences in their masses [5–9]. These have been applied to a diverse set of experiments ranging from metabolomic analyses to clinical proteomics [10–13]. Now their use is expanding into the field of food safety [14,15]. Along with other R&D laboratories working on such issues, we at ANSES—Fougères are using high-resolution mass spectrometry (HRMS) to develop analyses of residues of different classes of VMPRs in complex biological matrices and to screen them in the full scan mode.

In addition to several EU National Reference Laboratories (NRLs), our laboratory aims to develop a simple and rapid extraction method followed by trace-level identification of VMPRs in different complex biological matrices. In this program, as part of ANSES analytical research activity, we focus on the process(es) of identification of nontargeted veterinary drug metabolites in bio-origin products.

The methodology that we consider here for the near future would allow specialized laboratories to control via high-throughput screening a large number of regulated compounds through a quick, one-day/one-shot analysis to build consumers' confidence on the safety of the food placed into the market.

Unambiguous identification of all nontargeted drug metabolites, except those of simple known drug modifications (e.g., oxidation, reduction, and acetylation), in a given complex biological matrix, is still a challenging task for food testing laboratories. However, some challenges can be soon overcome with the following approach:

- Obtaining high-resolution mass measurement of VMPR analytes in full scan MS and MS/MS modes.
- Knowing elemental composition of $[M + H]^+$ or $[M - H]^-$ ions in full scan mode and product ions in MS/MS mode with mass error <2 ppm.
- Data processing through specialized software to find fingerprinting biomarkers.
- Synthesizing standards of suspected compounds.
- Comparison of mass spectral data of suspected samples with the standard compound.

First, this chapter focuses on the issues associated with the analysis of veterinary drug residues in various foods and the regulations that govern it. Then we present some examples implementing LC–HRMS using LTQ–Orbitrap instrumentation. These examples demonstrate its capability in resolving analytical issues, particularly through the molecular identification of unknown compounds that can be achieved only with HRMS.

7.2 ISSUES ASSOCIATED WITH VETERINARY DRUG RESIDUES AND EUROPEAN REGULATIONS

Veterinary drugs are used to treat various infectious diseases in food-producing animals, poultry, fish, and so on in order to gain more profit. Improper use of veterinary drugs by way of overdosage or not giving enough withdrawal time after treatment can result in the presence of drug residues in foods of animal origin intended for human consumption. As a result of stern safety requirements for food stuffs, the control of residues from veterinary medicinal products has become a subject of prime regulatory concern to the EU and other authorities. To protect human health, tolerance levels known as maximum residue limits (MRLs) in food products have been set by several countries around the world and recommended at the international level through the Codex Alimentarius (FAO/WHO) [16]. In the EU, procedures for setting MRLs are governed by a recent Regulation (EC) No. 470/2009 replacing the previous Regulation (EC) No. 2377/90 that dated from 1990 [1]. These limits support the Regulations governing food safety in terms of the residues of veterinary drugs in the tissues or fluids of production animals that may enter the human food chain. These MRLs are calculated based on toxicological data and account for safety coefficients that can be a factor of 10 or even 100, depending on the compound considered and its potential harmful effects on humans. Another Regulation (EC) No. 37/2010 lists these compounds [17]. It consists of two tables of substances—Table 1 listing all authorized chemical drug compounds and Table 2 listing those banned in the EU—together with the maximum permissible limits at which they can be present in various foods of animal/bio-origin. As a result, every Member State of the EU has a legal obligation to monitor VMPRs in foods. In principle, there are two types of regulatory monitoring programs for these residues. One concerns direct, on-site targeted compound monitoring organized at the point of production (animal or fish farms), slaughter (abattoirs), or product collection (dairies and egg hatcheries). In this case, the animals or their products are withheld by the competent veterinary services until it can be proven that they present no risk to public health. The other type of program consists in organizing a national residue control plan with the purpose of assessing the degree of compliance of overall national production, without excluding animals and their products from the market, but allowing the veterinary services to carry out rigorous inspections if the degree of compliance should fall. This type of monitoring is described at the EU community level in a Directive [18] that is in the process of revision in the framework of the regulations "on official controls performed to ensure the verification of compliance with feed and food law, animal health, and animal welfare rules" [19]. Directive 96/23/EC specifies the number of samples to be examined for each species of production animal (cows, pigs, horses, goats, sheep, poultry, game, fish, etc.) and their products (milk, eggs, and honey). It also specifies the different groups of veterinary medicinal products to be screened for (antimicrobials, anabolics, antiparasitics, etc.). In either type of monitoring, positive (i.e., noncompliant) or suspect samples must be properly separated from all negative (compliant) samples. For the meticulous implementation of these regulations and to curb the possibility of veterinary residue-tainted animal foods, proper identification and quantification tools are needed. Thus, the scientific and technical bases for these controls depend on the development, validation, and application of appropriate reliable analytical methods with excellent sensitivities for the criteria and requirements related to these regulations laid down in several European Commission Decision and Guidelines [20–22].

Bearing in mind the public health protection, it is logical to consider that both the regulations set for the control on food safety and the worrying perception that the consumers may have on the security of food tend to make proposals for building a faster, easier, but also broader and safer scope for this control. One of the cheapest but reliable routes for it will be using high-throughput methods designed to screen large numbers of harmful substances. It will maximize the food safety information obtained from a single sample and it will minimize the sets of samples to be collected.

7.3 CHOOSING A STRATEGY: TARGETED OR NONTARGETED ANALYSIS?

The most widely used MS technologies to confirm the presence of residues of veterinary drugs in routine/field laboratories are mainly QqQ LC–MS systems, hybrid quadrupole-ion trap (QTrap) LC–MS systems, and ion trap (IT) LC–MS systems. These systems are capable of analyses at a unit mass resolution (0.5–1 Da). They are generally dedicated to targeted quantitative confirmation analysis using SRM or MRM, providing high sensitivity and selectivity for the target compounds, but they are much less sensitive in full scan mode. The analytes of interest are selected before acquisition and the acquisition parameters are optimized so that only veterinary residues included in the "pretargeted" list are detected. While these are the currently prevailing methods, in recent years, some scientists have started to opt for high-resolution mass spectrometers in veterinary medicinal residue analysis by means of accurate mass measurement via ToF, Orbitrap, and FTICR mass analyzers, as mentioned in Section 7.1. Among these analyzers, FTICR has the highest resolving power, which mainly depends on the strength of the magnet. For example, a resolution of two million can be achieved with a 6 T magnet. However, FTICR is more expensive and incurs high maintenance compared with other mass analyzer systems. Alternately combining two different analyzers in a single instrument (hybridization), for example, Q–ToF, LTQ–Orbitrap, or Q–Exactive (Q–Orbitrap), increases the instrument performance by way of different possible scan modes. For comparison sake, ToF and Orbitrap analyzers have approximate resolving powers ranging from 30,000 for the ToF to >100,000 for the Orbitrap device. Figure 7.1 shows the

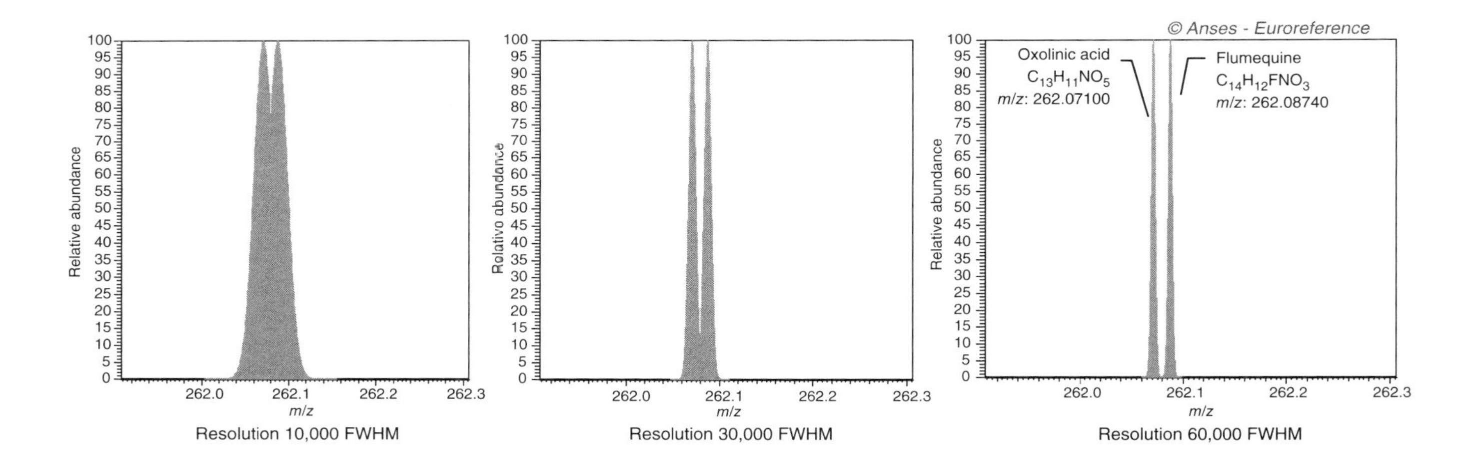

Figure 7.1. Simulated mass spectra of a mixture of isobaric compounds containing two antimicrobials, the oxolinic acid ($C_{13}H_{11}NO_5$; exact m/z ratio: 262.07100) and the flumequine ($C_{14}H_{12}FNO_3$; exact m/z ratio: 262.08740) at different resolving powers. Reproduced with permission from Anses - Euroreference Journal.

effect of the resolution of an instrument and its ability to discriminate between two nearly isobaric compounds. Regarding mass accuracy, mass accuracies ranging from 2 to 5 ppm can be achieved by ToF instruments with the external calibration method. It can even improve to <2 ppm with new generations of orbital trap instruments.

These high-resolution mass spectrometers allow nontargeted, full scan analyses. In such studies, retrospective analysis of the acquired data in full scan mode facilitates identification of suspect compounds from a theoretically unlimited number of analytes. Such methodologies can improve throughput by allowing the screening of multiple analytes simultaneously, eliminating the need for separate analyses for every analyte of interest. Because these HRMS instruments are becoming more common, it is worth discussing the factors that need to be considered when using them for targeted and nontargeted analyses.

7.3.1 Targeted Analysis Using HRMS

In this approach, development of chromatographic methods and optimization of mass spectrometer parameters for detection (source conditions: temperature, gas, voltage, etc.) in positive or negative ion modes (mode depends on ionization efficiency of the compounds) for different classes of standard veterinary medicinal products is necessary prior to real sample analysis. However, compromise is generally necessary to get voltages, temperatures, and gas values that allow the ionization of the maximum number of target compounds.

Extracts of biological matrices from foods, using previously developed methods, are subjected to LC–MS analysis under full scan high-resolution conditions. From full scan mass chromatograms, the first and most simple approach to find the target compounds is to extract the ion chromatogram (EIC) trace of the exact masses. In this analysis, the accurate masses of targeted $[M + H]^+$ ions or $[M - H]^-$ ions of the veterinary drugs are extracted from their MS raw chromatograms using a mass tolerance window of ±3–5 ppm. This narrow window avoids isobaric ion interference. The resulting extracted ion chromatogram shows a chromatographic peak of the requested exact mass, without interference from the other matrix components. The information in the mass spectrum associated with this peak (i.e., fragmentation and/or isotopic distribution) can be useful to confirm the presence of targeted analytes. In addition to mass spectral data, chromatographic retention times also support in the identification of veterinary drugs. For further confirmation, MS/MS analysis of suspected $[M + H]^+$ ions or $[M - H]^-$ ions can be performed. This, however, has obvious implications on the throughput of the overall methodology. These two events (full scan and MS/MS) can be implemented within a single run by using data-dependent acquisition (DDA). This combination allows the screening of analytes and their full confirmation at the same time. Indeed, using DDA allows acquiring MS/MS data extracted from ions selected from a full survey scan. The choice of the selected ions is made automatically either from several predefined criteria (signal intensity, etc.) or from a list of targeted masses. The first event, full scan, provides the screening step and the second event, that is, MS/MS analysis, gives the confirmatory step as the two product ions formed are unequivocal characteristics of a substance. In Directive 2002/657/EC, the concept of identification points (IP) was set to strengthen the MS confirmation of the presence of a substance. The need of a minimum of 3 IP (substance group B = MRL permitted veterinary drugs) or 4 IP (substance group A = banned veterinary drugs) was put into force. Thanks to HRMS technology, 2 IP for a precursor ion and 2.5 IP for a product ion could be achieved. So DDA experiment in HRMS leads to sufficient IP to confirm group A or group B substances.

7.3.2 Nontargeted Analysis Using HRMS: Screening for Unknown Compounds

Identification of untargeted analytes in biological sample matrices from high-resolution MS raw data is not an easy task due to the presence of an enormous number of endogenous matrix ions. This difficulty can be overcome by data processing through specialized software that can help in finding biomarkers or drug metabolites. These specialized software packages have the capability of comparing different MS raw data files. Generally, mass spectrometry-based drug metabolomic studies are carried out by comparing acquired MS data of both the control samples and the drug-incurred samples with the help of chemometric software tools. In the incurred samples, the statistically obtained differences in the mass-to-charge ratio (*m*/*z*) of the detected ions can assist in the identification of substances or metabolites. However, in this process, it must be verified that the observed *m*/*z* differences truly represent differences in the sample and are not simply due to analytical variations. But even with <3 ppm mass accuracy data, it is possible that a few dozen chemical formulas fit the specified mass-to-charge ratio. It then requires a lot of effort to deduce the reliable formula among all the likely candidates. In this pursuit, preliminary data mining can be implemented in order to choose compatible structures using the increasing number of databases that can be accessed from university web sites and elsewhere on the World Wide Web. The greater the instrument's mass resolution or the smaller the mass error, the fewer the compatible chemical formulas, which increases the confidence with which any unknown molecule can be identified. Furthermore, the suspected ions can be subjected to MS/MS analysis to obtain the fragmentation pattern with accurate mass and ring double bond equivalents that can help in the structure elucidation. Interpretation of product ion spectra of unknown analytes requires extensive knowledge and experience. In addition, manual interpretation is a time-consuming process and assigning of all product ions in a spectrum is difficult. In order to simplify the interpretation of the MS/MS spectra, some research groups are working on the development of algorithms to interpret high-resolution MS/MS spectra of unknown compounds or metabolites [23–25]. With these general targeted and nontargeted strategies in mind, we discuss two specific applications of LC–MS and LC–MS/MS to analyze the veterinary medicine degradation products or their metabolites in different sample matrices. The following examples provide a glimpse of the extraordinary variety of applications for which these instruments have been used.

7.4 APPLICATION NUMBER 1: IDENTIFICATION OF BRILLIANT GREEN AND ITS METABOLITES IN FISH UNDER HIGH-RESOLUTION MASS SPECTRAL CONDITIONS (TARGETED AND NONTARGETED APPROACHES)

To give a clear idea of the advances of analytical screening methods for detection of chemical traces in biological matrices, the first study we shall discuss, which was conducted at the ANSES—Fougères Laboratory, concerns pharmacologically active dyes that are not authorized in the EU due to their toxic properties. While this first example is not yet optimized for high throughput, it illustrates the importance of employing existing methods that are time consuming and require multiple steps, and streamlining them so that more analytes can be screened in each run. It also demonstrates the differences in using two different strategies: the targeted screening and the nontargeted approach. The dyes considered here were widely used in aquaculture and fisheries due to their low cost and efficiency in the prevention and/or treatment of bacterial or fungal infections in aquatic animals. Malachite green (MG) is one of the most popular and widely used of these compounds; it belongs to the triarylmethane family and is known to possess mutagenic, carcinogenic, teratogenic, and genotoxic properties [26]. Because of the potentially severe toxic effects of these dyes, most of the countries around the world restricted their use for the production of fish and shellfish intended for human consumption. Because of these restrictions, methods for their monitoring are regulated in the EU, and analytical techniques such as low-resolution tandem LC–MS are already widely used to screen farmed fish for this dye, and importantly, its metabolized product, leuco-malachite green (LMG). Recently, after suspicions were raised about the substitution of MG by another dye in farmed fish products imported from South America and Asia, European official control methods were extended to cover a second type of dye named gentian violet (also known as crystal violet (CV) in English-speaking countries). The Reference Laboratory at ANSES—Fougères thus developed a method for monitoring both malachite green and gentian violet and their respective metabolized leucobases LMG and LCV [27]. The analytes were isolated from the fish flesh matrix by liquid–liquid extraction with acetonitrile. Determination was performed using LC–MS/MS with positive electrospray ionization, using SRM mode, and monitoring two transitions for each compound. The chromatographic separation was performed on a reversed-phase HPLC C18 column, length 100 mm × diameter 2.1 mm, with 3.5 μm silica particles. The mobile phase consisted of ammonium formate buffer (0.05 M, pH 4.5) mixed with acetonitrile and was used in gradient mode. Very similar methods have now been in use for more than three years in almost all of the EU's official control laboratories. Very recently, the presence of brilliant green (BG) in certain fish farm products, a third dye belonging to the same family of triarylmethanes, has been suggested to possibly be in use in aquaculture farming, but has not yet been evidenced by the targeted methods currently used in official controls. The screening for the BG substance in aquaculture products was therefore studied. A new technique was proposed for detecting the metabolites of this compound in farmed fish flesh. At first, because of its similarity to other triarylmethane compounds, the hypothesis that the leucobase of brilliant green (LBG) would be found in trout treated with BG was considered. The presence of this compound had never been confirmed in previous studies [28]. We then screened the fish flesh to track the presence of other metabolites derived from the BG dye. The experiment involved a batch of trout treated with BG compared to a batch of BG-untreated trout. Samples of trout flesh were collected and subjected to a chemical extraction step prior to their analysis by high-resolution LC–MS using an LTQ–Orbitrap mass spectrometer, a hybrid MS with linear trap and orbital trap. The source of ionization was an electrospray ionization probe set in the positive ion mode. The instrument operated in full scan mode from m/z 100 to 1000 at a resolving power of 60,000 (full width at half maximum (FWHM)). Prior to the LC–MS analysis, the mass spectrometer was calibrated using the manufacturer's calibration solution (consisting of caffeine, the tetrapeptide MRFA®, and Ultramark®) in order to ensure reaching mass accuracies in the range of 1–3 ppm.

There were two different ways to process the resulting high-resolution LC–MS raw data: the targeted way and the nontargeted way. In the targeted approach, the suspected analyte ions, with their exact masses, were extracted from the total ion chromatogram(TIC) for their identification. The nontargeted approach consisted of using all the signals obtained from the samples of the same batch of trout treated with BG and comparing these with the signals obtained from the control batch. This comparison was conducted using Sieve® software.

In the first targeted approach, the presence of BG and of its leucobase LBG was identified in the treated fish flesh sample under high-resolution mass spectral conditions. This identification was processed by requesting extracted ion chromatograms of the targeted compounds, as shown in Figure 7.2. As can be seen in the figure, all the metabolites were separated in <15 min. The mass spectra of BG (theoretical mass M^+: 385.26382) and its leukobase LBG (theoretical mass MH^+: 387.27947) show <1.2 ppm in mass accuracy. In addition, the resulting isotopic patterns of the targeted compounds adequately matched with the theoretical isotopic patterns. Furthermore, the LBG $[M + H]^+$ ion was subjected to high-resolution MS/MS analysis and the resulting product ion spectrum was compared with MS/MS data obtained from its pure custom-made standard form derived by chemical synthesis, confirming the presence of LBG unambiguously.

In the second approach (nontargeted analysis) high-resolution LC–MS raw data of BG extracted from both treated and untreated trout samples were all processed through Sieve software for their comparison. For this, a set of six BG-treated trout fish samples and a set of six BG-untreated trout fish samples were used. The Sieve software extracted significant differences in signals between the groups of treated and untreated fishes, thus providing a list of ions (compounds) present in one group and absent in the other group. With this approach (Figure 7.2), we were able to first screen and subsequently confirm from all the high-resolution mass chromatograms obtained the presence of both BG and LBG. Second, we even extracted and characterized the presence of several other metabolized compounds, one of which was formally identified as desethyl LBG and another one, probably in a lower concentration, with a mass between 441.32759 and 441.33641, whose molecular structure has not yet been identified with enough certainty [27].

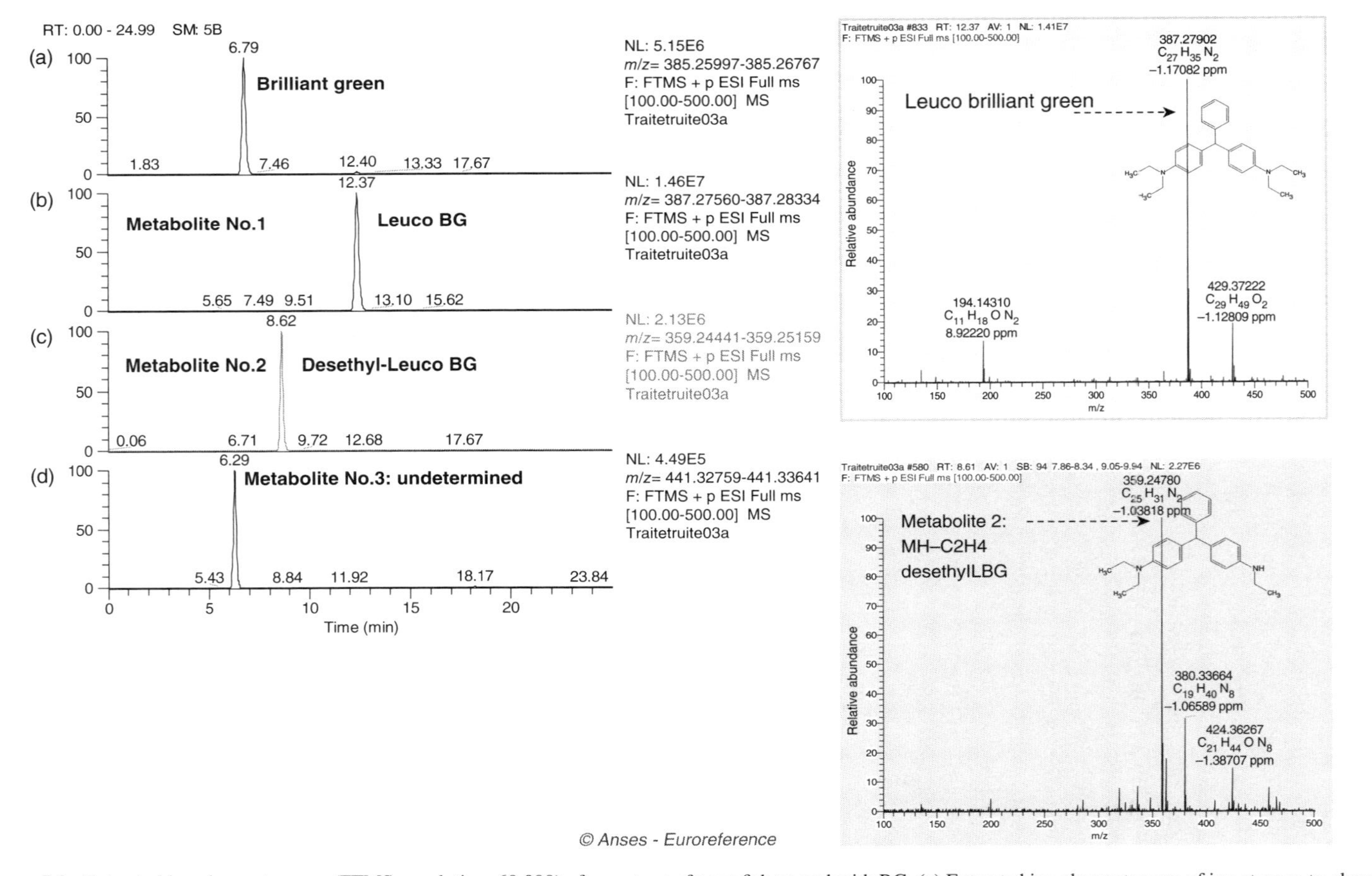

Figure 7.2. Extracted ion chromatograms (FTMS, resolution: 60,000) of an extract of trout fish treated with BG. (a) Extracted ion chromatogram of ion at mass-to-charge ratio (*m/z*) 385.26382, corresponding to BG. (b) Extracted ion chromatogram of ion at mass-to-charge ratio (*m/z*) 387.27947, corresponding to the leukobase of BG. (c) Extracted ion chromatogram of ion at mass-to-charge ratio (*m/z*) 359.24774, corresponding to desethyl leuko BG. (d) Extracted ion chromatogram of ion at mass-to-charge

7.5 APPLICATION NUMBER 2: TARGETED AND NONTARGETED SCREENING APPROACHES FOR THE IDENTIFICATION OF ANTIMICROBIAL RESIDUES IN MEAT

In another example, we were able to demonstrate experimentally the reliability of HRMS in the analysis of traces of veterinary medicinal products in foodstuffs of animal origin (Figure 7.3). This project was mainly focused on screening meat products for different families of antimicrobials that are authorized in the EU. Starting from a method we developed on an LC–MS/MS system and dedicated to multiantibiotic monitoring in meat tissues and in milk [29], this LC–HRMS method was implemented using an LTQ–Orbitrap mass spectrometer, a hybrid MS instrument with a linear trap hyphenated to an orbital trap [30].

A standard sample mixture of 60 antibiotics from different families was prepared at or close to their MRLs and then subjected to LC–HRMS analysis. From the resulting LC–HRMS raw data (i.e., TIC), we identified all the antibiotics, which were subjected to LC–HRMS analysis, by extracting the ion chromatogram of each antibiotic at their respective exact masses and the value of the retention time was recorded. These values

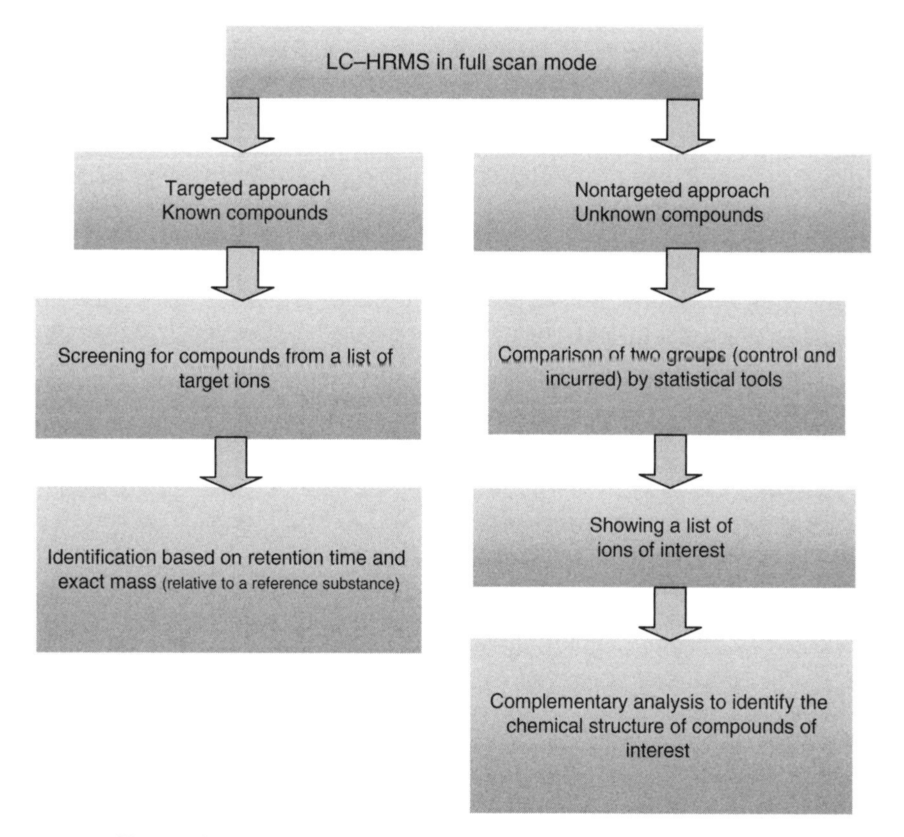

Figure 7.3. Comparison of a targeted approach against a nontargeted approach.

obtained from standard injections are used to create an Excel file containing each compound defined by its formula, the exact molecular weight, and the expected retention time. This list is the strategic heart of the automatic process of identification called ToxID®.

Using this software, a compound is positively identified from HRMS analysis when the following criteria are met simultaneously: measured retention time in accordance with the expected retention time, measured high-resolution accurate mass in accordance with the expected theoretical accurate mass with a tolerance of <5 ppm (criterion to be defined arbitrary), and peak intensity higher than an arbitrary threshold to be predefined (Figure 7.4). This method has been tested successfully on beef, pork, and poultry meat samples. However, the limitation of the method is in the efficient extraction step of those 60 antibiotic compounds from the matrices suitable at the MRL level. A single extraction procedure did not extract all compounds at concentration levels that could be high enough to be detected by the instrument. Thus, two different extractions had to be carried out in parallel for each sample to analytically cover the set of 60 antimicrobial compounds. Our next step in this work is to consider developing a comprehensive analytical method that may be able to detect not only multiple antimicrobial compounds but also multiple classes of veterinary drugs, including, if possible, antiparasitics, anti-inflammatories, anticoccidials, and tranquillizers. This particular approach for screening veterinary drug residues can be considered as a "posttarget screening" as the analytes are searched only after their mass has been acquired in the full scan mode.

Another approach we are working on is the nontargeted approach aimed at identifying metabolites of antimicrobials under high-resolution mass spectral conditions. One of the first studies for this approach was the identification of trace metabolites/degradation products of antibiotics in beef. In a preliminary test with a method involving microbiological analysis using agar diffusion, our beef sample showed positive microbial inhibition activity. We initially identified principal antibiotic components in the beef meat that displayed microbial inhibition by using MRM and high-resolution full scan MS. Second, we focused on identifying unknown metabolites of antibiotics using a nontargeted approach with high-resolution mass spectral conditions. In this approach, we compared high-resolution mass spectral data of a microbial inhibition-active beef muscle sample with a set of blank beef muscle samples taken from various batches of animals of the same species that had not been subjected to any intentional veterinary medication. For this, a set of four consecutive LC–MS analyses of the single microbial inhibition-active beef muscle sample and a single LC–MS analysis for each of the six different blank beef samples were carried out using high-resolution mass spectral conditions. The instrument was operated in full scan FTMS mode over a range of 10–1200 Da at a resolving power of 60,000 FWHM, in electrospray positive ion mode. LC separation was performed using a C18 column (125 mm × 3 mm) with 5 μm particle size. The mobile phase consisted of 1 mM HFBA in 0.5% formic acid solution and 0.5% formic acid in methanol/acetonitrile solution (50:50, v/v). The LC–MS raw data from the blank samples and the positive sample controlled were compared (Figure 7.5) with the help of Sieve software. Based on the Sieve

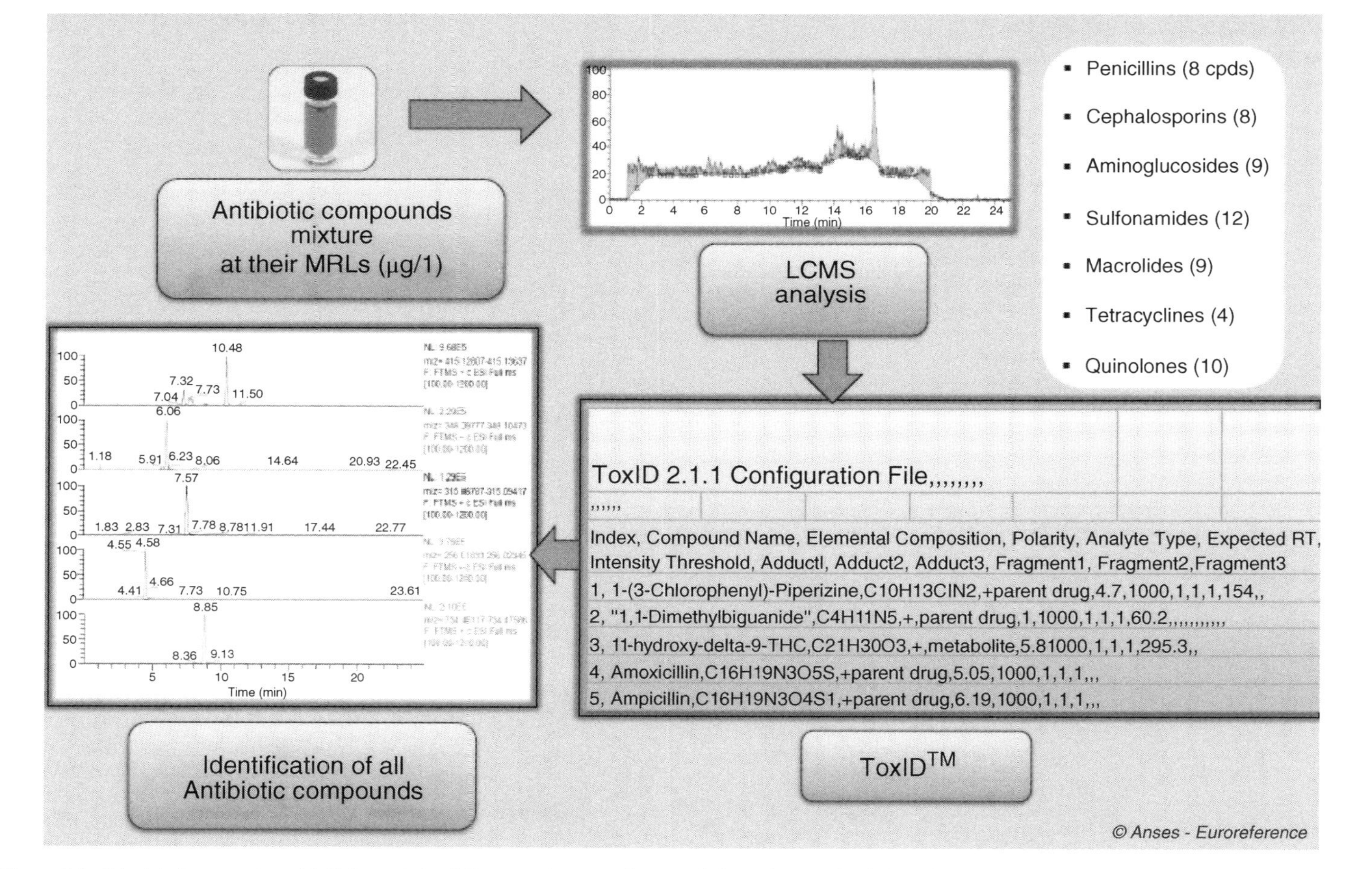

Figure 7.4. Principle for screening with high-resolution full scan and a targeted approach for a mixture of antimicrobial residues. Reproduced with permission from Anses - Euroreference Journal.

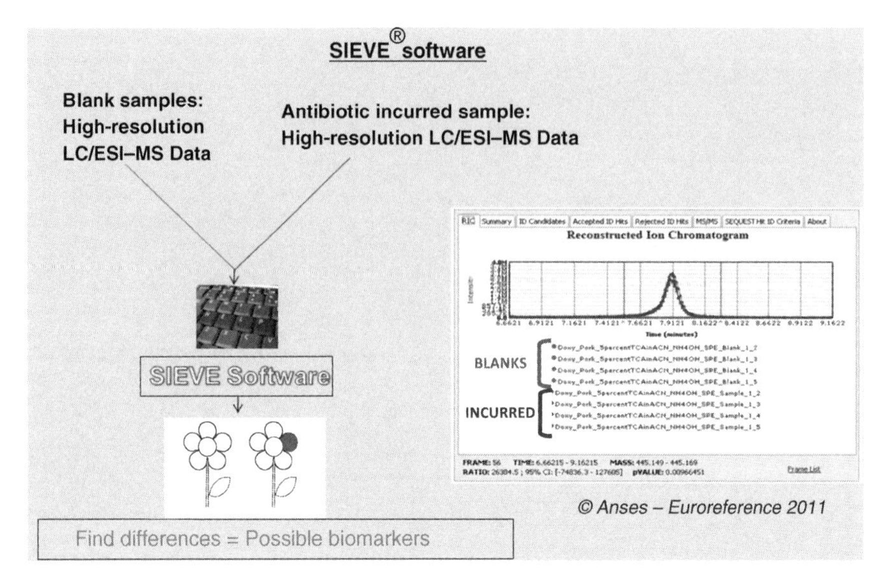

Figure 7.5. Principle of high-resolution full scan screening using a nontargeted approach with semi-quantitative differential expression analysis. Reproduced with permission from Anses - Euroreference Journal.

output results, we were able to identify some of the suspected ions (metabolites of drugs) in the microbial inhibition-active beef muscle sample. As an example, *N*4-acetylsulfadimethoxine, a metabolite of the sulfadimethoxine antimicrobial, was identified using this approach in a beef sample, in addition to the presence of sulfadimethoxine. Furthermore, MS/MS experiments of suspected $[M + H]^+$ ions were carried out in order to get structural information. Finally, some of the suspected analyte ionic structures were confirmed unambiguously by comparing the mass spectral data with reference standards synthetically prepared. At our laboratory, this type of analysis has since been repeated on muscle samples of various species like beef, pork, poultry, and rabbit. The detection of drug metabolites in animal tissues offers undoubted evidence of the administration of the parent drug to the animal.

The same method is now being applied to identify trace amounts of unknown antimicrobials in meat samples when these samples are screened showing positive microbial inhibition activity. In these samples, none of the known antibiotics, at least from those that were used at our laboratory, were identified by either MRM/SRM or high-resolution full scan MS.

This approach is time consuming for a unique sample to be screened thoroughly and is not successful for a certain number of samples. Nontargeted screening of ions of interest present only in the positive sample and not in the set of blank samples is highly difficult, especially when the concentration is not high enough for the instrument to get a significant signal (lack of sensitivity for residual traces of substances). If such techniques could be streamlined, however, the potential to screen for dozens of compounds in a single LC–HRMS analysis would significantly increase the throughput by reducing or eliminating the need for analyzing the same sample multiple times with methods that are sensitive to and selective for just few analytes per run.

7.6 CONCLUSIONS

The system for the surveillance of residues of chemical substances in biological matrices that form the basis of our diet is currently based on analytical methods that target molecules or classes of molecules. However, this targeted approach is beginning to face competition from innovative technologies in research laboratories developing future monitoring techniques. Indeed, the analytical power of high-resolution mass spectrometers, especially when used in full scan mode, makes it possible to envision an exclusive novel strategy for screening chemical residues in food products. Molecular identification by measuring the accurate mass would rapidly prioritize the identification of the metabolites and decomposition products of these substances. Researchers are paying attention to understanding the mechanisms at work in the search for target biomarkers of veterinary medicinal products, whether authorized with regulatory MRL or illegal and occurring through either misuse or fraud. These advances in analytical instrumentation, methodologies, and software packages will probably allow us to suggest future proposals for modification to the food safety surveillance system, with changes to the methods of chemical analysis from targeted to nontargeted modes as shown in the two examples presented in this chapter. These changes and the potential to widely open up these methods to include numerous other medicinal substances are, currently, quite limited not only by the vast variety of chemical properties of these compounds belonging to the different classes of veterinary medicinal products but also by the need for improvement of the sensitivity in the MS detection.

Despite these limitations, we can expect HRMS to become more prevalent, helping in more rapid and accurate analysis of nontarget analytes in a high-throughput style. It could further help enforce stringent regulations in connection with food safety and public health protection. In addition, the constant development of new algorithms for automatic and statistical interpretation of HRMS data will continue to facilitate the elucidation of the structure of unknown/untargeted veterinary medicinal product residues in food analysis.

REFERENCES

1. Regulation (EC) No. 470/2009 of the European Parliament and of the Council of 6 May 2009 laying down Community procedures for the establishment of residue limits of pharmacologically active substances in foodstuffs of animal origin. *Off. J. Eur. Union* **2009**, L152, 11–22.
2. US-FDA Code for Federal Regulations. Title 21: Food and Drugs, Chapter I: Food and Drug Administration, Department of Health and Human Services, Subchapter E: Animal Drugs, Feed and Related Products, Part 556: Tolerances for Residues of New Animal Drugs in Food.

3. De Brabander, H.; Noppe, H.; Verheyden, K.; VandenBussche, L.; Wille, K.; Okerman, L.; Vanhaecke, L.; Reybroeck, W.; Ooghe, S.; Croubels, S. Residue analysis: future trends from a historical perspective. *J. Chromatogr. A* **2009**, 1216, 7964–7976.
4. Verdon, E. Antibiotic residues in muscle tissues of edible animal products. In: L.M.L. Nollet and F. Toldra, editors. *Handbook of Muscle Tissues of Edible Animal Products*. CRC Press; **2009**, pp. 855–947.
5. Makarov, A.; Denisov, E.; Lange, O.; Horning, S. Dynamic range of mass accuracy in LTQ Orbitrap hybrid mass spectrometer. *J. Am. Soc. Mass Spectrom.* **2006**, 17, 977.
6. Peters, R.J.; Bolck, Y.J.; Rutgers, P.; Stolker, A.A.; Nielen, M.W. Multi-residue screening of veterinary drugs in egg, fish and meat using high-resolution liquid chromatography accurate mass time-of-flight mass spectrometry. *J. Chromatogr. A* **2009**, 1216, 8206–8216.
7. Kellmann, M.; Muenster, H.; Zomer, P.; Mol, H., Full scan MS in comprehensive qualitative and quantitative residue analysis in food and feed matrices: how much resolving power is required? *J. Am. Soc. Mass Spectrom.* **2009**, 20, 1464.
8. Erve, J.C.L.; Gu, M.; Wang, Y.; DeMaio, W.; Talaat, R.E. Spectral accuracy of molecular ions in an LTQ/Orbitrap mass spectrometer and implications for elemental composition determination. *J. Am. Soc. Mass Spectrom.* **2009**, 20, 2058.
9. Kaiser, N.K.; Quinn, J.P.; Blakney, G.T.; Hendrickson, C.L.; Marshall, A.G. A novel 9.4 Tesla FTICR mass spectrometer with improved sensitivity, mass resolution, and mass range. *J. Am. Soc. Mass Spectrom.* **2011**, 22, 1343–1351.
10. Sanders, M.A.; Shipkova, P.; Zhang, H.; Warrack, B.M. Utility of the hybrid LTQ-FTMS for drug metabolism applications. *Curr. Drug Metab.* **2006**, 7–5, 547–555.
11. Chen, C.H.W. Review of a current role of mass spectrometry for proteome research. *Anal. Chim. Acta* **2008**, 624, 16–36.
12. Han, J.; Danell, R.M.; Patel, J.R.; Gumerov, D.R.; Scarlett, C.O.; Speir, J.P.; Parker, C.E.; Rusyn, I.; Zeisel, S.; Borchers, C.H. Towards high-throughput metabolomics using ultrahigh-field Fourier transform ion cyclotron resonance mass spectrometry. *Metabolomics* **2008**, 4, 128–140.
13. Kang, H.J.; Yang, H.J.; Kim, M.J.; Han, E.S.; Kim, H.J.; Kwon, D.Y. Metabolomic analysis of meju during fermentation by ultra performance liquid chromatography–quadrupole–time of flight mass spectrometry (UPLC–Q–TOF MS). *Food Chem.* **2011**, 127, 1056–1064.
14. Dervilly-Pinel, G.; Weigel, S.; Lommen, A.; Chereau, S.; Rambaud, L.; Essers, M.; Antignac, J.P.; Nielen, M.; Le Bizec, B. Assessment of two complementary LC–HRMS metabolomics strategies for the screening of anabolic steroid treatment in calves. *Anal. Chim. Acta* **2011**, 700, 144–154.
15. Vanhaecke, L.; Van Meulebroek, L.; De Clerq, N.; Vanden Bussche, J. High resolution Orbitrap mass spectrometry in comparison with tandem mass spectrometry for confirmation of anabolic steroids in meat. *Anal. Chim. Acta* **2013**, 767, 118–127.
16. Codex veterinary drug residues in food. Available at http://www.codexalimentarius.org/standards/veterinary-drugs-mrls/en/
17. Regulation (EU) No. 37/2010 of 22 December 2010 on pharmacologically active substances and their classification regarding maximum residue limits in foodstuffs of animal origin. *Off. J. Eur. Union* **2009**, L15, 1–72.

18. European Council Directive 96/23/EC of 29 April 1996 on measures to monitor certain substances and residues thereof in live animals and animal products. *Off. J. Eur. Commun.* **1996**, L125, 10–32.
19. Regulation (EC) No. 882/2004 of the European Parliament and of the Council of 29 April 2004 on official controls performed to ensure the verification of compliance with feed and food law, animal health and animal welfare rules. *Off. J. Eur. Union* **2004**, L165, 1–141.
20. European Commission Decision No. 2002/657/EC of 12 August 2002 implementing Council Directive 96/23/EC concerning the performance of analytical methods and the interpretation of results. *Off. J. Eur. Commun.* **2002**, L221, 8–36.
21. Guidelines SANCO/2004/2726 rev4 of 4 December 2008 for the implementation of Decision 2002/657/EC, European Commission, Health & Consumer Protection Directorate-General, Directorate E-Safety of the Food Chain.
22. Guidelines from CRLs Residues of 20 January 2010 for the Validation of Screening Methods for Residues of Veterinary Medicines: Initial Validation and Transfer, European Commission, Health & Consumer Protection Directorate-General, Directorate E-Safety of the Food Chain.
23. Scheubert, K.; Hufsky, F.; Rasche, F.; Böcker, S. Computing fragmentation trees from metabolite multiple mass spectrometry data. *J. Comput. Biol.* **2011**, 18, 1383–1397.
24. Rasche, F.; Svatoš, A.; Maddula, R.K.; Böttcher, C.; Böcker, S. Identifying the unknowns by aligning fragmentation trees. *Anal. Chem.* **2011**, 83, 1243–1251.
25. Rasche, F.; Scheubert, K.; Hufsky, F.; Zichner, T.; Kai, M.; Svatoš, A.; Böcker, S. Identifying the unknowns by aligning fragmentation trees. *Anal. Chem.* **2012**, 84, 3417–3426.
26. Pierrard, M.A.; Kestemont, P.; Delaive, E.; Dieu, M.; Raes, M.; Silvestre, F. Malachite green toxicity assessed on Asian catfish primary cultures of peripheral blood mononuclear cells by a proteomic analysis. *Aquat. Toxicol.* **2012**, 114, 142–152.
27. Hurtaud-Pessel, D.; Couedor, P.; Verdon, E. Liquid chromatography tandem mass spectrometry method for the determination of dye residues in aquaculture products: development and validation. *J. Chromatogr. A* **2011**, 1218, 1632–1645.
28. Andersen, W.C.; Turnipseed, S.B.; Karbiwnyk, C.M.; Lee, R.H.; Clark, S.B.; Rowe, W.D.; Madson, M.R.; Miller, K.E. Multiresidue method for the triphenylmethane dyes in fish: malachite green, crystal (gentian) violet, and brilliant green. *Anal. Chim. Acta* **2009**, 637, 279–289.
29. Gaugain-Juhel, M.; Delepine, B.; Gautier, S.; Fourmond, M.P.; Gaudin, V.; Hurtaud-Pessel, D.; Verdon, E.; Sanders, P. Validation of a liquid chromatography–tandem mass spectrometry screening method to monitor 58 antibiotics in milk: a qualitative approach. *Food Addit. Contam.* **2009**, 26, 1459–1471.
30. Hurtaud-Pessel, D.; Jagadeshwar-Reddy, T.; Verdon, E. Developing a new screening method for the detection of antibiotic residues in muscle tissues in using liquid chromatography and high resolution mass spectrometry with a LC–LTQ–Orbitrap instrument. *Food Addit. Contam.* **2011**, 28, 10, 1340–1351.

CHAPTER

8

HIGH-THROUGHPUT ANALYSIS OF MYCOTOXINS

MARTA VACLAVIKOVA, LUKAS VACLAVIK, and TOMAS CAJKA

8.1 INTRODUCTION

Mycotoxins are well-known and abundant toxins that are widely considered to be the most important natural contaminants found in food and feed. Mycotoxins represent the low molecular weight organic compounds formed as secondary metabolites of microscopic, mostly saprophytic, filamentous fungi species, frequently referred to as molds. Under favorable environmental conditions, that is, when temperature and moisture are conducive, these fungi proliferate and may produce mycotoxins. Mycotoxins represent a group of compounds with diverse chemical structures, various biosynthetic origins, and a myriad of biological effects. An overview of selected mycotoxins is provided in Table 8.1. Although the definition of mycotoxins is relatively easy, their classification represents a more difficult and challenging task, especially because of the high number of different fungal species producing these natural toxins. Among many others, the most prominent fungal producers are toxicogenic molds of *Aspergillus*, *Penicillium*, *Fusarium*, *Claviceps*, and *Alternaria* fungi genera [1,2].

Because of the ubiquity of molds in the environment, abundance of soil and plant debris, and their dispersion by wind currents, insects, and rain, both these pathogenic organisms and their toxic secondary metabolites can be frequently found in foods and feeds. Mycotoxins are practically unavoidable because the growth of toxicogenic strains of molds cannot be completely eliminated under real-life conditions. However, it is important to minimize the conditions under which mycotoxins are formed, although this is not always feasible within common agricultural, market, and household practice. Mycotoxins are notoriously difficult to remove and the best method of control is prevention [3,4].

8.1.1 Legislation and Regulatory Limits

General public awareness of health risks related to mycotoxins is steadily growing. When present in foods or feeds at sufficiently high concentrations, toxic fungal metabolites can induce both acute and chronic adverse health effects in humans and

High-Throughput Analysis for Food Safety, First Edition.
Edited by Perry G. Wang, Mark F. Vitha, and Jack F. Kay.

Table 8.1. Overview of the Main Groups of Mycotoxin Representatives

Fungi Producer Species	Mycotoxins Group	Structure	Name of Mycotoxins (Abbreviation)	R-Structures	Molecular Formula	Relative Molecular Mass
Fusarium sp.	Trichothecenes A	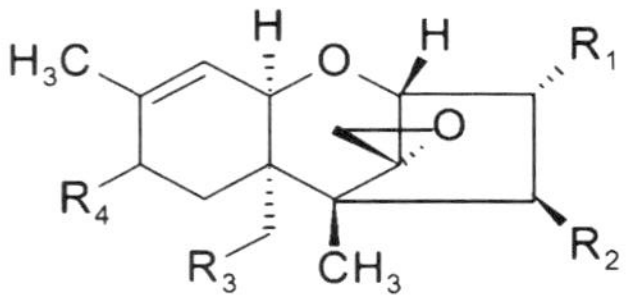	HT2 toxin (HT2)	R1=R2=OH; R3=OAc; R4=OCOi-Bu	$C_{22}H_{32}O_8$	424.2
			T2 toxin (T2)	R1=OH; R2=R3=OAc; R4=OCOi-Bu	$C_{24}H_{34}O_9$	466.2
			15-Acetoxyscirpenol (MAS)	R1=R2=OH; R3=OAc; R4=H	$C_{17}H_{24}O_6$	324.2
			Diacetoxyscirpenol (DAS)	R1=OH; R2=R3=OAc; R4=H	$C_{19}H_{26}O_7$	366.2
			Neosolaniol (NEO)	R1=OH; R2=R3=R4=OAc	$C_{19}H_{26}O_8$	382.2
			Verrucarol (VER)	R1=R4=H; R2=R3=OH	$C_{15}H_{22}O_4$	266.2
	Trichothecenes B	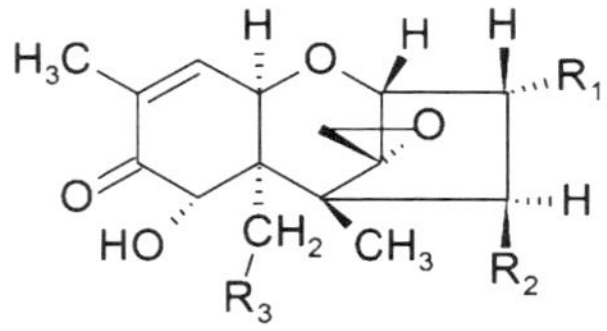	Deoxynivalenol (DON)	R1=R3=OH; R2=H	$C_{15}H_{20}O_6$	296.1
			Nivalenol (NIV)	R1=R2=R3=OH	$C_{15}H_{20}O_7$	312.1
			Fusarenon-X (FUS-X)	R1=R3=OH; R2=OAc	$C_{17}H_{22}O_8$	354.1
			3-Acetyldeoxynivalenol (3ADON)	R1=OAc; R2=H; R3=OH	$C_{17}H_{22}O_7$	338.1
			15-Acetyldeoxynivalenol (15ADON)	R1=OH; R2=H; R3=OAc	$C_{17}H_{22}O_7$	338.1
	Enniatins	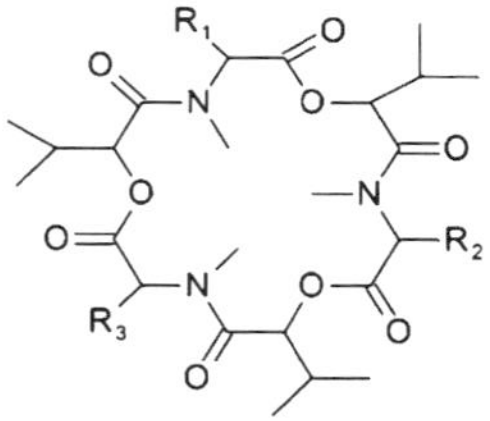	Enniatin A (ENN A)	R1=R2=R3=$CH_2C_6H_5$	$C_{36}H_{63}N_3O_9$	681.5
			Enniatin A1 (ENN A1)	R1=R2=R3=$CH(CH_3)CH_2CH_3$	$C_{35}H_{61}N_3O_9$	667.4
			Enniatin B (ENN B)	R1=$CH(CH_3)CH_2CH_3$, R2=R3=$CH(CH_3)_2$	$C_{33}H_{57}N_3O_9$	639.4
			Enniatin B1 (ENN B1)	R1=R2=R3=$CH(CH_3)_2$	$C_{34}H_{59}N_3O_9$	653.4
			Beauvericin (BEA)	R1=R2=R3=$CH_2C_6H_5$	$C_{45}H_{57}N_3O_9$	784.4

	Fumonisins	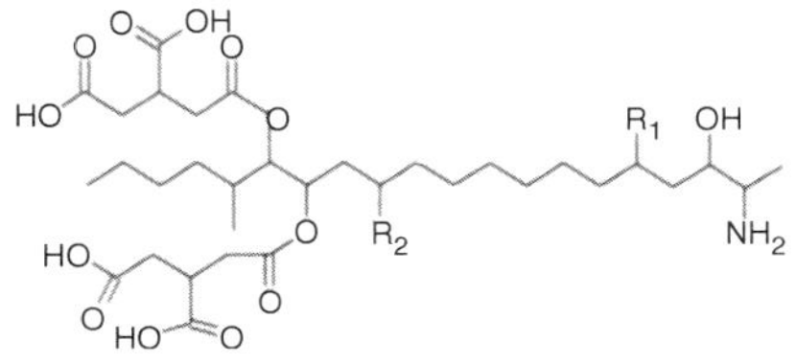	Fumonisin B1 (FB1)	R1=R2=OH	$C_{34}H_{59}NO_{15}$	721.4
			Fumonisin B2 (FB2)	R1=OH; R2=H	$C_{34}H_{59}NO_{14}$	705.4
			Fumonisin B3 (FB3)	R1=H; R2=OH	$C_{34}H_{59}NO_{14}$	705.4
	Others	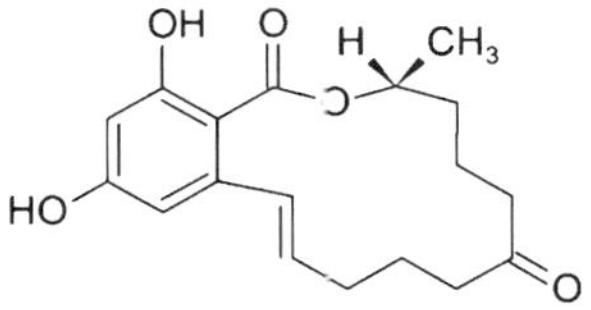	Zearalenone (ZON)		$C_{18}H_{22}O_5$	318.2
Aspergillus sp. and *Penicillium* sp.	Aflatoxins	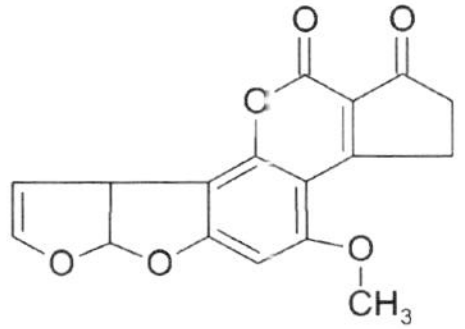	Aflatoxin B1 (AfB1)		$C_{17}H_{12}O_6$	312.1
		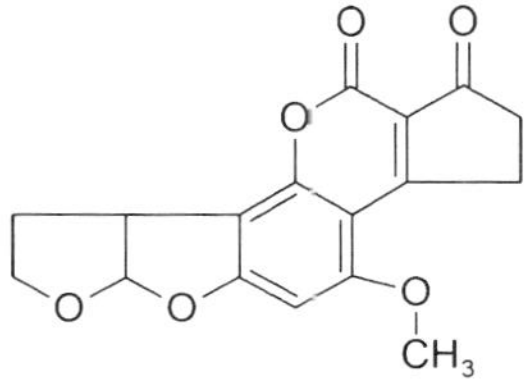	Aflatoxin B2 (AfB2)		$C_{17}H_{14}O_6$	314.1

(*continued*)

Table 8.1 (*Continued*)

Fungi Producer Species	Mycotoxins Group	Structure	Name of Mycotoxins (Abbreviation)	R-Structures	Molecular Formula	Relative Molecular Mass
		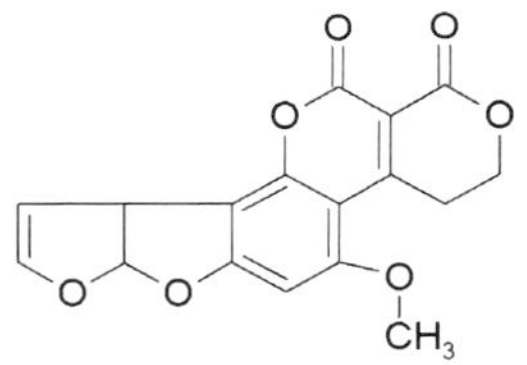	Aflatoxin G1 (AfG1)		$C_{17}H_{12}O_7$	328.1
		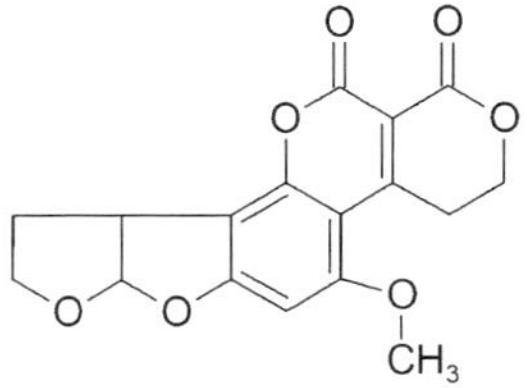	Aflatoxin G2 (AfG2)		$C_{17}H_{14}O_7$	330.1
		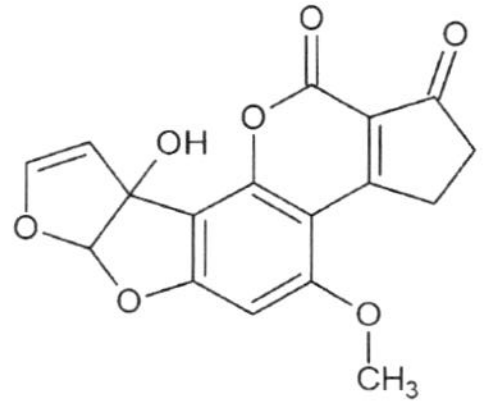	Aflatoxin M1 (AfM1)		$C_{17}H_{12}O_7$	328.1

	Ochratoxins	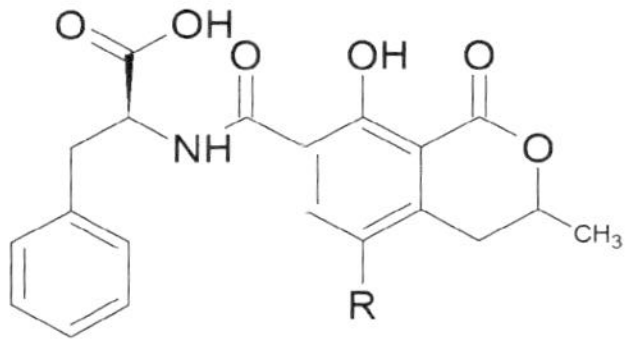	Ochratoxin A (OTA)	R=Cl	$C_{20}H_{18}NO_6Cl$	403.1
			Ochratoxin B (OTB)	R=H	$C_{20}H_{19}NO_6$	369.1
		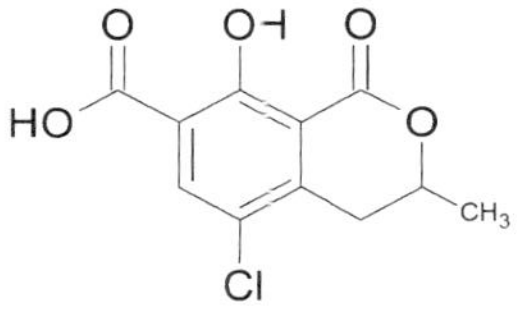	Ochratoxin α (OTα)		$C_{11}H_9O_5Cl$	256.0
	Others	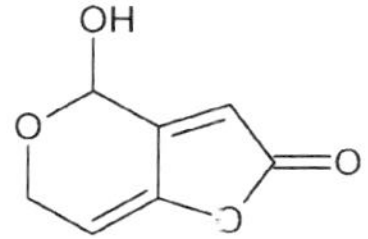	Patulin (PAT)		$C_7H_6O_4$	154.0
Alternaria sp.	Alternaria toxins	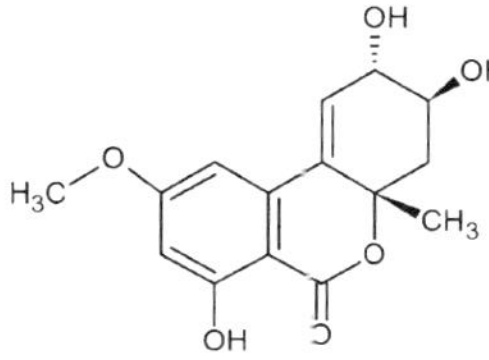	Altenuene (ALT)		$C_{15}H_{16}O_6$	292.1

(*continued*)

Table 8.1 (*Continued*)

Fungi Producer Species	Mycotoxins Group	Structure	Name of Mycotoxins (Abbreviation)	R-Structures	Molecular Formula	Relative Molecular Mass
		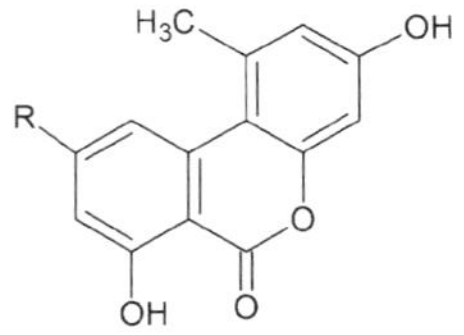	Alternariol (AOH)	R=OH	$C_{14}H_{10}O_5$	258.1
			Alternariol monomethyl ether (AME)	R=OCH_3	$C_{15}H_{12}O_5$	272.1
Claviceps sp.	Ergot alkaloids	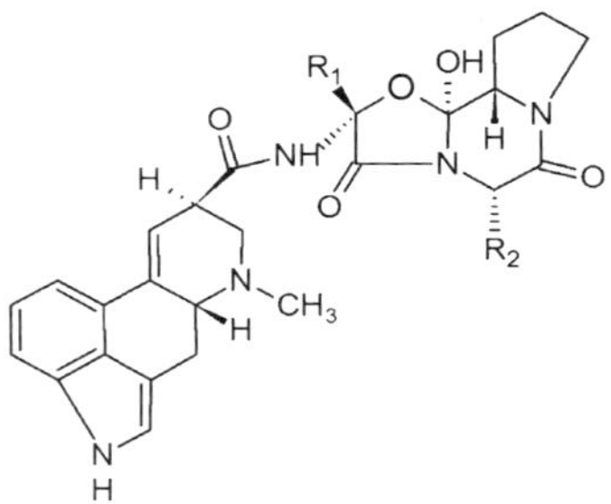	Ergosine	R1=CH_3; R2=$CH_2CH(CH_3)_2$	$C_{30}H_{37}N_5O_5$	547.3
			Ergotamine	R1=CH_3; R2=$CH_2C_6H_5$	$C_{33}H_{35}N_5O_5$	581.3
			Ergocornine	R1=R2=$CH(CH_3)_2$	$C_{31}H_{39}N_5O_5$	561.3
			Ergocristine	R1=$CH(CH_3)_2$; R2= $CH_2C_6H_5$	$C_{35}H_{39}N_5O_5$	609.3
		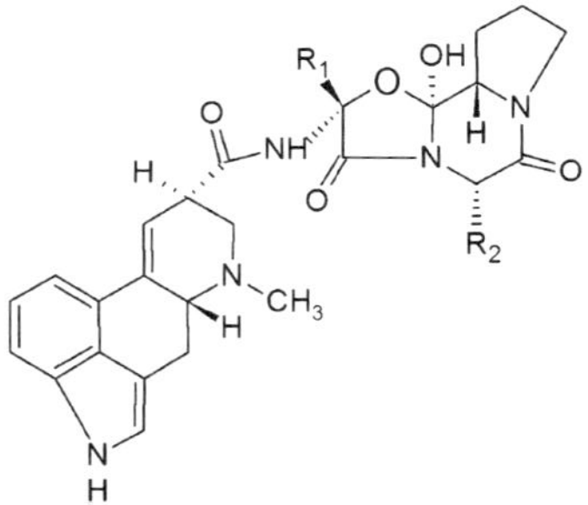	Ergosinine	R1=CH_3; R2=$CH_2CH(CH_3)_2$	$C_{30}H_{37}N_5O_5$	547.3
			Ergotaminine	R1=CH_3; R2=$CH_2C_6H_5$	$C_{33}H_{35}N_5O_5$	581.3
			Ergocornine	R1=R2=$CH(CH_3)_2$	$C_{31}H_{39}N_5O_5$	561.3
			Ergocristinine	R1=$CH(CH_3)_2$; R2= $CH_2C_6H_5$	$C_{35}H_{39}N_5O_5$	609.3

animals. While acute exposure to high concentrations can cause liver or kidney deterioration, chronic effects include carcinogenicity, cytotoxicity, nephrotoxicity, neurotoxicity, mutagenicity, teratogenicity, estrogenicity, and immune suppression [4,5].

Many international and governmental organizations, such as the World Health Organization (WHO), the Food and Agricultural Organization (FAO), the European Commission (EC), and the U.S. Food and Drug Administration (FDA), have recognized the occurrence of mycotoxins in food and feed as a serious health risk and have worked to establish and/or update respective maximum concentrations for these compounds [5]. In the past, numerous monitoring and toxicological studies were conducted to cope with problems related to occurrence of mycotoxins. Based on occurrence results, hygienic limits and/or tolerable daily intake (TDI) values were adopted for some related groups or individual mycotoxins. Out of ~500 currently known mycotoxins [6], only a few are recognized as major food safety hazards. From the food safety viewpoint, the most significant, the most often discussed, and thus the most frequently studied and controlled mycotoxins are aflatoxins, deoxynivalenol (DON), T2 and HT2 toxins, zearalenone (ZON), ochratoxin A (OTA), fumonisins, and patulin (PAT) [3]. Despite the serious acute and/or chronic toxic effects, there are relatively large gaps in legislation for different foodstuffs, especially compared with those for other toxicants such as pesticide residues, veterinary drugs (also referred to veterinary medicinal product residues (VMPRs)) and environmental contaminants. Additionally, the available legislation is not harmonized worldwide and varies significantly among respective countries [7]. It is noteworthy that the standardization of regulatory limits for mycotoxins is an extremely difficult task, as many factors have to be considered when making such decisions. In addition to scientific factors, such as risk assessment and analytical accuracy, economic and political factors arising from the commercial interests of each country and the constant need for an adequate food supply also play a role in the decision-making process.

When critically assessing the current regulatory systems, the EU probably has the most comprehensive, well-developed, and stringent legal limits worldwide. The EU regulation covers a wide range of various foodstuffs and raw materials (~50) intended for food production and direct consumption [8]. In addition to the EU, the following countries have at least partly established regulations for mycotoxins: Argentina, Australia, Bosnia and Herzegovina, Brazil, Canada, China, India, Japan, Mexico, New Zealand, Nigeria, Russia, South Africa, Switzerland, Turkey, and the United States. Aflatoxins are the only group of mycotoxins that are regulated in all of the above-mentioned countries. A summary of the worldwide regulatory limits on mycotoxins is available online [9]. Guidance on concentrations of representative mycotoxins (DON, ZON, OTA, fumonisins, and aflatoxins) in feedstuffs is also provided on the Internet [10].

8.1.2 Emerging Mycotoxins

In addition to mycotoxins already mentioned, there are many other compounds such as ergot alkaloids, alternaria toxins, beauvericin and enniatins, moniliformin, diacetoxyscirpenol, nivalenol (NIV), citrinin, sterigmatocystin, and phomopsin that occur in food and feed. Because both the evidence for toxicity and the occurrence of these compounds are increasing, they are currently the subjects of intensive research and monitoring studies. This occurrence trend is most probably linked to changes in climate conditions [11].

Mycotoxins occur in not only their native forms but also conjugated with peptides, carbohydrates, and/or sulfates. These so-called "masked mycotoxins" are formed after the metabolization of the original mycotoxins by plants, fungi, and mammals or during food processing. DON-3-glucoside (D3G), ZON-4-glucoside (Z4G), or masked fumonisins are in the forefront of conjugated mycotoxins research. In general, comprehensive information on the occurrence and toxicity of masked mycotoxins is not yet available. The main drawback of research on masked mycotoxins is the lack of both pure analytical standards and analytical methods needed for their accurate determination. The available data indicate that the native (parent) mycotoxins can be, at least to some extent, released from masked conjugates in the digestive tract, thus contributing to the overall exposure of both humans and animals to these toxins [12]. The issues related to masked mycotoxins were recently compiled in a comprehensive review by Berthiller et al. [13].

8.1.3 Analysis of Mycotoxins in High-Throughput Environment

Various analytical approaches have been developed and optimized to determine mycotoxins in food and feed; an overview is shown in Figure 8.1. Because the regulatory limits for certain compounds are set at very low (trace) concentrations, there is a need for highly sensitive analytical methods that are capable of detection, quantification, and confirmation of mycotoxins in complex matrices. Currently, the most frequently used technique for analysis of mycotoxins is liquid chromatography–mass spectrometry (LC–MS) utilizing various types of mass analyzers. Additionally, screening methods based on immunochemical techniques (e.g., enzyme-linked immunosorbent assay (ELISA)) or biosensors (e.g., protein chips and antibody/protein-coated electrodes) are also employed [14].

The term high-throughput analysis refers to an analytical procedure with minimal time requirements, which can be performed within minutes or which screens for a large number of compounds per unit time. As discussed in Chapter 1, high-throughput methods have to employ either no or only very simple sample preparation protocols followed by rapid, sensitive, and reliable detection and/or quantification steps suitable for all analytes of interest [15]. Keeping in mind certain limitations of analytical strategies applicable to mycotoxins, both biological (immunological) and instrumental methods can be effectively used for high-throughput analysis [5]. With the immunological methods, although the results may be obtained in minutes, the main drawbacks of their use are narrow scope in terms of number of analytes and typically poor accuracy. On the other hand, LC–MS-based analyses can provide accurate quantitative data for many target compounds. The number of analytes integrated in particular LC–MS-based methods can range from several (regulated) toxins to over 100 mycotoxins. The measurement of throughput of LC–MS methods is largely determined by the

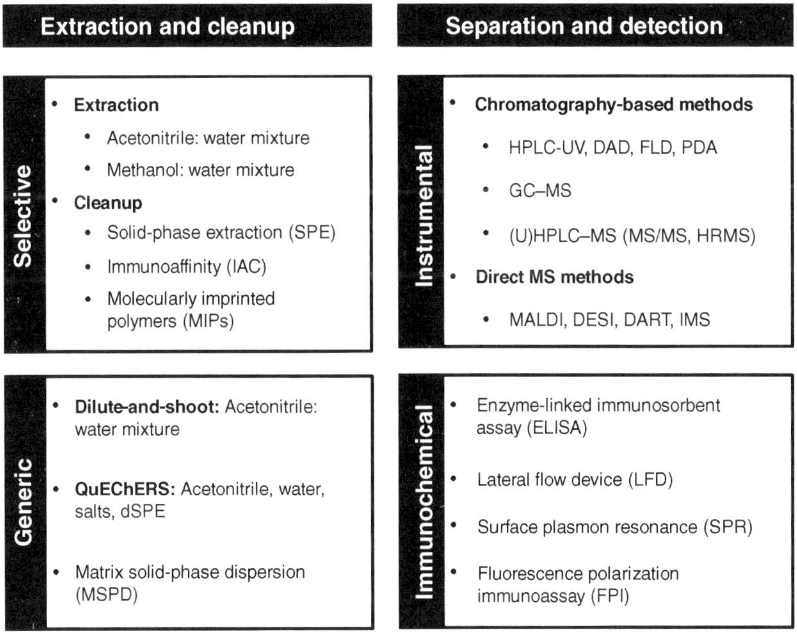

Extraction and cleanup

Selective

- **Extraction**
  - Acetonitrile: water mixture
  - Methanol: water mixture
- **Cleanup**
  - Solid-phase extraction (SPE)
  - Immunoaffinity (IAC)
  - Molecularly imprinted polymers (MIPs)

Generic

- **Dilute-and-shoot:** Acetonitrile: water mixture
- **QuEChERS:** Acetonitrile, water, salts, dSPE
- Matrix solid-phase dispersion (MSPD)

Separation and detection

Instrumental

- **Chromatography-based methods**
  - HPLC-UV, DAD, FLD, PDA
  - GC–MS
  - (U)HPLC–MS (MS/MS, HRMS)
- **Direct MS methods**
  - MALDI, DESI, DART, IMS

Immunochemical

- Enzyme-linked immunosorbent assay (ELISA)
- Lateral flow device (LFD)
- Surface plasmon resonance (SPR)
- Fluorescence polarization immunoassay (FPI)

Figure 8.1. The scheme and overview of analytical methods used in mycotoxins analysis.

complexity of the sample, and thus the extent of sample preparation needed prior to instrumental analysis can vary. Other important factors affecting the throughput of analysis are the time for chromatographic analyses followed by data processing and data evaluation.

In the following sections, topics relevant to high-throughput analysis of mycotoxins in food and feed, including sampling and sample preparation, are discussed. This chapter focuses primarily on workflows employing LC–MS, rapid immunological methods, and some nonchromatographic MS-based techniques, as these approaches represent the primary tools in this field.

8.2 SAMPLE PREPARATION

Sample preparation is a crucial step in the analysis of mycotoxins and creates a bottleneck in most analytical procedures [16]. Numerous sample preparation protocols that largely differ in overall time requirements have been described already, depending on the nature of the sample matrix and the type of detection and quantification. These processes typically involve (i) homogenization of the sample, (ii) extraction of target mycotoxins from the sample matrix, and (iii) cleanup of the crude sample extract with simultaneous preconcentration or dilution of analytes. In addition to sample preparation, sample collection (sampling) is another critical factor strongly influencing the results of a particular assay [4,5,17,18].

8.2.1 Sampling

The collection of representative samples is an important but often underappreciated phase in the analysis of mycotoxins. Because the distribution of mycotoxins in agricultural commodities is usually not homogeneous, an incorrect sampling procedure can cause extensive bias when determining the contamination of a particular commodity [5]. Whereas in the case of liquids it is often assumed that mycotoxins are evenly distributed, in fungus-contaminated solid samples (e.g., grains, nuts, or dried fruits) mycotoxins can occur in a few highly contaminated hot spots. The selection of an optimal strategy that enables proper collection of a representative sample is dependent on several factors, such as the properties of the sample matrix, type of packaging, and size of the sampling lot [16]. Sampling and homogeneity of the matrix become critical and extremely time-demanding, especially with regard to large samples. If sampling is performed in an improper way, low amounts of sample (1–5 g) that are being frequently used in rapid sample preparation procedures may lead to false negative results because local hot spots were missed or undersampled. In recent years, the design of official sampling procedures has become a significant concern to many national and international authorities, including the FDA, USDA, EC, and FAO [19,20]. This effort has resulted in the establishment of sampling methods that are believed to allow an objective assessment of contamination with regard to mycotoxins. Worldwide evaluation of these sampling protocols is still in progress.

8.2.2 Matrices of Interest

The studies dealing with monitoring mycotoxins and validating analytical methods for their determination are typically focused on matrices with the highest incidences of legislatively regulated compounds. Cereals, nuts, fruits, vegetables, and related products are in the forefront of interest due to their relatively high susceptibility to infestation by molds. According to the European Rapid Alert System for Food and Feed (RASFF), the greatest numbers of alerts are reported for the occurrence of aflatoxins and OTA in spices, nuts, cereals, and fruits. The presence of these toxins in named matrices is presumable and under strict control of producers, traders, and control authorities.

The monitoring studies are also performed for nonregulated compounds to fill gaps in knowledge on their occurrence in certain matrices to enable their eventual regulation. Examples are the lack of incidence data on T2 and HT2 toxins in oats and other cereals and on OTA in green coffee beans or licorice. The incidence of emerging mycotoxins in food, feed, and raw materials used for their production is also of significant concern. Recently, increased attention has been paid to dietary supplements that have gained high popularity among consumers. Dietary supplement products often contain extracts of various herbs and botanicals susceptible to fungal attack and may represent a significant source of consumers' exposure to mycotoxins [21].

In addition to food, feedstuffs represent an important matrix in the control of mycotoxins. Significant concerns are for silage, because raw materials employed for its production and by-products of food technologies are frequently contaminated. For instance, the occurrence of *Fusarium* mycotoxins was reported at high concentrations in samples of dried distiller's grains with solubles (DDGS). This material represents a valuable by-product of the ethanol production process and is used for feeding livestock. Last but not least, mycotoxins are also analyzed in nonfood/feed biological matrices, such as urine, blood, and feces to monitor their occurrence and metabolic transformations *in vivo* [22,23]. Given this range of target analytes and matrices, it is clear that no single extraction process will be optimal for all analytes in all matrices. Therefore, a range of extraction techniques have been developed, differing in their specificity, complexity, and speed. In Section 2.3, we discuss several common extraction protocols.

8.2.3 Extraction of Mycotoxins

The use of optimal extraction procedures is dictated by the physicochemical properties of the target mycotoxins and the matrix they are in. Similar to other contaminants discussed in this book, solid–liquid extraction (SLE) is the most frequently applied approach to extract mycotoxins from sample matrices. The choice of suitable extraction solvents is crucial to ensure sufficient recoveries, and thus accurate quantification. In procedures that aim to isolate only a single analyte or a small group of related mycotoxins, the composition of the extraction mixture can be adjusted for optimum recovery. Nevertheless, with regard to current trends aimed at the simultaneous determination of numerous mycotoxins, which largely differ in physicochemical properties, solvent mixtures allowing generic extraction of analytes are required. A number of extraction solvents, including methanol, chloroform, acetone, ethyl acetate, and acetonitrile, and their mixtures have already been employed for the extraction of mycotoxins. Among various solvent combinations, the mixture of acetonitrile and water in ratios ranging from 84:16 to 75:25 (v/v) represents the most efficient extraction solvent commonly used. Additionally, in order to improve recoveries of some acidic mycotoxins, formic or acetic acid is frequently added to the extraction mixture [4,5,17,18]. The generic extraction strategy called "dilute-and-shoot," which uses only pure solvents without any further purification, is nowadays commonly applied for the extraction of a wide range of mycotoxins [24]. The efficiency of extraction is usually improved by integrating shaking, sonication, or mixing into the extraction procedure. Alternatively, a combination of the above techniques is used.

Among various methods for the extraction of mycotoxins described in the literature, the QuEChERS (i.e., quick, easy, cheap, effective, rugged, and safe) protocol is probably one of the most relevant to high-throughput analysis. Since its original introduction for pesticide residues analysis [25], QuEChERS has already been used in numerous modifications to extract other chemical contaminants from various food matrices [26]. The QuEChERS procedure combines sample extraction from a mixture of an organic solvent (usually acetonitrile) with water and transfer of analytes into an organic layer with simultaneous separation of aqueous and organic phases induced by the addition of salts. The crude organic extract can be subsequently purified with the use of dispersive solid-phase extraction (dSPE) to remove undesired coextracts (e.g., sugars and/or fatty acids). The dSPE is based on the addition of the sorbent material to an aliquot of the sample extract to remove matrix interferences. The sorbent is subsequently separated from the extract bulk by centrifugation. Various sorbents, such as primary–secondary amine (PSA), silica gel, octadecylsilane-bonded silica gel (C_{18}), graphitized carbon black (GCB), or their combinations, are used for this purpose [26,27]. It is noteworthy that in the case of multitarget methods, the dSPE step is often omitted as it might otherwise decrease recoveries of some analytes [28]. The main advantages of QuEChERS over traditional extraction techniques are high sample throughput (15 versus 60 min per sample), use of small amounts of organic solvents (10 versus 25–100 ml), less glassware, and employment of relatively inexpensive laboratory equipment [26,27].

The use of QuEChERS in mycotoxin analysis was reported by several authors who applied this protocol mainly to cereals and cereal-derived products. In addition to these types of samples, wine, eggs, beer, fruits and vegetables, spices, oilseed, silage, milk, and meat were also matrices extracted for mycotoxins by employing a QuEChERS-type procedure (Table 8.2). Regarding the target analytes, the majority of studies focused on legislatively regulated mycotoxins and/or *Fusarium* mycotoxins. Several papers reported methods with broader scope, which in addition to the above analytes also included ergot alkaloids, alternaria toxins, and other mycotoxins produced by *Penicillium* and *Aspergillus* species. QuEChERS was also employed for simultaneous multiclass extraction of mycotoxins with other contaminants such as pesticides and veterinary drugs [24,29–31]. Table 8.2 provides an up-to-date overview of publications dealing with applications of QuEChERS to mycotoxin analysis and summarizes time demands of the extraction step [24,28–45].

The optimal QuEChERS-based extraction protocol largely depends on the type of matrix to be examined. Therefore, many modifications of the original QuEChERS design have been developed to fit particular sample types. The most important parameters of the QuEChERS method, which have a significant impact on recovery and other performance characteristics of the method, are the composition of extraction mixture, extraction time, type and amount of salts added, and the ratio between organic solvent volume and sample weight (matrix dilution factor, ml/g) [26]. Cereals and cereal-based products represent typical dry matrices that are frequently extracted for mycotoxins using QuEChERS. The matrix dilution factors applied to such samples are usually either 2.0 or 2.5 (i.e., 4 or 5 g of test sample and 10 ml of organic solvent). The volume of water used in published studies varied significantly and was in the range of 2–10 ml. The soaking of the sample matrix and/or prolonged extraction times were shown to be crucial in achieving sufficiently high recoveries of mycotoxins using QuEChERS-based extraction of cereals and similar dry samples [24,29]. However, longer extraction times ultimately result in diminished sample throughput. On the other hand, matrices with naturally high water content, such as vegetables, fruit, milk, beer, and wine, do not require soaking and can be processed at much higher throughput even without the addition of water (see Table 8.2). Regardless of the type of matrix, the extraction efficacy should always be assessed based on naturally contaminated reference materials rather than with the use of spiked samples. Improvement in recoveries of some problematic (acidic) analytes can be

Table 8.2. Overview of Recent QuEChERS Applications in the Analysis of Mycotoxins

Matrix	Analytes	Sample Amount	Extraction Mixture/ Extraction Time (min/ Sample)	Salts	Cleanup Procedure	Detection Technique/Run Time (min/Sample)	Reference
Feed, maize, honey, meat, egg, and milk	Multiple mycotoxins ($n=23$), plant toxins ($n=13$), pesticides ($n=136$), and veterinary drugs ($n=86$)	2.5 g	Water (7.5 ml) and acetonitrile with 1.0% acetic acid (10 ml)	$MgSO_4$ (4 g) and NaCl (1 g)	–	UPLC–MS/MS/20 HPLC–MS/MS/20	[29]
Wheat, maize, and millet	ADON, AOH, ALT, AME, DAS, DOM, DON, FUS-X, NIV, STC, and ZEA	2 g	Water (7.5 ml) and acetonitrile (10 ml)/15	$MgSO_4$ (4 g) and NaCl (1 g)	dSPE of 4 ml with PSA (200 mg) and $MgSO_4$ (600 mg)	DART–OrbitrapMS/ 0.2	[32]
Cereal-based commodities	3-ADON, 15-ADON, AFB1, AFB2, AFG1, AFG2, DAS, DON, FB1, FB2, FUS-X, HT2, NEO, NIV, OTA, T2, and ZON	5 g	Water (10 ml) and 0.5% acetic acid in acetonitrile (10 ml)/26	$MgSO_4$ (4 g) and NaCl (1 g)	Defatting by shaking with *n*-hexane	HPLC–MS/MS/24.5	[33]
Maize silage	Multiple mycotoxins ($n=27$)	10 g	Water (5 ml) with sodium acetate trihydrate (1.67 g) and 1% acetic acid in acetonitrile/13	$MgSO_4$ (4 g)	–	HPLC–MS/MS/44	[34]
Breakfast cereals and flour	15-ADON, DON, FUS-X, NIV, and ZON	5 g	Water (25 ml for cereals, 10 ml for flour) and acetonitrile (10 ml)/20	$MgSO_4$ (4 g) and NaCl (1 g)	dSPE of 6 ml with C_{18} (300 mg) and $MgSO_4$ (900 mg)	GC–MS (derivatization)/8	[35]
Wheat flour	DAS, DON, HT2, NIV, and T2	5 g	Methanol (8.5 ml) and acetonitrile (1.5 ml)/12	$MgSO_4$ (2 g) and NaCl (1 g)	–	HPLC–MS/25	[36]
Cereals	3-ADON, D3G, DON, FB1, FB2, FB3, FUS-X, HT2, NIV, T2, and ZON	4 g	Water (7.5 ml) with 0.1% formic acid and acetonitrile (10 ml)/8	$MgSO_4$ (4 g) and NaCl (1 g)	–	UHPLC–TOFMS/18	[37]
Flour, breakfast cereals, snacks, and bread	3-ADON, 15-ADON, ALT, AME, AOH, D3G, DAS, DON, enniatins, ergot alkaloids, FUS-X, HT2, NEO, NIV, T2, VOL, and ZON	4 g	Water (7.5 ml) and acetonitrile (10 ml)/11	$MgSO_4$ (4 g) and NaCl (1 g)	–	UHPLC–Orbitrap MS/18	[38]
Bread	DON, HT2, and T2	4 g	Water (7.5 ml) with 0.1% formic acid and acetonitrile (12.5 ml)/21	$MgSO_4$ (4 g) and NaCl (1 g)	–	UHPLC–Orbitrap MS/19	[39]

(continued)

Table 8.2 (*Continued*)

Matrix	Analytes	Sample Amount	Extraction Mixture/ Extraction Time (min/ Sample)	Salts	Cleanup Procedure	Detection Technique/Run Time (min/Sample)	Reference
Noodles	AFB1, AFB2, AFG1, and AFG2	2 g	Water (4 ml). methanol (5.1 ml), and acetonitrile (0.9 ml)/7	$MgSO_4$ (1.5 g) and NaCl (0.5 g)	–	HPLC–fluorescence detection (FLD)/ 23	[40]
Wine	OTA	3 ml	30 mM NaH_2PO_4 buffer (pH 7, 8 ml) and 5% formic acid in acetonitrile (10 ml)/5.3	$MgSO_4$ (4 g), NaCl (1 g), sodium citrate (1 g), and disodium citrate sesquihydrate (0.5 g)	–	HPLC–laser-induced fluorescence detection (LIF)/4	[41]
Eggs	AFB1, AFB2, AFG1, AFG2, BEA, CIT, enniatins, and OTA	2 g	Water with 1% acetic acid (2 ml) and methanol with 1% acetic acid (8 ml)	Na_2SO_4 (4 g) and sodium acetate (1 g)	–	UHPLC–MS/MS/ 6.5	[42]
Beer-based drinks	AFB1, AFB2, AFG1, AFG2, DON, FB1, FB2, FB3, HT2, NIV, OTA, PAT, T2, and ZON	10 ml	Acetonitrile (10 ml)/5.3	$MgSO_4$ (4 g), NaCl (1 g), sodium citrate (1 g), and disodium citrate sesquihydrate (0.5 g)	SPE cartridge (C_{18})	UHPLC–MS/MS/ 6.5	[27]
Cereals, cereal-based foods, cucumber, wine	AFB1, AFB2, AFG1, AFG2, HT2, OTA, T2, and pesticides ($n = 83$)	10 ml/5 g	Water (5 ml, cereals only) and 1% acetic acid in acetonitrile (10 ml)/ 60 min soaking (cereals only), 7	$MgSO_4$ (4 g) and NaCl (1.5 g)	–	UHPLC–MS/MS/13	[31]
Milk	AFB1, AFB2, AFG1, AFG2, AFM1, HT2, OTA, T2, and pesticides ($n = 42$)	10 ml	Acetonitrile with 1% acetic acid (10 ml)/7	$MgSO_4$ (4 g), NaCl (1 g), sodium citrate (1 g), and disodium citrate sesquihydrate (0.5 g)	–	UHPLC–MS/MS/ 6.5	[30]

Pear- and apple-based foods	PAT	10 g (puree, juice)/5 g	Water (10 ml, cereals, flakes, and concentrate only) and acetonitrile (10 ml)/20	$MgSO_4$ (4 g) and NaCl (1 g)	dSPE of 6 ml with PSA (400 mg), C_{18} (400 mg), and $MgSO_4$ (1200 mg)	HPLC–MS/MS/17.5	[43]
Popcorn	15-ADON, DON, FUS-X, NIV, and ZON	5 g	Water (10 or 20 ml for popped popcorn), saturated aqueous solution of Na_2CO_3 (5 ml), and acetonitrile (10 ml)/25	$MgSO_4$ (4 g) and NaCl (1 g)	Defatting by shaking with *n*-hexane, dSPE of 6 ml with C_{18} (300 mg), and $MgSO_4$ (900 mg)	GC–MS (derivatization)/16	[44]
Wheat, corn, rice, and noodles	3-ADON, 15-ADON, DAS, DON, FUS-X, HT2, NIV, and T2	1 g	Water (2 ml), acetonitrile (7.9 ml), and acetic acid (0.1 ml)/7	$MgSO_4$ (0.8 g) and NaCl (0.2 g)	–	HPLC–QTOFMS/23	[26]
Rice	CIT and OTA	10 g	Water (20 ml) and 1% acetic acid in acetonitrile (20 ml)/28	$MgSO_4$ (1.5 g) and sodium acetate (0.85 g)	dSPE of 20 ml with $MgSO_4$ (300 mg) and diatomaceous earth (Celite, 200 mg)	HPLC–photodiode array detection (PDA)/20	[45]
Fruit, cereals, spices, and oilseeds	Multiple mycotoxins ($n = 38$) and pesticides ($n = 288$)	2.5 g/10 g (fruit)	Water with 2% formic acid (10.0 ml) and acetonitrile (10 ml)/60 min soaking (dry matrices)/6.5	$MgSO_4$ (4 g) and NaCl (1 g)	dSPE of oilseed extracts (2 ml) with C18 (100 mg) and $MgSO_4$ (300 mg)	UHPLC–MS/MS/15.5	[24]
Barley	Multiple mycotoxins ($n = 32$)	2 g	Water with 0.1% formic acid (10 ml) and acetonitrile (10 ml)/21	$MgSO_4$ (4 g) and NaCl (1 g)	–	UHPLC–OrbitrapMS/18	[28]

3-ADON: 3-acetyldeoxynivalenol, 15-ADON: 15-acetyldeoxynivalenol, AFB1: aflatoxins B1, AFB2: aflatoxins B2, AFG1: aflatoxins G1, AFG2: aflatoxins G2, AFM1: aflatoxins M1, ALT: altenuene, AME: alternariol-methyl ether, AOH: alternariol, BEA: beauvericin, CIT: citrinin, D3G: deoxynivalenol-3-glucoside, DAS: diacetoxyscirpenol, DOM: deepoxy-deoxynivalenol, DON: deoxynivalenol, FB1: fumonisin B1, FB2: fumonisin B2, FB3: fumonisin B3, FUS-X: fusarenon-X, HT2: HT2 toxin, NEO: neosolaniol, NIV: nivalenol, OTA: ochratoxin A, PAT: patulin, STC: sterigmatocystin, T2: T2 toxin, VOL: verrucarol, ZON: zearalenone.

achieved by addition of various buffers and/or acids into the extraction mixture in order to adjust the pH and support the transfer of analytes into the organic layer. Regarding the type and optimal amount of salts, magnesium sulfate ($MgSO_4$) and sodium chloride (NaCl) at a ratio of 1:4 (w/w) have been used in most of applications.

8.2.4 Purification of Sample Extracts

The main goals of the cleanup step are to remove undesirable sample coextracts that may interfere with the analytes and to preconcentrate target analytes to allow acceptable sensitivity and selectivity to be achieved [5,17,18,27]. The most frequently employed sample purification approaches in the analysis of mycotoxins are using either solid-phase extraction (SPE) or immunoaffinity column (IAC) cleanup. Additionally, molecularly imprinted polymers (MIP) cleanup, matrix solid-phase dispersion (MSPD), or liquid–liquid partitioning of extract with *n*-hexane, acetone, or ethyl acetate have been used in this field, but to a limited extent [14].

The SPE cleanup technique is based on the partitioning of analytes and the sample matrix between mobile and stationary phases. SPE is performed by passing the sample extract through a disposable cartridge containing sorbents with bound phases (with C_{18} being the most used) and various adsorbents, such as charcoal, Florisil, or Celite. SPE in three different modes can be generally applied in food analysis: (i) selective extraction, (ii) selective washing, and (iii) selective elution [4,46]. In the selective extraction mode, the SPE cartridge retains mycotoxins and allows impurities to pass through the cartridge. In a subsequent step, target analytes are released from the SPE stationary phase using a suitable solvent. In the sample washing mode, both analytes and impurities are first retained on the SPE sorbent bed, the interfering matrix components are further rinsed out using strong enough solvent, and analytes are finely eluted with other but stronger solvents. In the selective elution mode, interfering impurities are retained by the stationary phase of the SPE cartridge and the target mycotoxins are allowed to pass through the column. No washing and elution steps are further required. Retention and elution of analytes and impurities are strongly dependent on properties of the stationary bed and elution/wash solvents, which are commonly designed for specific usage of analyte–matrix–solvent combinations. The description of sorbent and solvent selectivity is thoroughly described in a review by Lucci et al. [46]. The respective SPE modes are illustrated in Figure 8.2. It is apparent that for achieving the highest possible sample throughput, the matrix removal SPE mode is the most desirable as it enables the sample extract cleanup to be performed in a single step. It is worth noting that the MycoSep SPE cartridges, which are currently the most frequently used in mycotoxin analysis, are operated in the matrix removal mode [18].

While SPE offers cleanup and preconcentration for a broad range of analytes, IAC provides a higher selectivity and specificity to target analyte(s). After application of the sample extract to the IAC, mycotoxins are selectively bound to antibodies (either monoclonal or polyclonal) that are immobilized in the cyanobromide-activated sepharose gel [47] present in the cartridge. IAC combines the sample cleanup and sample concentration modes. Components of the sample matrix that do not interact with antibodies are gradually eluted from the column during application of the extract.

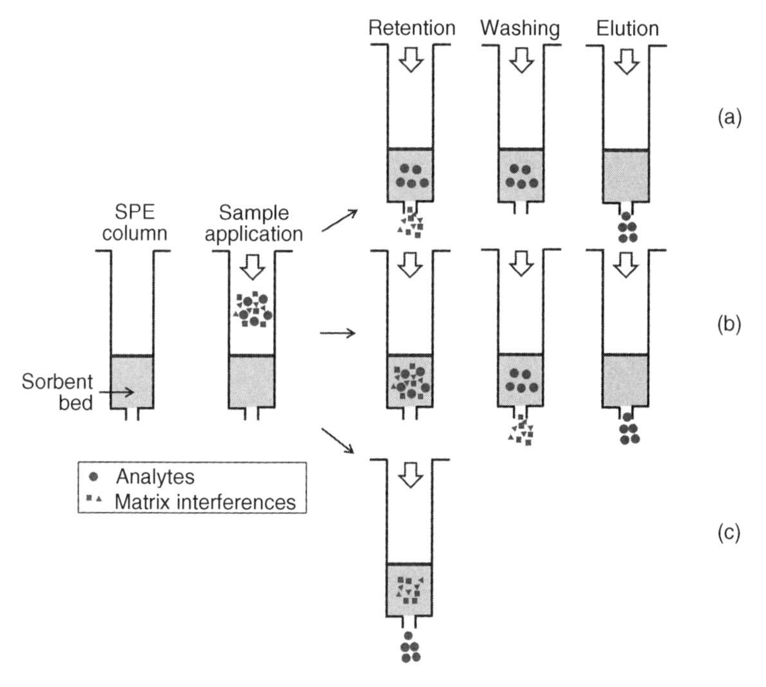

Figure 8.2. A scheme of SPE column modes and functionality. (a) Selective extraction mode. (b) Selective washing mode. (c) Selective elution mode.

The analytes are then eluted by a small volume of pure organic solvent (methanol or acetonitrile), which disrupts mycotoxin–antibody bonds by protein denaturation. The most comprehensive studies on this topic have been published by Turner et al. [18] and Rahmani et al. [4].

Although both multifunctional IAC and SPE columns are commercially available for all of the main regulatory significant mycotoxins, these columns are still not developed for a wide range of mycotoxin groups. Moreover, the high cost and relatively long time of analysis caused by large loading and elution solvent volumes make these cleanup strategies poorly suited for rapid sample preparation within a high-throughput environment. In addition to the dSPE cleanup strategy that has already been described and that fulfills the requirements needed for high throughput, the dilute-and-shoot strategy has also been employed in conjunction with LC–MS for high-throughput multimycotoxin analysis. In this case, the poorer sensitivity caused by higher amounts of coextracted matrix compounds, which hamper the ionization of target analytes, is the price paid for generic extraction, no purification steps, and improved sample throughput [24,37,44,48].

8.3 SEPARATION AND DETECTION OF MYCOTOXINS

There are several chromatographic techniques available for the analysis of mycotoxins, including gas chromatography (GC), thin layer chromatography (TLC), and high- or ultrahigh-performance liquid chromatography (HPLC and UHPLC). GC was frequently used for this purpose in the 1990s. However, the obvious drawbacks of GC-based methods relate to the need for time-consuming sample preparation and derivatization of analytes, which have led to their reduced use in mycotoxin analysis [49]. On the other hand, LC coupled to either conventional detectors [ultraviolet (UV) detector, diode array detector (DAD), fluorescence detector (FLD), and photodiode array detector (PDA)] or mass spectrometers is currently the most frequently applied separation technique. Note that conventional detectors are selective only for a limited number of toxins, and are thus less versatile than MS detection. LC-based methods are also used to confirm results of novel rapid screening techniques [47,50].

8.3.1 Liquid Chromatography–Mass Spectrometry-Based Methods

Currently, the analysis of mycotoxins relies largely on LC separation employing reversed-phase (RP) columns in combination with MS using different mass analyzers. Such methods represent the reference and definitive protocols for mycotoxin analysis [18]. Although most of the published LC–MS-based workflows have focused on simultaneous determination of structurally related mycotoxins in single food/feed matrices, several studies have described successful integration of analysis of multiple nonrelated mycotoxins into a single determinative LC–MS method. This was made possible by substantial advances in MS instrumentation that resulted in sufficiently sensitive and selective high-throughput broad-scope mycotoxin analysis. In these LC–MS methods, various combinations of LC operated in either high-pressure (HP) or ultrahigh-pressure (UHP) mode with low-resolution (LR) tandem MS or high-resolution (HR) MS have been used.

The most recently published studies aiming at rapid analysis of multiple mycotoxins using LC–MS are Refs [24,28,29,48,51–53]. In some of these studies, mycotoxins were analyzed simultaneously with other food contaminants or natural toxins, such as pesticides, plant toxins, marine toxins, and/or veterinary drugs [24,29,52]. The average number of mycotoxins analyzed by these methods was between 30 and 40. Typical groups of mycotoxins for which analytical standards are commercially available (e.g., trichothecenes, enniatins, fumonisins, aflatoxins, ergot alkaloids, alternaria toxins, ZON, OTA, and PAT) were tested. The only exceptions were multimycotoxin methods described by Sulyok et al. [48], Abia et al. [51], and Varga et al. [53], who developed procedures capable of simultaneous analysis of 106, 320, and 191 mycotoxins and other toxic or potentially toxic fungal secondary metabolites, respectively. Not only cereals, nuts, and related products were used for evaluation of the method recoveries, but other important matrices such as baby foods, fruits, seeds, spices, honey, milk, eggs, meat, alcoholic and nonalcoholic beverages, soybeans, and cheese were also included.

Regarding the separation step, (U)HPLC systems are usually applied for multimycotoxins analysis. An ongoing development in UHPLC instrumentation allows separation to be performed under substantially higher pressures using chromatographic columns with a sub-2 μm stationary-phase particle size, which generally result in narrower chromatographic peaks and lower overall run times. The effective LC separation of multiple analytes requires the proper selection of both chromatographic columns and composition of mobile phases and a careful optimization of chromatographic parameters such as temperature, mobile-phase gradient, pH, and composition of buffers. In practice, chromatographic columns using RP-C_{18} stationary phases are almost universally applied. The most comprehensive multitarget LC–MS method dealing with the analysis of 320 toxic and potentially toxic mycotoxins was developed and published by Abia et al. [51]. This methodology amends methods published previously by Sulyok et al. [48,54]. For the chromatographic separations used here [51,53], an HPLC system employing an RP-C_{18} column was used. For sufficiently sensitive analyses for all compounds, two separate chromatographic runs had to be performed in positive (ESI(+)) and negative (ESI(−)) electrospray modes using a triple quadrupole linear ion trap (QLIT) MS instrument. The time needed for LC–MS analysis of all 320 analytes was 41 min per sample. A similar strategy was also employed in a study by Lacina et al. [24], who analyzed 38 mycotoxins together with 288 pesticides. This analysis was also subdivided into two consecutive runs with run times of 15.5 min each. In a study by Mol et al. [29], two 20 min UHPLC–MS/MS methods were applied for the determination of mycotoxins and natural toxins ($n = 36$), pesticides ($n = 136$), and veterinary drugs ($n = 86$) in both positive and negative ionization modes. Herrmann et al. [52] performed simultaneous analysis of 36 mycotoxins together with some drugs, pesticides, and other chemical contaminants, representing in total 127 target analytes. This analysis was again subdivided into two separate runs. Each was accomplished within 22 min and resulted in a total analysis time of 44 min per sample. Two separate runs in ESI (+) and ESI(−) modes are commonly applied in multitarget analyses where triple quadrupole (QqQ) or QLIT are used as mass analyzers. This is necessary because of the high number of simultaneously eluted analytes that differ in terms of their optimal ionization modes. Because the polarity switching is not rapid enough to enable simultaneous acquisition in both ionization modes when employing common LR-MS instruments, the only viable solution to achieve acceptable LODs is to separate the analytes into positive and negative ionization mode runs. For example, aflatoxins (ionizing in ESI(+)) and trichothecenes (ionizing in ESI(−)), which represent highly important regulated toxins, cannot be easily separated with C_{18} columns and therefore typically overlap or coelute.

The LC parameters of the above LC–MS methods were more or less similar. The dimensions of the most frequently employed UHPLC columns were 100 or 150×2.1 mm with 1.7 or 1.8 μm particle sizes [24,29,53] or 150×4.6 mm with 5 μm particle sizes for HPLC analysis [48,51]. In one case, a shorter column (50×2.1 mm, with 1.8 μm particle sizes) was applied in UPLC–MS/MS analysis [52]. The column temperatures ranged from 25 to 55 °C. The majority of methods employed acidified ammonium formate (1–5 mM) and acidified methanol in ESI (+) ionization mode for the mobile phase, while aqueous ammonium acetate (5 mM) and methanol were used in ESI(−) mode.

Multianalyte methods developed specifically for the determination of mycotoxins usually have quite similar parameters, as described in the above applications. These were recently summarized in a review by Hajslova et al. [55]. The state-of-the-art trends focus on the development of high-throughput methods with generic sample preparation and low detection limits of a broad range of food contaminants. In all of these studies, the LR-MS represented by a QqQ is the most prevalent MS option in mycotoxin analysis for selective detection and confirmation of analytes. Using the detection/confirmation strategy based on monitoring two MRM transitions (one precursor ion → two product ions, or first precursor ion → one product ion and second precursor ion → one product ion) for each analyte, the requirements for analyte identification established by official documents such as Commission Decision 2002/657/EC [56] and the SANCO/12495/2011 document [57] can be fulfilled.

8.3.2 High-Resolution Mass Spectrometry in Mycotoxins Analysis

In addition to tandem MS, HR-MS analyzers have also been applied to quantitative, semiquantitative, and nontargeted screening analyses of multimycotoxins. Despite their ability to simultaneously detect and confirm multiple analytes, the HR-MS techniques have not yet been extensively used for multimycotoxin analysis. Additionally, current EU legislation requires certain conditions to be fulfilled when confirming positive findings with HR-MS. Confirmatory analysis must provide at least two characteristic masses (*m/z*) acquired at HR-MS conditions for a target analyte to fulfill the requirement for confirmation [56]. Unfortunately, achieving two ions with significant intensity is often difficult for certain analytes, especially when they are present at trace concentrations. From this perspective, hybrid HR-MS instruments capable of operating in the MS/MS mode to provide fragmentation mass spectra with accurate mass represent a new possibility for simultaneous analysis and confirmation of mycotoxins in food and feed.

The pioneering use of HR-MS techniques (utilizing a time-of-flight (TOF) analyzer) in mycotoxin analysis was described by Tanaka et al. [58], Mol et al. [29], and Zachariasova et al. [37]. Tanaka published an LC–TOFMS method with atmospheric pressure chemical ionization (APCI) for simultaneous determination of trichothecenes, aflatoxins, and ZON in corn, wheat, cornflakes, and biscuits. The disadvantage of this method was the additional SPE cleanup that had to be employed resulting in slightly decreased throughput of the entire workflow. Zachariasova et al. [37] employed UHPLC coupled to TOF and Orbitrap mass analyzers to examine 11 major *Fusarium* mycotoxins (fumonisins, DON, 3-ADON, NIV, HT2, T2, ZON, D3G, and fusarenon-X) in cereals. Two alternative sample preparation procedures based on either modified QuEChERS extraction or aqueous acetonitrile extraction were used prior to instrumental analysis. The UHPLC–TOFMS chromatograms of DON are shown in Figure 8.3. Based on these results, it was concluded that both technologies are applicable for mycotoxin detection, but the approach using TOFMS required some additional cleanup strategy to achieve sufficient sensitivity for the target analytes [37]. In a comparative study by Mol et al. [29], the UHPLC–TOFMS method was shown to be a generic tool in multiresidue and contaminants analysis compatible with the MS/MS approach regardless of sample preparation. Hybrid quadrupole/time-of-flight (QTOF) instrumentation was applied in a study by Sirhan et al. [26], who determined trichothecene mycotoxins in wheat, corn, rice, and

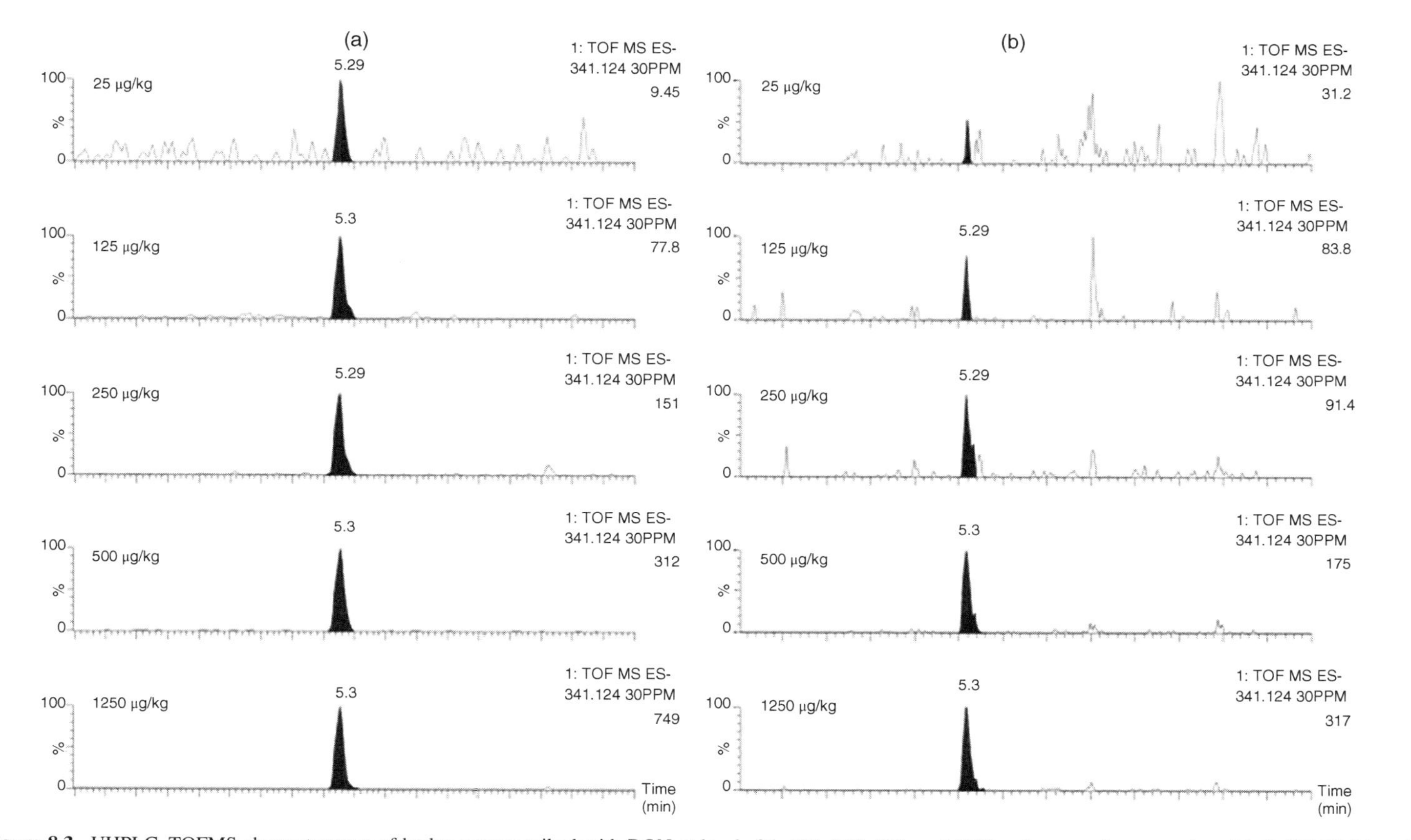

Figure 8.3. UHPLC–TOFMS chromatograms of barley extract spiked with DON at levels 25, 125, 250, 500, and 1250 μg/kg; sample preparation: (a) QuEChERS-based method and (b) crude extract-based method; extraction window was 30 ppm [37]. *Source*: Ref. [37], Figure 1, p. 56. Reproduced with permission of Elsevier Science Ltd.

noodles. Another study by Polizzi et al. [59] investigated the occurrence of mycotoxins in air, dust, wallpaper, and silicone materials using both LC–MS/MS and LC–QTOFMS techniques. Application of QTOF technology was also described by Veprikova et al. [60], who used it for identification of masked glycosylated forms of T2 and HT2 toxins.

The most comprehensive HR-MS studies devoted to the application of UHPLC–Orbitrap MS technology in multimycotoxin analysis were published by Herebian et al. [61], Zachariasova et al. [37,62], Rubert et al. [28], and De Dominicis et al. [63]. All tested 32 mycotoxins as the main representatives of *Fusarium*, *Claviceps*, *Aspergillus*, *Penicillium*, and *Alternaria* fungi. In the study of Herebian et al. [61], the HPLC–ESI–MS/MS and microcapillary-HPLC–LTQ/Orbitrap MS instruments were critically assessed for their use in cereal examination. Based on analyses of the undiluted acetonitrile:water extracts, it was concluded that HR-MS is also a time-saving method useful for the suggested purpose. Zachariasova et al. [37,62] published the use of Orbitrap MS technology for the analysis of mycotoxins in cereals and beer. Both studies were focused on comparing two HR-MS instruments (TOF and Orbitrap MS) and their possible applicability for the fully validated screening and quantitative methods. In both cases, Orbitrap MS instrumentation was shown to offer superb sensitivity without the need for lengthy sample preparation protocols. This particular instrumentation was also used in a validation study aimed at regulated mycotoxins in wheat/barley flours, crisp bread, and other bakery ingredients [8,63]. The increasing interest in HR-MS for nontargeted screening of masked forms of mycotoxins and various metabolites was also demonstrated in several publications [64–66]. HR-MS was shown to be applicable as a detection tool for potentially harmful compounds, for which analytical standards were not available.

8.4 NO-SEPARATION MASS SPECTROMETRY-BASED METHODS

In addition to MS-based applications employing separation of the sample extract, some rapid no-separation techniques, such as matrix-assisted laser desorption MS, ambient ionization MS, and ion mobility spectrometry, have also been used to analyze mycotoxins. Examples of applications of these techniques are provided in the following sections.

8.4.1 Matrix-Assisted Laser Desorption Ionization–Mass Spectrometry

The principles of matrix-assisted laser desorption ionization–mass spectrometry (MALDI–MS) have been described in other chapters. Although not widely employed, several applications of MALDI–MS aimed at analysis of mycotoxins have been published. With regard to the need for internal standardization to allow quantification, MALDI–MS has been used mainly for qualitative analysis. MALDI is also a useful tool in characterization and classification of toxigenic fungi and mycotoxin-related proteomics.

In a study by Elosta et al. [67], a thorough optimization of positive-mode MALDI coupled to TOFMS was performed to allow sensitive determination of DON, NIV, and ADONs in SPE-purified acetonitrile–water extracts of barley and malt. The use of sodium azide matrix provided good reproducibility and relatively low limits of detection ranging from 0.6 to 0.9 μg/ml. The authors also explored the capability of the method to quantify DON in naturally contaminated malt based on external calibration. The results for DON obtained by MALDI–TOFMS and the reference HPLC–MS/MS method were 507 ± 9 and 780 ± 124 μg/kg, respectively.

Work published by Catharino et al. [68] described the MALDI–TOFMS protocol for screening aflatoxins (AFB1, AFB2, AFG1, and AFG2) in peanuts at concentrations as low as 50 fmol. The use of an ionic liquid matrix (triethylamine–α-cyano-4-hydroxycinnamic acid solution in methanol) enabled the acquisition of interference-free mass spectra. The target mycotoxins were isolated from the samples by a procedure based on extraction with an aqueous–methanol solution containing potassium chloride and chloroform and purified with the use of $CuSO_4$ and diatomaceous earth (Celite).

An interesting MALDI–TOFMS approach to qualitative analysis of gliotoxin was reported by Davis et al. [69], who developed a single-pot derivatization strategy using sodium borohydride-mediated reduction of gliotoxin followed by immediate alkylation of exposed thiols by reaction with 5′-iodoacetamidofluorescein to yield a stable product, diacetamidofluorescein-gliotoxin, of molecular mass 1103.931 Da ($[M+H]^+$ ion). Unlike free gliotoxin, this product was readily detectable by MALDI–TOFMS at concentrations above 530 fmol. Although demonstrated only for the analysis of *Aspergillus fumigatus* culture supernatants, the above strategy may also be applicable to analysis of extracts of food and feed.

Marchetti-Deschmann et al. [70] used MALDI–TOFMS to classify closely related *Fusarium* species responsible for *Fusarium* head blight disease of crops based on the analysis of intact spores. The spore suspensions were directly embedded into a MALDI matrix without laborious sample cleanup or enrichment steps and the surface-associated compounds were analyzed by MALDI–TOFMS. These mass spectra were used to develop partial least-squares discriminant analysis (PLSDA) models for sample classification. The authors demonstrated the potential to build a database for accurate *Fusarium* species identification and for fast response in the case of infection in the cornfield. In another study, MALDI–TOFMS was used to identify resistance-associated proteins in response to *Aspergillus flavus* infection under drought stress [71]. MALDI–MS is also frequently used for detection and identification of conjugates of mycotoxins with proteins, which can also be used in analytical applications, such as immunogens for production of selective antibodies [72,73] or as biomarkers of intoxications with mycotoxins [74].

8.4.2 Ambient Ionization Mass Spectrometry

Novel ambient desorption ionization techniques such as direct analysis in real time (DART) and desorption electrospray ionization (DESI) hold great potential in high-throughput analysis of food. Various techniques and principles have been described in other chapters. Only a few applications were described for the analysis of mycotoxins in food and feed with the use of ambient ionization MS.

The most comprehensive study dealing with the high-throughput ambient MS analysis of mycotoxins in cereals was performed by Vaclavik et al. [32], who used a DART ion source coupled with an Orbitrap mass spectrometer. In the first step, the DART ionization efficiency of various mycotoxins was investigated. Of the 24 tested mycotoxins, 11 target analytes could be efficiently ionized by the DART technology. Only poor ionization of major trichothecenes A (T2 and HT2) and some aflatoxins (AFB1 and AFB2) was achieved by DART. The ionization of OTA and other mycotoxins such as ergot alkaloids, fumonisins, and D3G was not possible under the experimental conditions employed. The samples of test cereals were processed by a modified QuEChERS extraction procedure and, due to relatively high ion signal fluctuations, analytes were quantified by means of matrix-matched standards with addition of isotope-labeled internal standards (Figure 8.4). The data generated by DART–MS analysis of certified reference materials were in good agreement with those obtained by a UHPLC–TOFMS method. The method was shown to be applicable for high-throughput detection of DON and ZON at limits established in the EU for unprocessed wheat and maize.

Another study described rapid DART–TOFMS analysis of DON in beer samples following immunoaffinity cleanup and sample preconcentration [75]. In a paper focused on UHPLC–MS analysis of multiple mycotoxins in beer, DART–Orbitrap MS fingerprinting was employed to document the purification effect achieved by acetonitrile-induced precipitation of some matrix components [62]. The application of the DESI ionization technique coupled with an ion-trap mass spectrometer was demonstrated for the determination of mycotoxins in a review by Maragos et al. [76]. Fumonisin B1 (0.2 ng) was deposited on the surface of maize kernels and, after drying, was easily detected as $[M+H]^+$ ion by DESI–MS. Moreover, after subjecting the DESI-analyzed kernels to a germination test, 9 of 10 were found viable. Such results document the nondestructive nature of the DESI technique.

8.4.3 Ion Mobility Spectrometry

Ion mobility spectrometry (IMS) is an analytical technique that has gained widespread use in many applications dealing with the detection of contaminants due to its excellent sensitivity and rapid operation. Its main advantages include low detection limits, rapid response, simplicity, portability, and relatively low cost. IMS is a gas-phase ion separation technique in which ion mobility measurement is based on the drift velocities of ions in an electric field at ambient pressure. The technique is similar to TOFMS except that it operates under atmospheric pressure [77,78].

The IMS approach has been applied to the detection of mycotoxins in only a few studies. The first study focused on determining aflatoxins B1 and B2 in pistachios by means of corona discharge IMS [78]. In another study, the mycotoxin ZON and its metabolites α-zearalenol (α-ZOL), β-zearalenol (β-ZOL), and α-zearalanol (α-ZAL) were analyzed by means of a novel high-field asymmetric waveform ion mobility spectrometry (FAIMS) method coupled with electrospray ionization (ESI). In comparison with ordinary ESI–MS performance parameters, significantly lower detection limits were obtained [79]. Khalesi et al. [80] described the IMS determination of OTA

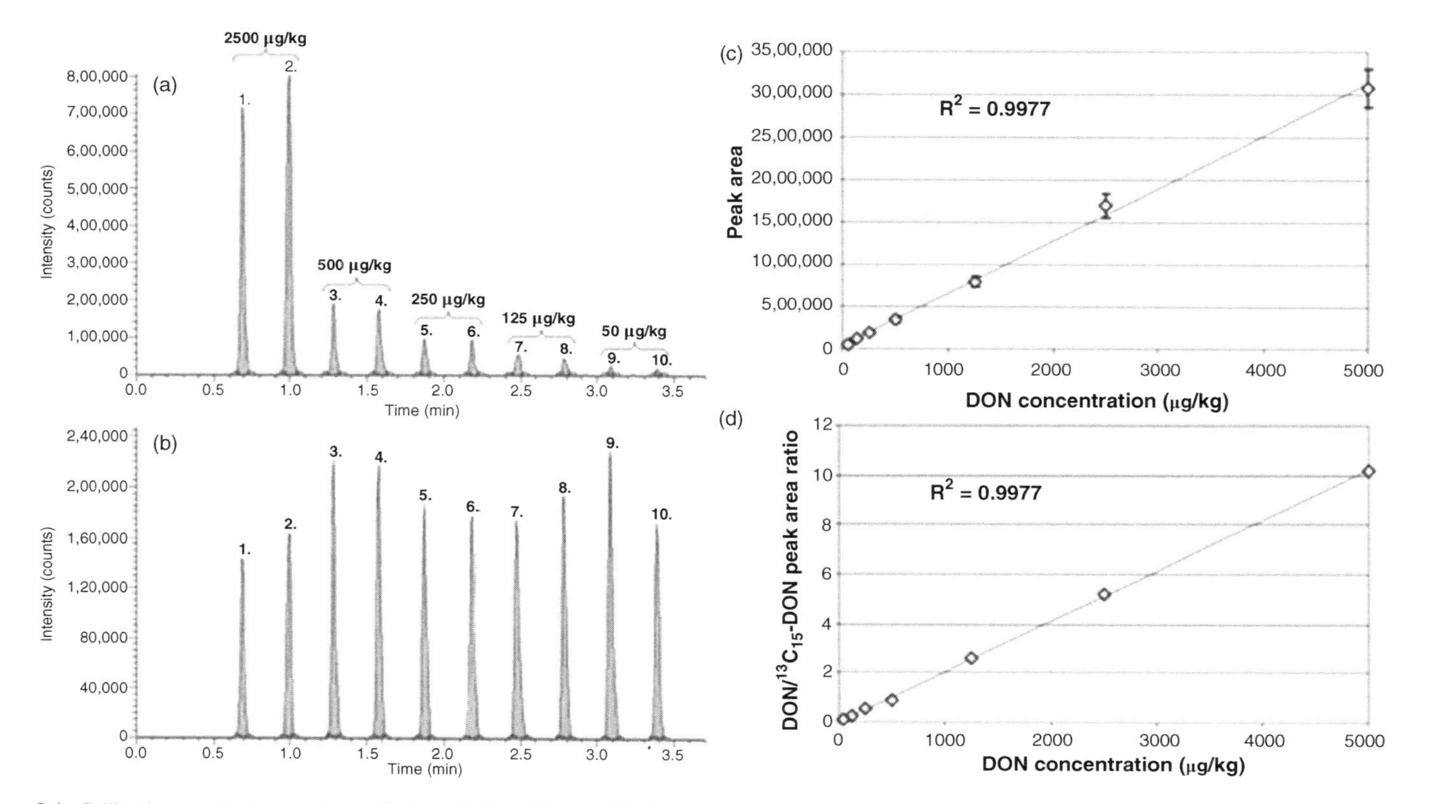

Figure 8.4. Calibrations employing matrix-matched standards and isotope dilution. (a) Extracted target ion record: DON (m/z 331.0943 ± 4 ppm); concentration in the range of 50–2500 μg/kg. (b) Extracted target ion record: $^{13}C_{15}$-DON (m/z 346.1446 ± 4 ppm); concentration 500 μg/kg. (c) External calibration curve. (d) Isotope dilution calibration curve. Error bars are standard deviations calculated from three repeated injections [32]. *Source*: Ref. [32], Figure 3, p. 1956. Reproduced with permission of Elsevier Science Ltd.

in licorice root after sodium bicarbonate (0.13 M) and methanol (9:1, v/v) extraction and immunoaffinity cleanup. A detection limit as low as 0.01 ng/g of OTA in matrix was reported.

8.4.4 Immunochemical Methods

Immunochemical screening assays represent an important group of high-throughput tools for analyzing mycotoxins in various biological matrices, including food and feed. These techniques are characterized by rapid sample preparation and minimal time of analysis [17]. Because of their high selectivity provided by specific antibodies, their relative simplicity, and field portability, immunochemical methods are widely employed in industry and for purposes of agricultural control to obtain instant information on contamination with mycotoxins [81,82]. The predominant immunochemical techniques are based on ELISA, lateral flow devices (LFD), and surface plasmon resonance (SPR) technology [83]. Similar to the previously discussed techniques, the trends in this field are toward the development of rapid multimycotoxin screening methods with improved detection limits, decreased matrix effects, and simplified operation [81,83]. Several comprehensive reviews have recently been published by Zheng et al. [15], Goryacheva et al. [81] and Maragos et al. [84]. The following sections provide an overview of current applications of and future trends in immunochemical methods.

8.4.4.1 Enzyme-Linked Immunosorbent Assay

The microtiter plate ELISA is the most frequently applied rapid method for the analysis of mycotoxins. Both direct and indirect ELISA kits are commercially available for a variety of mycotoxins. The ELISA kits are usually intended for the analysis of aflatoxins, fumonisins, trichothecenes, OTA, and ZON in cereals (maize, wheat, and oats), nuts, milk or cheese (AFM1), and feed. The majority of studies employing ELISA are aimed at monitoring mycotoxins in raw materials and food products. Additionally, new synthetic antigens and monoclonal antibodies for other mycotoxins, such as citrinin, are continuously being developed [85]. Other new polyclonal antibodies and ELISA kits for determination of tenuazonic acid in flour [86], trichothecene mycotoxin verrucarin A in indoor environments [87], and aflatoxins in herbal medicine products [88] have been recently introduced.

The main disadvantage of ELISA tests is the existence of antibody cross-reactivity to matrix or structurally related mycotoxins, which can produce overestimation or false positive results. Therefore, LC–MS-based confirmation of positive results obtained by ELISA is often performed. Although there is good agreement between data generated by ELISA and instrumental techniques for some matrices (cereals and rice), this trend cannot be generalized. To provide more accurate results, each lot of ELISA kits should be characterized by the producer in terms of cross-reactivity and recovery and this respective information should be provided to the users and declared on the product [89]. Currently, no ELISA kits that enable simultaneous determination of multiple mycotoxins are available.

8.4.4.2 Membrane-Based Immunoassays

Noninstrumental immunoassays based on antimycotoxin antibody principles are LFD, dipstick tests, and flow-through assays. In these assays, antigens or antibodies are immobilized on carrier membranes prepared from polyvinylidene difluoride, nylon, or nitrocellulose. Based on the appearance of colored lines on analysis strips, qualitative, semiquantitative, and in some cases quantitative results can be obtained by membrane immunoassays. Concentrations of mycotoxins then correlate with the intensity of the color. Test kits enabling both qualitative and quantitative analyses are commercially available for routinely controlled analytes, such as *Fusarium* mycotoxins (DON, ZON, T2 toxin, and fumonisins), aflatoxins, and OTA in many matrices. Additionally, new antimycotoxin monoclonal antibodies have also been developed, such as those for ZON or total fumonisins [90,91].

To document that a particular assay is fit-for-purpose, several studies focused on comparison between data obtained and those generated by conventional ELISA or LC–MS techniques. Most recently, the concentrations of both DON and 3-ADON have been assessed by both ELISA and LFD assays [92]. Although the data obtained with LFDs were in agreement with ELISA at most of the concentrations tested, in some cases, the recoveries of LFDs were outside the range of EU requirements (70–120%). In a study by Liu J. et al. [93], the accuracy of a new quantitative LFD for DON determination in durum wheat, semolina, and pasta was verified by parallel LC–MS/MS analyses. The assay was shown to be capable of simple, rapid, cost-effective, and robust on-site screening or remote quantitative analysis for ZON at concentrations fulfilling the worldwide legislation requirements.

Great attention has also recently been paid to the development of reliable multi-target dipsticks. For instance, a study describing semiquantitative determination of multiple mycotoxins in wheat, oats, and maize by multiplex indirect dipstick immunoassay was published in 2012 [8]. In this study, two application reports on the use of commercial dipsticks for simultaneous determination of DON, ZON, T2/HT2 toxins and fumonisins FB1, FB2, and FB3 in cereals were described. A methanol and water mixture used for the extraction of samples demonstrated recoveries in the range of 73–109% for all tested mycotoxins in all examined matrices (wheat, oats, and maize). The complete sample preparation and extraction was performed within 10 min and the dipstick analysis was performed in ~30 min. The reliability of these assays was confirmed by LC–MS analysis. The rate of false positive results, which can be caused by cross-reactivity of structural analogs, was below 13%.

The development of these types of devices is still in progress. In particular, the use of nanotechnologies and nanomaterials for preparation and construction of assays has been applied and published. For instance, a quantitative LFD for measuring of OTA in maize and wheat was developed [94], in which a ready-to-use device with antibodies labeled with gold nanoparticles was applied. Similar establishment was also published for the detection of aflatoxins B1 in food [95], but in this particular case a monoclonal antibody immobilized on nanoparticles with a silver core and a gold shell as a detection reagent was used. The assay was evaluated with the use of naturally contaminated rice, wheat, sunflower, cotton, chili peppers, and almonds. A good correlation was obtained between results obtained with a commercially available ELISA. Additionally, magnetic nano-gold particles were also applied in microsphere-based lateral flow immune-dipsticks for the detection of AFB2 in food [96].

8.4.4.3 *Surface Plasmon Resonance*

SPR represents a relatively new analytical technique that has gained increasing popularity due to its rapid, real-time, and highly selective and sensitive determination of analytes. Various applications of SPR in biochemistry, clinical diagnosis, and food analysis have already been described and several reviews describing the principles and benefits have been published [97]. SPR is an optical phenomenon used to measure changes on the surface of thin metal films under conditions of total internal reflection [81]. It allows direct detection of analytes without any labeling of interactants. As in the case of other immunoassay-based methods, SPR sensors have also been developed exclusively for mycotoxins of regulatory interest such as aflatoxins, trichothecenes, ZON, fumonisins, and OTA. Some of these tests are also commercially available [97]. There is a trend in the use of SPR technique to develop and validate multisensors for detection and quantification of numerous mycotoxins in a single analysis. This was achieved by van der Gaag et al. [98], who introduced a multiple SPR sensor for simultaneous determination of AFB1, DON, ZON, and FB1. This unique device was constructed from four flow cells containing four types of antimycotoxin antibodies. The evaluation of a prototype of the multiplex microimmunoassay quantification sensor for DON and ZON was published by Dorokhin et al. [99]. The limits of detection achieved in this study were 84 and 68 μg/kg for DON and 64 and 40 μg/kg for ZON in maize and wheat, respectively.

8.4.4.4 *Fluorescence Polarization Immunoassay*

In fluorescence polarization (FP) immunoassays, an analyte labeled with fluorophore (fluorescein) competes with free analyte for specific antibody-binding sites in solution, while fluorescence polarization of the fluorescein label is measured. An FP immunoassay has been successfully used for the determination of DON, ZON, and OTA in wheat, corn, and some food samples [17]. In a study by Bondarenko et al. [100], the influence of various fluorescent-based tracers on sensitivity of the assay for the determination of ZON and OTA was examined. The LODs (15 and 10 μg/kg for ZON and OTA, respectively) and acceptable recoveries ranging from 84 to 97% were obtained. The development and application of new FP immunoassay has recently been published for simultaneous quantitative analysis of T2 and HT2 toxins in contaminated wheat samples [101]. In this particular study, the synthesis of four fluorescein-labeled T2 or HT2 toxin tracers was carried out and their binding responses with seven monoclonal antibodies were evaluated. Using extraction with a methanol:water mixture (90:10, v/v), it was possible to obtain an average recovery of 96% and a LOD as low as 8 μg/kg for the sum of the toxins. The assay allowed quantitation of target analytes within 10 min.

8.5 CONCLUSIONS

Monitoring, control, and prevention of occurrence of mycotoxins in agricultural raw materials, food, and feed represent an important task related to quality and safety of the technological production of food and feed and to human health. With increasing number of mycotoxins of interest, there is an ongoing need for developing rapid and robust analytical strategies for analysis of these hazardous compounds in a wide range of matrices. Both instrumental (LC–MS-based techniques) and immunochemical methods (ELISA, LFD, SRM, etc.) can fulfill requirements for detectability, selectivity, and throughput. These techniques ensure accurate and reliable data applicable for further food/feed risk assessments. Both types of procedures have their advantages and disadvantages. Generally speaking, the application of sophisticated UHPLC–MS/MS instrumentation is a cutting-edge methodology for the simultaneous multimycotoxin analysis in a wide range of matrices. Considering the sensitivity of MS coupled with rapid and simple sample preparation of dilute-and-shoot or QuEChERS strategies, this combination enables development of screening methods for rapid monitoring (several minutes) of a wide range of contaminants. On the other hand, the noninvasive and easy-to-handle methods such as immunoassays, dipsticks, and biosensors offer much less costly but still sufficiently accurate strategies, which are also able to determine mycotoxins in a relatively short time. Moreover, these approaches can potentially be used on-site in industrial or agricultural settings. In the case of MS-based techniques, future trends and challenges can be seen in the incorporation of HR-MS instrumentation into routine determination of mycotoxins and, in the case of immuno-based methods, in the increase in the number of matrices and target mycotoxin combinations.

ACKNOWLEDGMENTS

M.V. and L.V. acknowledge the support by an appointment to the Research Participation Program at the Center for Food Safety and Applied Nutrition administered by the Oak Ridge Institute for Science and Education through an interagency agreement between the U.S. Department of Energy and the U.S. Food and Drug Administration. The authors wish to thank Timothy H. Begley and Jeanne I. Rader for their helpful discussions and comments.

REFERENCES

1. Bennett, J.W.; Klich, M. Mycotoxins. *Clin. Microbiol. Rev.* **2003**, 16, 497–516.
2. Murphy, P.A.; Hendrich, S.; Landgren, C.; Bryant, C.M. Food mycotoxins: an update. *J. Food Sci.* **2006**, 71, R51–R65.
3. Magan, N.; Aldred, D. Post-harvest control strategies: minimizing mycotoxins in the food chain. *Int. J. Food Microbiol.* **2007**, 119, 131–139.
4. Rahmani, A.; Jinap, S.; Soleimany, F. Qualitative and quantitative analysis of mycotoxins. *Compr. Rev. Food Sci. Food Saf.* **2009**, 8, 202–251.

5. Krska, R.; Schubert-Ulrich, P.; Molinelli, A.; Sulyok, M.; MacDonald, S.; Crews, C. Mycotoxin analysis: an update. *Food Addit. Contam.* **2008**, 25, 152–163.
6. Nielsen, K.F.; Smedsgaard, J. Fungal metabolite screening: database of 474 mycotoxins and fungal metabolites for dereplication by standardised liquid chromatography–UV–mass spectrometry methodology. *J. Chromatogr. A* **2003**, 1002, 111–136.
7. Mol, H.G.J.; Van Dam, R.C.J.; Zomer, P.; Mulder, P.P.J. Screening of plant toxins in food, feed and botanicals using full/scan high-resolution (Orbitrap) mass spectrometry. *Food Addit. Contam. A* **2011**, 28, 1405–1423.
8. Lattanzio, V.M.T.; Nivarlet, N.; Lippolils, V.; Della Gatta, S.; Huet, A.-C.; Delahaut, P.; Granier, B.; Visconti, A. Multiplex dipstick immunoassay for semi-quantitative determination of *Fusarium* mycotoxins in cereals. *Anal. Chim. Acta* **2012**, 718, 99–108.
9. European Mycotoxins Awareness Network. Available at http://www.mycotoxins.org/ (accessed August, 2013).
10. Know Mycotoxins Website. Available at http://www.knowmycotoxins.com/ (accessed August 20, 2013).
11. Kokkonen, M.; Ojala, L.; Parikka, P.; Jestoi, M. Mycotoxin production of selected *Fusarium* species at different culture conditions. *Int. J. Food Microbiol.* **2010**, 143, 17–25.
12. Dall'Erta, A.; Cirlini, M.; Dall'Asta, M.; Del Rio, D.; Galaverna, G.; Dall'Asta, Ch. Masked mycotoxins are efficiently hydrolyzed by human colonic microbiota releasing their aglycones. *Chem. Res. Toxicol.* **2013**, 26, 305–312.
13. Berthiller, F.; Crews, C.; Dall'Asta, Ch.; De Saeger, S.; Haesaert, G.; Karlovsky, P.; Oswald, I.P.; Seefelder, W.; Speijers, G.; Stroka, J. Masked mycotoxins: a review. *Mol. Nutr. Food Res.* **2013**, 57, 165–186.
14. Capriotti, A.L.; Caruso, G.; Cavaliere, C.; Foglia, P.; Samperi, R.; Lagana, A. Multiclass mycotoxin analysis in food, environmental and biological matrices with chromatography/mass spectrometry. *Mass Spectrom. Rev.* **2012**, 31, 466–503.
15. Zheng, M.Z.; Richard, J.L.; Binder, J. A review of rapid methods for the analysis of mycotoxins. *Mycopathologia* **2006**, 161, 261–273.
16. Shephard, G.S.; Berthiller, F.; Burdaspal, P.A.; Crews, C.; Jonker, M.A.; Krska, R.; MacDonald, S.; Malone, R.J.; Maragos, C.; Sabino, M.; Solfrizzo, M.; Van Egmond, H. P.; Whitaker, T.B. Developments in mycotoxin analysis: an update for 2010–2011. *World Mycotoxin J.* **2012**, 5, 3–30.
17. Koppen, R.; Koch, M.; Siegel, D.; Merkel, S.; Maul, R.; Nehls, I. Determination of mycotoxins in foods: current state of analytical methods and limitations. *Appl. Microbiol. Biotechnol.* **2010**, 86, 1595–1612.
18. Turner, N.W.; Subrahmanyam, S.; Piletsky, S.A. Analytical methods for determination of mycotoxins: a review. *Anal. Chim. Acta* **2009**, 632, 168–180.
19. European Commission. Commission Regulation (EC) No. 401/2006 laying down the methods of sampling and analysis for the official control of the levels of mycotoxins in foodstuff. *Off. J. Eur. Union* **2006**, L70, 12–34.
20. U.S. FDA Compliance Program Guidance Manual. Chapter 07: Molecular biology and natural toxins, 7307.001, 2007. Available at www.fda.gov/downloads/Food/Guidance-ComplianceRegulatoryInformation/ComplianceEnforcement/ucm073294.pdf. (accessed on Aug 20, 2013).

21. Di Mavungu, J.D.; Monbaliu, S.; Scippo, M.L.; Maghuin-Rogister, G.; Schneider, Y.J.; Larondelle, Y.; Callebaut, A.; Robbens, J.; Van Peteghem, C.; De Saeger, S. LC–MS/MS multi-analyte method for mycotoxin determination in food supplements. *Food Addit. Contam.* **2009**, 26, 885–895.
22. Song, S.Q.; Ediage, E.N.; Wu, A.B.; De Saeger, S. Development and application of salting-out assisted liquid/liquid extraction for multi-mycotoxin biomarkers analysis in pig urine with high performance liquid chromatography/tandem mass spectrometry. *J. Chromatogr. A* **2013**, 1292, 111–120.
23. Warth, B.; Sulyok, M.; Fruhmann, P.; Mikula, H.; Berthiller, F.; Schuhmacher, R.; Hametner, Ch.; Abia, W.A.; Adam, G.; Fröhlich, J.; Krska, R. Development and validation of a rapid multi-biomarker liquid chromatography/tandem mass spectrometry method to assess human exposure to mycotoxins. *Rapid Commun. Mass Spectrom.* **2012**, 26, 1533–1540.
24. Lacina, O.; Zachariasova, M.; Urbanova, J.; Vaclavikova, M.; Cajka, T.; Hajslova, J. Critical assessment of extraction methods for the simultaneous determination of pesticide residues and mycotoxins in fruits, cereals, spices and oil seeds employing ultra-high performance liquid chromatography–tandem mass spectrometry. *J. Chromatogr. A* **2012**, 1262, 8–18.
25. Anastassiades, M.; Lehotay, S.J.; Stajnbaher, D.; Schenck, F.J. Fast and easy multiresidue method employing acetonitrile extraction/partitioning and "dispersive solid-phase extraction" for the determination of pesticide residues in produce. *J. AOAC Int.* **2003**, 86, 412–431.
26. Sirhan, A.Y.; Tan, G.H.; Wong, R.C.S. Simultaneous detection of type A and type B trichothecenes in cereals by liquid chromatography coupled with electrospray ionization quadrupole time of flight mass spectrometry. *J. Liq. Chromatogr. Relat. Technol.* **2012**, 35, 1945–1957.
27. Tamura, M.; Uyama, A.; Mochizuki, N. Development of a multi-mycotoxin analysis in beer-based drinks by a modified QuEChERS method and ultra-high-performance liquid chromatography coupled with tandem mass spectrometry. *Anal. Sci.* **2011**, 27, 629–635.
28. Rubert, J.; Dzuman, Z.; Vaclavikova, M.; Zachariasova, M.; Soler, C.; Hajslova, J. Analysis of mycotoxins in barley using ultra high liquid chromatography high resolution mass spectrometry: comparison of efficiency and efficacy of different extraction procedures. *Talanta* **2012**, 99, 712–719.
29. Mol, H.G.J.; Plaza-Bolanos, P.; Zolmer, P.; de Rijk, T.C.; Stolker, A.A.M.; Mulder, P.P.J. Toward a generic extraction method for simultaneous determination of pesticides, mycotoxins, plant toxins, and veterinary drugs in feed and food matrixes. *Anal. Chem.* **2008**, 80, 9450–9459.
30. Aguilera-Luiz, M.M.; Plaza-Bolanos, P.; Romero-Gonzalez, R.; Vidal, J.L.M.; Frenich A.G. Comparison of the efficiency of different extraction methods for the simultaneous determination of mycotoxins and pesticides in milk samples by ultra-high performance liquid chromatography–tandem mass spectrometry. *Anal. Bioanal. Chem.* **2011**, 399, 2863–2875.
31. Romero-Gonzalez, R.; Frenich, A.G.; Vidal, J.L.M.; Prestes, O.D.; Grio, S.L. Simultaneous determination of pesticides, biopesticides and mycotoxins in organic products applying a quick, easy, cheap, effective, rugged and safe extraction procedure and ultra-high performance liquid chromatography–tandem mass spectrometry. *J. Chromatogr. A* **2011**, 1218, 1477–1485.

32. Vaclavik, L.; Zachariasova, M.; Hrbek, V.; Hajslova, J. Analysis of multiple mycotoxins in cereals under ambient conditions using direct analysis in real time (DART) ionization coupled to high resolution mass spectrometry. *Talanta* **2010**, 82, 1950–1957.
33. Desmarchelier, A.; Oberson, J.-M.; Tella, P.; Gremaud, E.; Seefelder, W.; Mottier, P. Development and comparison of two multiresidue methods for the analysis of 17 mycotoxins in cereals by liquid chromatography electrospray ionization tandem mass spectrometry. *J. Agric. Food Chem.* **2010**, 58, 7510–7519.
34. Rasmussen, R.R.; Storm, I.M.L.D.; Rasmussen, P.H.; Smedsgaard, J.; Nielsen, K.F. Multi-mycotoxin analysis of maize silage by LC–MS/MS. *Anal. Bioanal. Chem.* **2010**, 397, 765–776.
35. Cunha, S.C.; Fernandes, J.O. Development and validation of a method based on a QuEChERS procedure and heart-cutting GC–MS for determination of five mycotoxins in cereal products. *J. Sep. Sci.* **2010**, 33, 600–609.
36. Sospedra, I.; Blesa, J.; Soriano, J.M.; Manes, J. Use of the modified quick easy cheap effective rugged and safe sample preparation approach for the simultaneous analysis of type A- and B-trichothecenes in wheat flour. *J. Chromatogr. A* **2010**, 1217, 1437–1440.
37. Zachariasova, M.; Lacina, O.; Malachova, A.; Kostelanska, M.; Poustka, J.; Godula, M.; Hajslova, J. Novel approaches in analysis of *Fusarium* mycotoxins in cereals employing ultra-performance liquid chromatography coupled with high resolution mass spectrometry. *Anal. Chim. Acta* **2010**, 662, 51–61.
38. Malachova, A.; Dzuman, Z.; Veprikova, Z.; Vaclavikova, M.; Zachariasova, M.; Hajslova, J. Deoxynivalenol, deoxynivalenol-3-glucoside, and enniatins: the major mycotoxins found in cereal-based products on the Czech market. *J. Agric. Food Chem.* **2011**, 59, 12990–12997.
39. Monaci, L.; De Angelis, E.; Visconti, A. Determination of deoxynivalenol, T-2 and HT-2 toxins in a bread model food by liquid chromatography–high resolution-Orbitrap-mass spectrometry equipped with a high-energy collision dissociation cell. *J. Chromatogr. A* **2011**, 1218, 8646–8654.
40. Sirhan, A.Y.; Tan, G.H.; Wong, R.C.S. Method validation in the determination of aflatoxins in noodle samples using the QuEChERS method (quick, easy, cheap, effective, rugged and safe) and high performance liquid chromatography coupled to a fluorescence detector (HPLC–FLD). *Food Control* **2011**, 22, 1807–1813.
41. Arroyo-Manzanares, N.; Garcia-Campana, A.M.; Gamiz-Gracia, L. Comparison of different sample treatments for the analysis of ochratoxin A in wine by capillary HPLC with laser-induced fluorescence detection. *Anal. Bioanal. Chem.* **2011**, 401, 2987–2994.
42. Frenich, A.G.; Romero-Gonzales, R.; Gomez-Perez, M.L.; Vidal, J.L.M. Multi-mycotoxin analysis in eggs using a QuEChERS-based extraction procedure and ultra-high-pressure liquid chromatography coupled to triple quadrupole mass spectrometry. *J. Chromatogr. A* **2011**, 1218, 4349–4356.
43. Desmarchelier, A.; Mujahid, C.; Racault, L.; Perring, L.; Lancova, K. Analysis of patulin in pear- and apple-based foodstuffs by liquid chromatography electrospray ionization tandem mass spectrometry. *J. Agric. Food Chem.* **2011**, 59, 7659–7665.
44. Ferreira, I.; Fernandes, J.O.; Cunha S.C. Optimization and validation of a method based in a QuEChERS procedure and gas chromatography–mass spectrometry for the determination of multi-mycotoxins in popcorn. *Food Control* **2012**, 27, 188–193.

45. Hackbart, H.C.S.; Prietto, L.; Primel, E.G.; Garda-Buffon, J.; Badiale-Furlong, E. Simultaneous extraction and detection of ochratoxin A and citrinin in rice. *J. Braz. Chem. Soc.* **2012**, 23, 103–109.

46. Lucci, P.; Pacetti, D.; Núñez, O.; Frega, N.G. Current trends in sample treatment techniques for environmental and food analysis. In: Calderon, L., editor. *Chromatography: The Most Versatile Method of Chemical Analysis*. InTech; **2012**, pp. 127–164.

47. Songsermsakul, P.; Razzazi-Fazeli, E. A review of recent trends in applications of liquid chromatography–mass spectrometry for determination of mycotoxins. *J. Liq. Chromatogr. Relat. Technol.* **2008**, 31, 1641–1686.

48. Sulyok, M.; Krska, R.; Schuhmacher, R. Application of an LC–MS/MS based multi-mycotoxin method for the semi-quantitative determination of mycotoxins occurring in different types of food infected by moulds. *Food Chem.* **2010**, 408–416.

49. Rodriguez-Carrasco, Y.; Berrada, H.; Font, G.; Manes, J. Multi-mycotoxin analysis in wheat semolina using an acetonitrile-based extraction procedure and gas chromatography–tandem mass spectrometry. *J. Chromatogr. A* **2012**, 1270, 28–40.

50. Cigić, I.K.; Prosen, H. An overview of conventional and emerging analytical methods for the determination of mycotoxins. *Int. J. Mol. Sci.* **2009**, 10, 62–115.

51. Abia, W.A.; Warth, B.; Sulyok, M.; Krska, R.; Tchana, A.N.; Njobeh, P.B.; Dutton, M.F.; Moundipa, P.F. Determination of multi-mycotoxin occurrence in cereals, nuts and their products in Cameroon by liquid chromatography tandem mass spectrometry (LC–MS/MS). *Food Control* **2013**, 31, 438–453.

52. Herrmann, A.; Rosen, J.; Jansson, D.; Hellenas, K.-E. Evaluation of a generic multi-analyte method for detection of >100 representative compounds correlated to emergency events in 19 food types by ultrahigh-pressure liquid chromatography–tandem mass spectrometry. *J. Chromatogr. A* **2012**, 1235, 115–124.

53. Varga, E.; Glauner, T.; Berthiller, F.; Krska, R.; Schuhmacher, R.; Sulyok, M. Development and validation of a (semi-)quantitative UHPLC–MS/MS method for the determination of 191 mycotoxins and other fungal metabolites in almonds, hazelnuts, peanuts and pistachios. *Anal. Bioanal. Chem.* **2013**, 405, 5087–5104.

54. Sulyok, M.; Krska, R.; Schuhmacher, R. Application of a liquid chromatography–tandem mass spectrometric method to multi-mycotoxin determination in raw cereals and evaluation of matrix effects. *Food Addit. Contam.* **2007**, 24, 1184–1195.

55. Hajslova, J.; Zachariasova, M.; Cajka, T. Analysis of multiple mycotoxins in food. In: Zweigenbaum, J., editor. Humana Press; **2011**, pp. 233–258.

56. European Commission (EC). Commission Decision (2002/657/EC) of 12 August 2002. Implementing Council Directive (96/23/EC) concerning the performance of analytical methods and the interpretation of results. *Off. J. Eur. Commun.* **2002**, L221, 8–36.

57. European Commission (EC). Document No. SANCO/12495/2011: Method validation and quality control procedures for pesticide residues analysis in food and feed, Available at http://ec.europa.eu/food/plant/plant_protection_products/guidance_documents/docs/qualcontrol_en.pdf. (accessed August 20, 2013).

58. Tanaka, H.; Takino, M.; Sugita-Konishi, Y.; Takana, T. Development of a liquid chromatography/time-of-flight mass spectrometric method for the simultaneous determination of trichothecenes, zearalenone and aflatoxins in foodstuffs. *Rapid. Commun. Mass Spectrom.* **2006**, 20, 1422–1428.

59. Polizzi, V.; Delmulle, B.; Adams, A.; Moretti, A.; Susca, A.; Picco, A.M.; Rosseel, Y.; Kindt, R; Van Bocxlaer, J.; De Kimpe, N.; Van Peteghem, C.; De Saeger, S. JEM spotlight: fungi, mycotoxins and microbial volatile organic compounds in mouldy interiors from water-damaged buildings. *J. Environ. Monitor.* **2009**, 11, 1849–1858.

60. Veprikova, Z.; Vaclavikova, M.; Lacina, O.; Dzuman, Z.; Zachariasova, M.; Hajslova, J. Occurrence of mono- and di-glycosylated conjugates of T-2 and HT-2 toxins in naturally contaminated cereals. *World Mycotoxin J.* **2012**, 5, 231–240.

61. Herebian, D.; Zühlke, S.; Lamshöft, M.; Spiteller, M. Multi-mycotoxin analysis in complex biological matrices using LC–ESI/MS: experimental study using triple stage quadrupole and LTQ-Orbitrap. *J. Sep. Sci.* **2009**, 32, 939–948.

62. Zachariasova, M.; Cajka, T.; Godula, M.; Malachova, A.; Veprikova, Z.; Hajslova, J. Analysis of multiple mycotoxins in beer employing (ultra)-high-resolution mass spectrometry. *Rapid Commun. Mass Spectrom.* **2010**, 24, 3357–3367.

63. De Dominicis, E.; Commissati, I.; Suman, M. Targeted screening of pesticides, veterinary drugs and mycotoxins in bakery ingredients and food commodities by liquid chromatography–high-resolution single-stage Orbitrap mass spectrometry. *J. Mass Spectrom.* **2012**, 47, 1232–1241.

64. Cirlini, M.; Dall'Asta, Ch.; Galaverna, G. Hyphenated chromatographic techniques for structural characterization and determination of masked mycotoxins. *J. Chromatogr. A* **2012**, 1255, 145–152.

65. Kostelanska, M; Dzuman, Z.; Malachova, A.; Capouchova, I.; Prokinova, E.; Skerikova, A.; Hajslova, J. Effects of milling and baking technologies on levels of deoxynivalenol and its masked form deoxynivalenol-3-glucoside. *J. Agric. Food Chem.* **2011**, 59, 9303–9312.

66. Zachariasova, M.; Vaclavikova, M.; Lacina, O.; Vaclavik, L.; Hajslova, J. Deoxynivalenol oligoglycosides: new "masked" *Fusarium* toxins occurring in malt, bear and breadstuff. *J. Agric. Food Chem.* **2012**, 60, 9280–9291.

67. Elosta, S.; Gajdosova, D.; Hegrova, B.; Havel, J. MALDI TOF mass spectrometry of selected mycotoxins in barley. *J. Appl. Biomed.* **2007**, 5, 39–47.

68. Catharino, R.R.; de Azevedi Marques, L.; Silva Santos, L.; Baptista, A.S.; Gloria, E.M.; Calori-Dominguez, M.A.; Facco, E.M.P.; Eberlin, M.N. Aflatoxin screening by MALDI-TOF mass spectrometry. *Anal. Chem.* **2005**, 77, 8155–8157.

69. Davis, C.; Gordon, N.; Muphy, S.; Singh, I.; Kavanagh, K.; Carberry, S.; Doyle, S. Single-pot derivatization strategy for enhanced gliotoxin detection by HPLC and MALDI–TOF mass spectrometry. *Anal. Bioanal. Chem.* **2011**, 401, 2519–2529.

70. Marchetti-Deschmann, M.; Winkler, W.; Dong, H.; Lohninger, H.; Kubicek, C.P.; Allmaier, G. Using spores for *Fusarium* spp. classification by MALDI-based intact cell/spore mass spectrometry. *Food Technol. Biotechnol.* **2012**, 50, 334–342.

71. Wang, T.; Zhang, E.; Chen, X.; Li, L.; Liang, X. Identification of seed proteins associated with resistance to pre-harvested aflatoxins contamination in peanuts (*Arachis hupogaea L*). *BMC Plant Biol.* **2010**, 10, 267.

72. Cervino, C.; Knopp, D.; Weller, M.G.; Niessner, R. Novel aflatoxins derivatives and protein conjugates. *Molecules* **2007**, 12, 641–653.

73. Fernandez-Arguelles, M.T.; Costa-Fernandez, J.M.; Pereiro, R.; Sanz-Medel, A. Simple bio-conjugation of polymer-coated quantum dots with antibodies for fluorescence-based immunoassays. *Analyst* **2008**, 133, 444–447.

74. Kim, E.J.; Jeong, S.H.; Cho, J.H.; Ku, H.O.; Pyo, H.M.; Kang, H.G.; Choi, K.H. Plasma haptoglobin and immunoglobulins as diagnostic indicators of deoxynivalenol intoxication. *J. Vet. Sci.* **2008**, 9, 257–266.
75. Hajslova, J.; Vaclavik, L.; Poustka, J.; Schurek, J. *Analysis of Deoxynivalenol in Beer: Application Notebook*. Jeol USA, Inc.; **2008**.
76. Maragos, C.M.; Busman, M. Rapid and advanced tools for mycotoxin analysis: a review. *Food Addit. Contam.* **2010**, 27, 688–700.
77. Holopainen, S.; Nousiainen, M.; Anttalainen, O.; Sillanpaa, M.E.T. Sample-extraction methods for ion-mobility spectrometry in water analysis. *Trend Anal. Chem.* **2012**, 37, 124–134.
78. Sheibani, A.; Tabrizchi, M.; Ghaziaskar, H.S. Determination of aflatoxins B1 and B2 using ion mobility spectrometry. *Talanta* **2008**, 75, 233–238.
79. McCooeye, M.; Kolakowski, B.; Boison, J.; Mester, Z. Evaluation of high-field asymmetric waveform ion mobility spectrometry mass spectrometry for the analysis of the mycotoxin zearalenone. *Anal. Chim. Acta* **2008**, 627, 112–116.
80. Khalesi, M.; Sheikh-Zeinoddin, M.; Tabrizchi, M. Determination of chratoxin A in licorice root using inverse ion mobility spectrometry. *Talanta* **2011**, 83, 988–993.
81. Goryacheva, I.Y.; De Saeger, S.; Eremin, S.A.; Van Peteghem, C. Immunochemical methods for rapid mycotoxin detection: evolution from single to multiple analyte screening: a review. *Food Addit. Contam.* **2007**, 24, 1169–1183.
82. Posthuma-Trumpie, G.A.; Korf, J.; van Amerongen, A. Lateral flow (immuno)assay: its strengths, weaknesses, opportunities and threats: a literature survey. *Anal. Bioanal. Chem.* **2009**, 393, 569–582.
83. Krska, R.; Becalski, A.; Braekevelt, E.; Koerner, T.; Cao, X.L.; Dabeka, R.; Godefroy, S.; Lau, B.; Moisey, J.; Rawn, D.F.K.; Scott, P.M.; Wang, Z.; Forsyth, D. Challenges and trends in the determination of selected chemical contaminants and allergens in food. *Anal. Bioanal. Chem.* **2012**, 402, 139–162.
84. Maragos, C.M. Biosensors for mycotoxin analysis: recent developments and future prospects. *World Mycotoxin J.* **2009**, 2, 221–238.
85. Li, Y.N.; Wang, Y.Y.; Guo, Y.H. Preparation of synthetic antigen and monoclonal antibody for indirect competitive ELISA of citrinin. *Food Agric. Immunol.* **2012**, 23, 145–156.
86. Yang, X.X.; Liu, X.X.; Wang, H.; Xu, Z.L.; Shen, Y.D.; Sun, Y.M. Development of an enzyme-linked immunosorbent assay method for detection of tenuazonic acid. *Chin. J. Anal. Chem.* **2012**, 40, 1347–1352.
87. Gosselin, E.; Denis, O.; Van Cauwenberge, A.; Conti, J.; Vanden Eynde, J.J.; Huygen, K.; De Coninck, J. Quantification of the trichothecene Verrucarin-A in environmental samples using an antibody-based spectroscopic biosensor. *Sens. Actuators B Chem.* **2012**, 166, 549–555.
88. Shim, W.B.; Kin, K.; Ofori, J.A.; Chung, Y.C.; Chung, D.H. Occurrence of aflatoxins in herbal medicine distributed in South Korea. *J. Food Prot.* **2012**, 75, 1991–1999.
89. Tangni, E.K.; Motte, J.C.; Callebaut, A.; Pussemier, L. Cross-reactivity of antibodies in some commercial deoxynivalenol test kits against some fusariotoxins. *J. Agric. Food Chem.* **2010**, 58, 12625–12633.
90. Liu, G.; Han, Z.; Nie, D.; Yang, J.H.; Zhao, Z.H.; Zhang, J.B.; Li, H.P.; Liao, Y.C.; Song, S.Q.; De Saeger, S.; Wu, A.B. Rapid and sensitive quantitation of zearalenone in food and feed by lateral flow immunoassay. *Food Control* **2012**, 27, 200–205.

91. Molinelli, A.; Grossalber, K.; Krska, R. A rapid lateral flow test for the determination of total type B fumonisins in maize. *Anal. Bioanal. Chem.* **2009**, 395, 1309–1316.
92. Aamot, H.U.; Hofgaard, I.S.; Brodal, G.; Elen, O.; Jestoi, M.; Klemsdal, S.S. Evaluation of rapid test kits for quantification of deoxynivalenol in naturally contaminated oats and wheat. *World Mycotoxin J.* **2012**, 5, 339–350.
93. Liu, J.; Zanardi, S.; Powers, S.; Suman, M. Development and practical application in the cereal food industry of a rapid and quantitative lateral flow immunoassay for deoxynivalenol. *Food Control* **2012**, 26, 88–91.
94. Anfossi, L.; D'Arco, G.; Baggiani, C.; Giovannoli, C.; Giraudi, G. A lateral flow immunoassay for measuring ochratoxin A: development of a single system for maize, wheat and durum wheat. *Food Control* **2011**, 22, 1965–1970.
95. Liao, J.Y.; Li, H. Lateral flow immunodipstick for visual detection of aflatoxin B-1 in food using immuno-nanoparticles composed of a silver core and a gold shell. *Microchim. Acta* **2010**, 171, 289–295.
96. Tang, D.; Sauceda, J.C.; Lin, Z.; Ott, S.; Basova, E.; Goryacheva, I.; Biselli, S.; Lin J.; Niessner, R.; Knopp, D. Magnetic nanogold microspheres-based lateral-flow immunodipstick for rapid detection of aflatoxin B-2 in food. *Biosens. Bioelectron.* **2009**, 25, 514–518.
97. Li, Y.; Liu, X.; Lin, Z. Recent developments and applications of surface plasmon resonance biosensors for the detection of mycotoxins in foodstuffs. *Food Chem.* **2012**, 132, 1549–1554.
98. van der Gaag, B.; Spath, S.; Dietrich, H.; Stigter, E.; Boonzaaijer, G.; van Osenbruggen, T.; Koopal, K. Biosensors and multiple mycotoxin analysis. *Food Control* **2003**, 14, 251–254.
99. Dorokhin, D.; Haasnoot, W.; Franssen, M.C.R.; Zuilhof, H.; Nielen, M.W.F. Imaging surface plasmon resonance for multiplex microassay sensing of mycotoxins. *Anal. Bioanal. Chem.* **2011**, 400, 3005–3011.
100. Bondarenko, A.P.; Eremin, S.A. Determination of zearalenone and ochratoxin A mycotoxins in grain by fluorescence polarization immunoassay. *J. Anal. Chem.* **2012**, 67, 790–794.
101. Lippolis, V.; Pascale, M.; Valenzano, S.; Pluchinotta, V.; Baumgartner, S.; Krska, R.; Visconti, A. A rapid fluorescence polarization immunoassay for the determination of T-2 and HT-2 toxins in wheat. *Anal. Bioanal. Chem.* **2011**, 401, 2561–2571.

INDEX

A
Acceptable daily intake (ADI), 170, 171
Acetic acid, 98, 139, 243–245
Acetonitrile, 7, 22, 33, 34, 54, 121, 123, 150, 151, 241, 243, 245, 253
Acrylamide, 8, 94–99
  determination using GC-MS, 98
  determination using LC–MS/MS, 98
    chromatogram, 99
  formation, in food processing, 98
  levels in foodstuffs, 95
  techniques for analysis, 98
Adipose tissue, 118
Administrative MRLs (AMRLs), 170
Adsorption processes, 23
Aflatoxins, 233–234, 237, 253
Alkyl-bonded silica columns, 173
Alternaria fungi, 231, 252
  toxins, 235, 237
Ambient desorption/ionization techniques, 38–62, 253
  applications, 55, 61, 65
    relevant to food safety and quality, 56–60
Ammonia, 52, 94
Analog-to-digital converter (ADC), 176
Analyte retention factors, 17
Analytical quality control (AQC) system, 76, 86
  internal quality control, 86
    method performance verification in routine use, 86
  proficiency testing, 86, 87
Analytical workflow, elements, 80
  extraction efficiency, 81
  sample preparation, 80, 81
  sample processing, effects, 81
Animal fat, 117
Anthocyanidins, 9
Antimicrobial residues
  in meat, 223–227
  principle for screening, 225
  targeted/nontargeted screening approaches, 223–227
APCI. *See* Atmospheric pressure chemical ionization (APCI)
AQC guidelines, 76, 78, 90
Arginine, 109
ASAP. *See* Atmospheric pressure solids analysis probe (ASAP)
Aspergillus, 231, 233, 242, 252, 253
Atmospheric pressure chemical ionization (APCI), 26, 44, 45, 53, 100, 102, 106, 250
Atmospheric pressure solids analysis probe (ASAP), 45, 46, 55, 60
  ion source, and ionization process, 46
Atrazine, 9, 56, 59, 133, 147
Audits, 75
Australian government statutory authority, 171
Australian Pesticides and Veterinary Medicines Authority (APVMA), 171
Automated extraction systems, 8
Automated SPE workstations, 8
Automation of weighing and preparing standard solutions, 5
  QuEChERS, 6
  SweEt, 6
Azo-dyes, 105

High-Throughput Analysis for Food Safety, First Edition.
Edited by Perry G. Wang, Mark F. Vitha, and Jack F. Kay.

B
Beauvericin, 232, 237, 245
Belgium, dioxins in pork and milk products, 1
Bensulfuron-methyl, 118, 122, 137, 149
Benzene, 1, 10, 119
Benzimidazoles, 193
Benzyl butyl phthalate, 103, 105
Bioaccumulation, 103
Bioactivity-based methods, 2
Bioflavonoids, 9
Biological techniques, 172, 238, 256–258
Biosensors, 194, 238, 259
Bis(2-ethylhexyl)phthalate, 103
Bisphenol A (BPA), 101
Bovine spongiform encephalopathy, 1
Brilliant green (BG), 220
  extracted ion chromatograms, 222
  high-resolution LC–MS, 221
  mass spectra, 221
  and metabolites in fish
    identification under HRMS conditions (*See* Nontargeted approache; Targeted approache)
Brominated flame retardants, 19
Bruker micrOTOF system, 182
Butyl benzyl phthalate (BBP), 120
β-Zearalenol (β-ZOL), 254

C
Canadian Food Inspection Agency (CFIA), 170
Capillary electrophoresis, 10
Carcinogen, 94, 117, 119, 220, 237
Centers for Disease Control and Prevention (CDC), 1, 2
Chemical contamination, 90, 93, 95, 97, 109
Chemical ionization (CI), 18
Chemical pollutants, 117
  in animal fat, 118
China, 1
  melamine milk crisis, 3, 23, 94
  pork samples, analysis, 152
  regulations for mycotoxins, 237
  risk assessment of environmental pollution, 162
Chloramphenicol, 168, 170, 174, 187, 193, 194
Chloropropanols, 100
Chlortoluron, 118, 122, 137, 149
Chromatograms, 20, 99, 104, 110, 138, 139, 185, 196, 197, 200, 222, 251
Chromatography
  based approaches, 15
  performance, 22
  separation of VDs, 174–175
CI. *See* Chemical ionization (CI)
Citrinin, 237
Claviceps fungi, 252
  toxicogenic molds of, 231
Code of federal regulations (CFR), 96-97
Codex Alimentarius Commission (CAC), 2, 76, 95, 171, 172, 215
  standards, recommendations for, 2
Column
  Florisil, 121
  geometry, 16, 174
  length, 16, 220
  Megabore, 17
  particle-based, 23
  with a small internal diameter, 16
Confirmatory analysis, 79, 191, 250
Confirmatory methods, 2, 78, 79, 199
  validation and EU regulation, 79
Control of records, 75
Core–shell columns, 24
Correlation coefficients, 28–29
Crystal violet (CV), 220
C-shaped storage trap, 178
Cyanuric acid, 23, 94, 109
  chromatogram, 110

D
DAPCI. *See* Desorption atmospheric pressure chemical ionization (DAPCI)
DAPPI. *See* Desorption atmospheric pressure photoionization (DAPPI)
Darcy's law, 22
DART ionization. *See* Direct analysis in real-time (DART) ionization
DART–Orbitrap MS fingerprinting, 254
Data-dependent acquisition (DDA) mode, 177, 218
Decision limit (CCα), 171, 192, 195, 196
Deoxynivalenol (DON), 54, 232, 237, 243, 245, 253–255, 257, 258
  calibrations, and isotope dilution, 255

matrix-matched standards/isotope dilution, 255
Desciprofloxacin, 185
DESI. *See* Desorption electrospray ionization (DESI)
Desorption atmospheric pressure chemical ionization (DAPCI), 45, 46, 59
Desorption atmospheric pressure photoionization (DAPPI), 46, 47, 60
Desorption electrospray ionization (DESI), 41–43, 53–54, 253
geometry-independent, 43
ion source, and ionization process, 42
mass analyzers, suitable to deal with, 53
matrix effects, 54
optimization, 47–48
for food-related DESI applications, 42
quantification, 54–55
sensitivity, 42
source parameters, 48–50
transmission mode (TM) DESI, 42
Detection capability (CCβ), 88–89, 171, 192, 195, 196
Diacetoxyscirpenol, 58, 232, 237, 245
Dibutyl phthalate (DBP), 104, 105, 120
Dichloromethane, 52, 103, 122, 138, 139
1,3-Dichloropropan-2-ol (1,3-DCP), 100
Dicyandiamide, 108
Diethyl phthalate (DEP), 104, 105, 120
Diethylstilbestrol, 168
Dimethyl phthalate (DMP), 104, 105, 120
Dioctyl phthalate (DOP), 120
Diode array detector (DAD), 183, 192, 248
Dioxins, 1
Direct analysis in real-time (DART) ionization, 44–45, 53–54, 253, 254
applicability of DART-MS to chicken meat metabolomics for, 62
authenticity assessment of extra virgin olive oil, 61
efficient tool for rapid determination of lipids/ionizable impurities, 62, 254
mass analyzers, suitable to deal with
TOF/Orbitrap mass analyzers, for fungicides, 53
matrix effects, 54
multivariate analysis, 62
optimization, 47–48
of partition-based sample cleanup, 62
quantification, 54–55
source parameters, 50–53
dopants, 50, 52–53
geometry, 50
ionization gas parameters, 50–52
voltages, 50
Dispersive solid-phase extraction (dSPE), 241–243, 245, 247
Diterpene glycosides, 9, 61
Docosahexaenoic acid (DHA), 5, 6
Document control, 74
DON. *See* Deoxynivalenol (DON)
DON-3-glucoside (D3G), 238, 243, 245, 250, 254
2,4-D pesticide, 118
Dried distiller's grains with solubles (DDGS), 241
Drug discovery, 3
Drug residues, 3, 76, 79, 193, 213, 215, 224
physicochemical analyses, 213
Dynamic range enhancement (DRE), 176

E

Eicosapentaenoic acid (EPA), 5, 6, 119, 120
electronic Laboratory Information Management System (LIMS), 75
Electron ionization (EI), 18
Electrospray ionization (ESI), 26, 41, 98, 103, 122, 183, 187, 190, 198, 249, 254
Electrostatic interactions, 23
Endosulfan, 20, 117, 122, 124, 148, 159, 161
Enniatins, 232, 237, 243, 244
Environmental contaminants, 2, 19
Environmental pollutants, 117
long-term hazards, 117
residues, 117
Environmental protection, 117
Enzyme-linked immunosorbent assay (ELISA), 77, 238, 256, 257, 259
disadvantage of, 256
LC–MS-based confirmation, 256
Ergot alkaloids, 236, 237, 242, 243, 248, 254
EU. *See* European Union (EU)

The European Food Safety Authority (EFSA), 1, 95, 171
European Medicines Agency (EMA), 171
European Pharmacopoeia, 171
European Union (EU), 1
analytical methods, minimum performance criteria, 171, 172
food safety framework, 168
human health of VD residues, regulated controls, 169
characterization of quinolones regulated by, 192
Commission Regulation (EU) 37/2010, 169
legislative documents, 169
MRL values, 170, 171, 189, 196, 199
principle of zero tolerance, 170
legislation on contaminants, 95
substances banned, 168
Exactive^{TM} Orbitrap, 178
Extracted ion chromatograms (EIC), 181, 218, 221–223
Extractive electrospray ionization (EESI), 43, 55, 57, 61
ion source, and ionization process, 43
QTOF mass spectra of grapes, 61
tolerance to sample matrix, 43
Extraction efficiency, 81
comparison of, 150

F
False negative, 3, 89
False positive, 3
Fast gas chromatography, 16
applications, 19
The Federal Food, Drug, and Cosmetic Act (FFDCA), 95
Federal government agencies, 2
Fertilizers, 93
Fipronil, 118
Flavor compounds, 19
Flumequine, mass spectra, 217
Fluorescence
detector, 248
liquid chromatography (LC), 168
Fluorescence polarization (FP) immunoassays, 258
Food analysis, 2, 15, 17, 55, 59, 227, 246, 258
Food and Agriculture Organization of the United Nations (FAO), 2, 170, 215
Food and Drug Administration (FDA), 1, 94
Food authenticity, 2
Food-borne pathogen, 2
Food-containing residues, with antimicrobial activity, 167
Food contamination, 16, 93, 94
accidental, 94
nonintentionally added substances (NIAS), 94
undesirable packaging contaminants, 94
Food control applications, 181
comparison studies, 195–201
confirmation/quantification methods, 191–195
screening applications, 181–191
Food monitoring program, 95
Food-producing animal drugs, 168
notice of compliance (NOC) for, 170
Food safety
classifications, 3
Food Safety Modernization Act (FSMA), 1, 95
U. S./Canada, 168
Food Standard Agency (FSA), 93
Formic acid, 98
Fourier transform ion cyclotron resonance (FTICR) mass spectrometers, 214, 216
Fourier transform orbital trap (FTMS–Orbitrap), 214
France, tainted coca-cola in, 1
Fumonisins, 233, 237, 245, 254, 256–258
Furan, 100, 101
Fusarium mycotoxins, 241, 257

G
Gas chromatography (GC), 15, 172, 247
steps, 16
Gas chromatography–mass spectrometry (GC–MS), 2, 4, 15–21, 78, 117, 124
Gastric intestinal disturbances, 167
Gel permeation chromatography (GPC), 103, 117, 121, 124, 136, 138, 151, 153

GEMS/Food Programme, 95
Graphitized carbon black (GCB), 242
Growth-promoting agents (GPAs), 169

H

Headspace analysis, 10
Headspace gas chromatography–mass spectrometry (HS-GC–MS), 10
  equilibration times, 38
  HS-MS (e-nose), for volatile compounds analysis, 38
Headspace solid-phase microextraction (HS-SPME), 19
  features, 37–38
  incubation and extraction times, 38
  MS e-nose, in food authenticity studies, 38
Heat-induced food processing contaminants, 97–101
Helium, 17, 44, 50–52, 124
Heptafluorobutyric acid (HFBA), 108, 175, 224
Heptafluorobutyrylimidazole (HFBI), 100
Hexythiazox, 118, 122, 137, 149
Higher energy collision dissociation (HCD), 178, 198
High-field asymmetric waveform ion mobility spectrometry (FAIMS) method, 254
High-performance liquid chromatography (HPLC), 7, 21, 22, 24, 101, 105, 191, 192, 195, 220, 243, 244, 249, 253
High-resolution/accurate mass analyzers, 18
High-resolution mass spectrometry (HRMS), 25, 80, 168, 173, 195–201, 214, 218, 250–252
  comparison studies, 195–201
  food control applications
    extracted MS/MS ion chromatograms, enrofloxacin milk sample, 185
    screening applications, 181–182, 185–186, 189–190
  hybrid quadrupole-time of flight (QqTOF), 168
  metabolites in fish, 220–222
  nontargeted analysis using, 219, 223, 226
  Orbitrap-MS analyzer, 168
    detection of veterinary drug residues in food samples, 187–188
  targeted analysis using, 218–219, 223
  time of flight (TOF) MS, 168
    detection of veterinary drug residues in food samples, 183–184
  veterinary drugs (VDs), 192–195
  veterinary medicinal products, in foodstuffs, 223
High-resolution TOFMS (HRTOFMS), 19
High screening capacities, 4
High-speed high-resolution/accurate mass analyzers, 18
High-speed TOFMS (HSTOFMS), 19
High-throughput analysis, 61
  for food safety, 3
High throughput concept, 3
High-throughput definition, 4
High-throughput drug analysis, 3
High-throughput screening, 3
High/ultrahigh-performance liquid chromatography (HPLC/UHPLC), 248
HPLC–QLIT–MS/MS
  using SRM mode for multiscreening of VDs, 191
HPLC–QqQ–MS/MS system, 195
HPLC–TOF–MS method, 196
HRMS. *See* High-resolution mass spectrometry (HRMS)
HT2 toxins, 252
Hybrid QqTOF- MS analyzer
  application in field of VDs, 177
  MS/MS mode, 191
Hybrid quadrupole-ion trap (QTrap) LC–MS systems, 216
Hybrid quadrupole/time-of-flight (QTOF), 250
Hydrogen bonding, 23
Hydrophilic interaction liquid chromatography (HILIC), 7, 22, 23, 108
  food analyses, 23
Hydrophilic stationary phase, 22
Hygienic limits, 237

I

Identification points (IPs), 171
Igacure, 101, 102

Immunoaffinity column (IAC), 246
Immunochemical methods, 172, 238, 256–258
  enzyme-linked immunosorbent assay, 238, 256
  fluorescence polarization (FP) immunoassays, 258
  membrane-based immunoassays, 257–258
  surface plasmon resonance, 258
Injection techniques, 16
Insecticides, 121
  in honey, 9
  organophosphorus, 49
  persistent organic pollutants, 118
Internal quality control (IQC), 76
  method performance verification in routine use, minimum requirements, 86
Ionization efficiencies, 16, 23, 26, 47, 50, 174, 218, 254
Ion mobility spectrometry (IMS), 254
  determination of OTA, 254
Ion trap (IT) LC–MS systems, 173, 189–191, 194, 216
Isobaric interferences, 178
  in food samples, 53
Isopropyl thioxanthone (ITX), 94
ISO 17025 quality standard, 73
IT technology
  detection of veterinary drug residues, 190
  identification and confirmation of VD residues in food samples, 189, 194
  improvements in, 179

J
Japan, 1
  government "positive list" to regulate, 1
  MHLW establish positive list with MRLs for pesticides, 170, 171
  pesticides in contaminated foods, 1
  regulations for industrial use of PFOS, 1
  regulations for mycotoxins, 237
Joint FAO/WHO Expert Committee on Food Additives (JECFA), 170

K
Kjeldahl method, 94

L
Labeling accuracy, 2
Lateral flow devices (LFD), 256
LC coupled with Q–MS and QqQ–MS/MS
  chromatograms for enroloxacin, 196
  LRMS limitations, 168
  scanning modes, 194–195
LC-TOF-MS method, 192
  for screening and quantification of VD residues, 185
Legislation. *See also* regulation
  for adulterated food, 97
  Code of Federal Regulations (CFR), 96, 97
  Commission Regulation (EC) No. 1881/2006, 96
  Directive 96/23/EC specifies, 215
  657/2002/EC document, 1
  on food contaminants, 95
  foodstuffs, 237
    toxicants, 237
  instrumentation and software, 97
  mycotoxins, 231, 237
  Rapid Alert System for Food and Feed (RASFF), 95, 240
  Regulation 10/2011, 96
  Regulation 2003/460/EC, 96
  Regulation No. 450/2009, 96
  toxicants, 237
  veterinary drug, 168–172
    Commission Decision 2002/657/EC, 171
    Council Directive 96/23/EC, 169
    Directive 2001/82/CE, 169
    Regulation 2377/90/EC, 169
    Regulation (EC) No. 470/2009, 169
Leucobase of brilliant green (LBG), 220, 221
  mass spectra, 221
Leuco-malachite green (LMG), 220
Limits of detection (LODs), 33, 84, 118, 121, 139, 148, 184, 187, 190, 258
Lincomycin, 193
Linear ion traps (LITs), 173, 178
  improvements in technology, 179
  use of third quadrupole (Q3) as, 180
Linear-orbital trap (LTQ-Orbitrap), 173, 215
  detection of veterinary drug residues in food samples, 187–188

Liquid chromatography–mass spectrometry (LC–MS), 2, 15, 21–22, 78, 97, 117, 124, 168, 173, 238, 248–249
Liquid extraction surface analysis, 5, 9–10
Liquid–liquid extraction (LLE), 5, 8
LODs. *See* Limits of detection (LODs)
Low-pressure gas chromatography (LP–GC), 17
  advantages, 17
Low-resolution mass spectrometry analyzers (LRMS)
  detectors, 178, 199
  limitations, 168

M

Macrolides, 193
Malachite green (MG), 220
Malicious contamination of food, 105–111
Masked mycotoxins, 238
Mass accuracy, 25, 176, 178, 218
Mass spectrometry (MS), 15, 24
  ambient desorption/ionization methods, 38–41
  based techniques, 2, 10
  calibration, role of weighting factors for, 28–30
  matrix effects, 26–28
  nontargeted analysis, 26
  targeted analysis, 24–26
Mass-to-charge ratio (*m/z*), 219, 221
Mass window setting, 21
Matrix-assisted laser desorption/ionization-mass spectrometry (MALDI-MS), 30, 252–253
  applications relevant to food safety and quality, 33–34
  in high-throughput analysis of food, 36–37
  instrumentation, 30–31
  optimization of key parameters, 31
    laser parameters, 35
    matrix, 32
    sample preparation, 32, 35
  principles, 30–31
Matrix effects, 26
Matrix solid-phase dispersion (MSPD), 5, 118, 120, 246
Maximum residue limits (MRLs), 84, 85, 169, 170, 183–184, 187, 192, 193, 215, 219, 224, 227
  CODEX Alimentarius, guidelines, 171, 172, 215
  in food products, 215
  Regulation (EC) No. 37/2010, 215
  Regulation (EC) No. 470/2009, 215
  veterinary drug, 171
Melamine, 23, 93, 109
  chromatogram, 110
Membrane-based immunoassays, 257–258
Methanol, 7, 22, 23, 33, 34, 36, 58, 98, 100, 122, 224, 247, 253, 256–258
Metsulfuronmethyl, 118
Microextraction by packed sorbent (MEPS), 5, 9
Microwave assisted extractions (MAE), 118, 119
Minimum required performance limit (MRPL), 79, 169, 170, 184
  defined, 169
  values established for, 170
Molecularly imprinted polymers (MIP), 246
Moniliformin, 237
3-Monochloropropane-1,2-diol (3-MCPD), 100, 101
Monolithic columns, 23, 24
Monomers, 94
MRLs. *See* Maximum residue limits (MRLs)
MRPL. *See* Minimum required performance limit (MRPL)
MS/MS mode, 189, 218
MS/MS transitions, 19
MSPD. *See* Matrix solid-phase dispersion (MSPD)
Multiclass/multiresidue analyses, 120–122
  actual sample analysis, 157–161
  analytical methods, 124, 136
    GC–MS/MS with EI source, 124–135
    GC–MS with NCI source, 124–137
    GPC cleanup, 124
    LC–MS/MS, 124, 136
  experiment, 122
  extraction solvent, selection of, 138–139
  GPC cleanup conditions, selection of, 136, 138
    GPC chromatogram
      blank sample, 138

Multiclass/multiresidue analyses (*Continued*)
standard fortified sample, 139
instruments, 122
linear range/LOD/LOQ, 140–149, 152
pesticide and VD residue analyses, 181, 182
precisions, 152, 157
qualitative/quantitative determination, 136
reagents, 122
recoveries, 152, 157
sample cleanup, comparison of, 151–156
sample extraction methods, comparison of, 150–151
accelerated solvent extraction, 150
homogeneous extraction, 150
oscillation extraction, 150
sample preparation, 123
standard solutions, preparation, 122–123
Multiple reaction monitoring (MRM), 18, 24, 98, 213. *See also* selected reaction monitoring (SRM)
Multiresidue methods (MRMs), 77, 81, 120
quantitative, 78, 80
screening, 181
Mycotoxins, 2, 9, 186, 231–236
analysis in high-throughput environment, 238–239
analytical methods, for analysis, 239
chromatographic techniques, 247
countries partly established regulations for, 237
definition of, 231
emerging, 237–238
in food and feed, 237
health risks, public awareness, 231
legislation and regulatory limits, 231–237
masked, 238
matrices, 240
NO-separation mass spectrometry-based methods
ambient ionization mass spectrometry, 253–254
enzyme-linked immunosorbent assay, 256
fluorescence polarization (FP) immunoassays, 258
immunochemical methods, 256
ion mobility spectrometry (IMS), 254–256
matrix-assisted laser desorption ionization–mass spectrometry, 252–253
membrane-based immunoassays, 257–258
surface plasmon resonance, 258
QuEChERS applications, 243–245
sample preparation, 239
dispersive solid-phase extraction (dSPE), 241–242
extraction of, 241–246
matrices of interest, 240–241
purification of sample extracts, 246–247
sampling, 240
use of QuEChERS, 242
separation/detection, 247
high-resolution mass spectrometry (*See* High-resolution mass spectrometry (HRMS))
liquid chromatography–mass spectrometry-based methods, 238, 248–250

N

Nano-ESI-MS analysis, 9, 10
Nanoparticles, 10, 257
National Oceanic and Atmospheric Administration (NOAA), 2
Negative chemical ionization (NCI), 18, 117, 122, 124, 136
NIR spectroscopy, 10
Nitrofurans, 168, 170, 193
Nitrogen content, of processed food, 94
Nitroimidazoles, 193
Nivalenol (NIV), 57, 232, 237, 245
N-nitrosamines, 102–103
Nongovernmental organizations (NGOs), 170
Nonintentionally added substances (NIAS), 94, 95
Nontargeted analysis, 26, 214, 219–221
aimed at identifying metabolites of antimicrobials, 224
residues in meat, 223–227
principle of high-resolution full scan screening using, 226
vs. targeted approach, 26, 216, 223
veterinary drugs, 189

Notice of compliance (NOC), 170
Nuclear magnetic resonance (NMR), 10, 109

O

Ochratoxins, 9, 235, 237, 245
Omega-6 fatty acid, 9
Optimal QuEChERS-based extraction protocol, 242
Orbitrap-MS analyzer, 25, 53, 168, 177, 214, 216, 221, 223, 252
  determination of norfloxacin and its isobaric interference, 179
  mycotoxin applications, 250–252, 254
  veterinary drugs applications, 177–179, 186, 193–194, 197
Organochlorine pesticides, 16, 118
Oxociprofloxacin, 185
Oxolinic acid, 217

P

Packaging migrants, 101–105
  bisphenol A, 101
  chemical migration, 102
  detection levels, 101
  faster QuECh-ERS method, 101
  irgacure, 101
  ITX, 101
  LC–MS/MS chromatogram, 104
  legislation, 101
  N-nitrosamines, 102
  photoinitiators in food, 102
  phthalates, 102, 105
  sample preparation, 101
  TRP-ITX, 101
PAHs. *See* Polycyclic aromatic hydrocarbons (PAHs)
Partial least-squares discriminant analysis (PLSDA) models, 253
Particle size, 10, 22
Partitioning theory, 23
Patulin (PAT), 237
PCBs. *See* Polychlorinated biphenyls (PCBs)
Penicillins, 9, 193
Penicillium fungi, 231, 242, 252
Perfluorinated contaminants (PFCs), 103, 105, 106
Perfluorooctane sulfonate (PFOS), 1, 103, 106
Perfluorooctanesulfonic acid (PFOA), 103
Persistent organic pollutants (POPs), 118–119, 157
Pesticide residue analysis, 2
Pesticides
  residue analysis, 2, 4, 19, 78, 117, 121, 186, 241
  residues in foods, screening methods for validation, 79
PFCs. *See* Perfluorinated contaminants (PFCs)
Phomopsin, 237
Photoinitiators, 94
Photoionization, 47
Phoxim, 118
Phthalate bis-2-ethylhexyl ester (DEHP), 120
Phthalate esters (PAEs), 102, 117, 119, 120, 157
Phthalates, 1, 94, 102, 105
Pistachios, 1
Plant toxins, 186
Plasticizers, 94, 103
Polar organic mobile phase, 23
Polybrominated diphenyl ethers, 120
Polychlorinated biphenyls (PCBs), 9, 19, 96, 117, 119–120, 125–135, 157
Polycyclic aromatic hydrocarbons (PAHs), 6, 19, 58, 96, 117, 119, 157
Poly(methyl methacrylate) (PMMA), 47
Polytetrafluoroethylene (PFTE), 47, 49
Polyvinylchloride (PVC), 103
Positive chemical ionization (PCI), 18
Pressurized liquid extraction, 7–8
Primary–secondary amine (PSA), 242
Product ion scan (PIS), 173
Proficiency testing, 76, 86, 87
  ISO/IEC 17043, 86
  laboratory performance, dependent on, 87
Propanil, 118
Protein precipitation (PPT), 5
Proteomics, applications, 24

Q

QqQ LC–MS systems, 216, 217
Quadrupole-time of flight mass spectrometry (QqTOF-MS), 18, 168, 177, 191, 195
  confirmation of VDs in milk samples, 185
    MS/MS ion chromatograms, 185
  methodologies used for the detection of veterinary drug residues, 183–184

Quadrupole IT (QIT), 178
detection of veterinary drug residues, 190
two-dimensional, 179
Quadrupole-linear ion trap (QqLIT), 173, 178, 249
Qualitative screening methods, 76–78
biochemical methods, 77
biological methods, 77
confirmatory methods, 78, 79
physicochemical methods, 77
selectivity of mass spectrometry-based methods, 78
validation (*See* validation)
Quality control (QC), 75, 76, 86–87
Quality manual, 74
Quality systems, 73
advantages of implementing, 73
core elements, 73
audits, 75
control of records, 75
document control, 74
internal quality control, 76
manual, 74
method performance criteria, 76
procedures, 74
roles and responsibilities, 74
staff competency, 75
system design, 73, 74
validation of methodology, 75
design, 73, 74
QuEChERS method, 5, 6, 19, 27, 56–60, 101, 118, 119, 181, 241, 242, 250, 251, 254
applications in the analysis of mycotoxins, 243–245
approaches, 181
extraction method, 19
procedure, modified, 54
Quinolones, 193

R

Rapid Alert System for Food and Feed (RASFF), 95
Recovery, 27, 152
Red wine, 9
Regulations, 1–4, 80, 215–216. *See also* Legislation
Relative isotopic abundance (RIA), 77, 189, 202
Relative standard deviations (RSDs), 118, 140–149, 198
Residue analysis. *See* Pesticides; Veterinary drugs (VDs)
Resolution
chromatography, 17, 22
mass spectrometry, 18, 25, 175–176, 178, 186, 216–217, 221, 224
Retention time reproducibility, 17
Reversed-phase liquid chromatography (RPLC), 23, 173
RIA. *See* Relative isotopic abundance (RIA)
RSDs. *See* Relative standard deviations (RSDs)

S

Salmonella, in peanuts and pistachios, 1
Salting out LLE (SALLE), 5
Sample capacity, 17
Sample cleanup comparison, 151, 246
Sample preparation, 9, 19, 80, 239
techniques, advanced, 5
Sample throughput, 4, 15, 17, 19, 181, 242, 246, 247
Sample volumes, 9, 17
Sampling, 240
Scheduled selected reaction monitoring (sSRM) algorithm, 2
Screening capacity, 4
Screening detection limit (SDL), 79
Screening method, 3, 76, 171, 181, 199
Screening target concentration (STC), 88
Selected ion monitoring (SIM) mode, 18, 189, 213
Selective reaction monitoring (SRM) modes, 24, 173, 213. *See also* Multiple reaction monitoring (MRM)
Selectivity, 78
Sensitivity, 16, 18, 24, 35, 49, 53, 102, 176, 186, 196, 246, 254, 259
Sieve® software, 221, 224
Size exclusion chromatography, 10
"Soft" ionization techniques, 18
Sol–gel process, 23
Solid–liquid extraction (SLE), 118, 241
Solid-phase extraction (SPE), 5, 8, 27, 100, 118, 119, 120, 241, 246
cleanup technique, 246
column modes, 247

Solid-phase microextraction (SPME), 8–9, 27, 101, 119
Soxhlet extraction 118, 119
Spoilage markers, 2
Staff competency, 75
State government agencies, 2
Stationary phase, 19, 23
Sterigmatocystin, 237
Steroids, 201
Sudan dyes, 93, 96, 106, 108
  contamination in Europe, 93
  in food as contaminants, 107
Sulfonamides, 168, 193
  trace analysis, 9
Surface plasmon resonance (SPR) technology, 256, 258
Swedish extraction technique (SweET), 5, 6

T
Taiwan, phthalates in drinks and foods, 1, 3
Targeted approach, 189, 218–219, 221
  identification of antimicrobial residues in meat, 223–227
  vs. non-targeted, 26, 216, 223
  principle for screening with high-resolution full scan and, 225
Tetracyclines, 168, 193
  residues, in calves, 170
Thin layer chromatography (TLC), 42, 45, 247
Time-to digital converter (TDC), 176
TOF-MS. See Time-of-flight mass spectrometry
Time-of-flight mass spectrometry (TOF-MS), 18, 78, 168, 214, 216, 250–252, 254
  application in the field of mycotoxins, 250–252
  application in field of veterinary drugs, 175–176, 183–184, 192, 194, 197
  limits of detection, 185
Tolerable daily intake (TDI) values, 237
Toluene, 47
Total ion chromatogram (TIC), 221
Toxic effects
  direct/indirect, 167
Toxicogenic strains, 231
ToxID®, 224
Tranquilizers, 193
Transesterification, 19
Traveling wave-based radio frequency-only stacked ring ion guide (TWIG), 176
Trichlorphon, 118
Trichothecenes, 232
Tridecafluoroheptanoic acid (TFHA), 108
Triethylamine (TEA), 173
Turbulent flow chromatography, 7
Tylosin, 170

U
UHPLC–LTQ–Orbitrap–MS, 186, 189
  analysis, 199
UHPLC–TOF–MS method, 193, 195, 251
Ultrahigh-performance liquid chromatography (UHPLC), 10, 21, 22, 173, 174, 181–182, 186, 192–193, 195, 197, 248–249
United Kingdom
  benzene in carbonated drinks, 1
  bovine spongiform encephalopathy in beef, 1
  tests by dairies, 3
United States
  federal laws, 2
  salmonella in peanuts and pistachios, 1
United States Department of Agriculture (USDA), 2, 94
United States Food and Drug Administration (USFDA), 167
Unit-resolution instruments, 18
UV-absorbing analytes, 47

V
Validation, 75, 79, 81–85 216
  commodity groups and representative commodities, 82, 83
  compliance with ISO 17025, 75
  method validation parameters and criteria, 84
  qualitative screening multiresidue methods
    for pesticide residues in foods, 79, 80
    SANCO document, 85
    for veterinary drug residues in foods, 87
      the Community Reference Laboratories Guidelines, 77

Validation (*Continued*)
determination of specificity/selectivity and detection capability, 88
determination of the applicability, 89
establishment of a cutoff level and calculation, 88, 89
EU legislation covering method, 87, 88
HRMS technologies, 80
Valine, 109
VDs. *See* Veterinary drugs (VDs)
Veterinary drugs (VDs), 1, 167, 172–174, 186, 189, 191, 192. *See also* Veterinary medicinal product residues; Veterinary medicinal products
chromatographic separation, 174–175
and European regulations, 168, 215–216
food-producing, infectious diseases, 215
in food samples, 168, 183, 187, 190
high-resolution mass spectrometers, 175, 213, 214, 216, 237
hybrid QqQLIT platform, 179, 216
ion trap (IT) LC–MS systems, 216
LIT scanmodes, 180
LRMS detectors, 178
LTQ–Orbitrap, 216
Orbitrap–MS analyzer, 177–178
Q–Exactive (Q–Orbitrap), 216
QqTOF–MS analyzer, 177, 216
(QTrap) LC–MS systems, 216
technologies, advantages, 173–174
TOF–MS analyzer, 175–176
legislation, 168–172
metabolites, 214
multiresidue screenings of, 182
U.S./Canada, 168
Veterinary medicinal product residues (VMPRs), 213, 214, 237. *See also* Veterinary drugs
Veterinary medicinal products (VMPs), 167, 169, 213, 215, 216, 218, 227. *See also* Veterinary drugs
Viscosity, 22
VMPRs. *See* Veterinary medicinal product residues (VMPRs)
VMPs. *See* Veterinary medicinal products (VMPs)
Volatile acids, 98
Volatility, 16, 38, 55, 174

W
Weighting factors, 28-29
World Health Organization (WHO), 2, 215, 237
World Wide Web, 219

Z
Zearalenone (ZON), 57, 233, 237, 245, 256–258
ZON-4-glucoside (Z4G), 238

CHEMICAL ANALYSIS

A SERIES OF MONOGRAPHS ON ANALYTICAL CHEMISTRY
AND ITS APPLICATIONS

Series Editor
MARK F. VITHA

Vol. 1 **The Analytical Chemistry of Industrial Poisons, Hazards, and Solvents**. *Second Edition.* By the late Morris B. Jacobs
Vol. 2 **Chromatographic Adsorption Analysis**. By Harold H. Strain (*out of print*)
Vol. 3 **Photometric Determination of Traces of Metals**. *Fourth Edition*
Part I: General Aspects. By E. B. Sandell and Hiroshi Onishi
Part IIA: Individual Metals, Aluminum to Lithium. By Hiroshi Onishi
Part IIB: Individual Metals, Magnesium to Zirconium. By Hiroshi Onishi
Vol. 4 **Organic Reagents Used in Gravimetric and Volumetric Analysis**. By John F. Flagg (*out of print*)
Vol. 5 **Aquametry: A Treatise on Methods for the Determination of Water**. *Second Edition* (*in three parts*). By John Mitchell, Jr. and Donald Milton Smith
Vol. 6 **Analysis of Insecticides and Acaricides**. By Francis A. Gunther and Roger C. Blinn (*out of print*)
Vol. 7 **Chemical Analysis of Industrial Solvents**. By the late Morris B. Jacobs and Leopold Schetlan
Vol. 8 **Colorimetric Determination of Nonmetals**. *Second Edition.* Edited by the late David F. Boltz and James A. Howell
Vol. 9 **Analytical Chemistry of Titanium Metals and Compounds**. By Maurice Codell
Vol. 10 **The Chemical Analysis of Air Pollutants**. By the late Morris B. Jacobs
Vol. 11 **X-Ray Spectrochemical Analysis**. *Second Edition.* By L. S. Birks
Vol. 12 **Systematic Analysis of Surface-Active Agents**. *Second Edition.* By Milton J. Rosen and Henry A. Goldsmith
Vol. 13 **Alternating Current Polarography and Tensammetry**. By B. Breyer and H.H. Bauer
Vol. 14 **Flame Photometry**. By R. Herrmann and J. Alkemade
Vol. 15 **The Titration of Organic Compounds** (*in two parts*). By M. R. F. Ashworth
Vol. 16 **Complexation in Analytical Chemistry: A Guide for the Critical Selection of Analytical Methods Based on Complexation Reactions**. By the late Anders Ringbom
Vol. 17 **Electron Probe Microanalysis**. *Second Edition.* By L. S. Birks
Vol. 18 **Organic Complexing Reagents: Structure, Behavior, and Application to Inorganic Analysis**. By D. D. Perrin
Vol. 19 **Thermal Analysis**. *Third Edition.* By Wesley Wm. Wendlandt
Vol. 20 **Amperometric Titrations**. By John T. Stock
Vol. 21 **Reflctance Spectroscopy**. By Wesley Wm. Wendlandt and Harry G. Hecht
Vol. 22 **The Analytical Toxicology of Industrial Inorganic Poisons**. By the late Morris B. Jacobs
Vol. 23 **The Formation and Properties of Precipitates**. By Alan G. Walton
Vol. 24 **Kinetics in Analytical Chemistry**. By Harry B. Mark, Jr. and Garry A. Rechnitz
Vol. 25 **Atomic Absorption Spectroscopy**. *Second Edition.* By Morris Slavin
Vol. 26 **Characterization of Organometallic Compounds** (*in two parts*). Edited by Minoru Tsutsui
Vol. 27 **Rock and Mineral Analysis**. *Second Edition.* By Wesley M. Johnson and John A. Maxwell
Vol. 28 **The Analytical Chemistry of Nitrogen and Its Compounds** (*in two parts*). Edited by C. A. Streuli and Philip R. Averell
Vol. 29 **The Analytical Chemistry of Sulfur and Its Compounds** (*in three parts*). By J. H. Karchmer
Vol. 30 **Ultramicro Elemental Analysis**. By Güther Toölg
Vol. 31 **Photometric Organic Analysis** (*in two parts*). By Eugene Sawicki
Vol. 32 **Determination of Organic Compounds: Methods and Procedures**. By Frederick T. Weiss
Vol. 33 **Masking and Demasking of Chemical Reactions**. By D. D. Perrin

Vol. 34 **Neutron Activation Analysis**. By D. De Soete, R. Gijbels, and J. Hoste
Vol. 35 **Laser Raman Spectroscopy**. By Marvin C. Tobin
Vol. 36 **Emission Spectrochemical Analysis**. By Morris Slavin
Vol. 37 **Analytical Chemistry of Phosphorus Compounds**. Edited by M. Halmann
Vol. 38 **Luminescence Spectrometry in Analytical Chemistry**. By Mark F. Vitha, S. G. Schulman, and T. C. O'Haver
Vol. 39. **Activation Analysis with Neutron Generators**. By Sam S. Nargolwalla and Edwin P. Przybylowicz
Vol. 40 **Determination of Gaseous Elements in Metals**. Edited by Lynn L. Lewis, Laben M. Melnick, and Ben D. Holt
Vol. 41 **Analysis of Silicones**. Edited by A. Lee Smith
Vol. 42 **Foundations of Ultracentrifugal Analysis**. By H. Fujita
Vol. 43 **Chemical Infrared Fourier Transform Spectroscopy**. By Peter R. Griffiths
Vol. 44 **Microscale Manipulations in Chemistry**. By T. S. Ma and V. Horak
Vol. 45 **Thermometric Titrations**. By J. Barthel
Vol. 46 **Trace Analysis: Spectroscopic Methods for Elements**. Edited by Mark F. Vitha
Vol. 47 **Contamination Control in Trace Element Analysis**. By Morris Zief and James W. Mitchell
Vol. 48 **Analytical Applications of NMR**. By D. E. Leyden and R. H. Cox
Vol. 49 **Measurement of Dissolved Oxygen**. By Michael L. Hitchman
Vol. 50 **Analytical Laser Spectroscopy**. Edited by Nicolo Omenetto
Vol. 51 **Trace Element Analysis of Geological Materials**. By Roger D. Reeves and Robert R. Brooks
Vol. 52 **Chemical Analysis by Microwave Rotational Spectroscopy**. By Ravi Varma and Lawrence W. Hrubesh
Vol. 53 **Information Theory as Applied to Chemical Analysis**. By Karl Eckschlager and Vladimir Stepanek
Vol. 54 **Applied Infrared Spectroscopy: Fundamentals, Techniques, and Analytical Problemsolving**. By A. Lee Smith
Vol. 55 **Archaeological Chemistry**. By Zvi Goffer
Vol. 56 **Immobilized Enzymes in Analytical and Clinical Chemistry**. By P. W. Carr and L. D. Bowers
Vol. 57 **Photoacoustics and Photoacoustic Spectroscopy**. By Allan Rosencwaig
Vol. 58 **Analysis of Pesticide Residues**. Edited by H. Anson Moye
Vol. 59 **Affity Chromatography**. By William H. Scouten
Vol. 60 **Quality Control in Analytical Chemistry**. *Second Edition.* By G. Kateman and L. Buydens
Vol. 61 **Direct Characterization of Fineparticles**. By Brian H. Kaye
Vol. 62 **Flow Injection Analysis**. By J. Ruzicka and E. H. Hansen
Vol. 63 **Applied Electron Spectroscopy for Chemical Analysis**. Edited by Hassan Windawi and Floyd Ho
Vol. 64 **Analytical Aspects of Environmental Chemistry**. Edited by David F. S. Natusch and Philip K. Hopke
Vol. 65 **The Interpretation of Analytical Chemical Data by the Use of Cluster Analysis**. By D. Luc Massart and Leonard Kaufman
Vol. 66 **Solid Phase Biochemistry: Analytical and Synthetic Aspects**. Edited by William H. Scouten
Vol. 67 **An Introduction to Photoelectron Spectroscopy**. By Pradip K. Ghosh
Vol. 68 **Room Temperature Phosphorimetry for Chemical Analysis**. By Tuan Vo-Dinh
Vol. 69 **Potentiometry and Potentiometric Titrations**. By E. P. Serjeant
Vol. 70 **Design and Application of Process Analyzer Systems**. By Paul E. Mix
Vol. 71 **Analysis of Organic and Biological Surfaces**. Edited by Patrick Echlin
Vol. 72 **Small Bore Liquid Chromatography Columns: Their Properties and Uses**. Edited by Raymond P. W. Scott
Vol. 73 **Modern Methods of Particle Size Analysis**. Edited by Howard G. Barth
Vol. 74 **Auger Electron Spectroscopy**. By Michael Thompson, M. D. Baker, Alec Christie, and J. F. Tyson
Vol. 75 **Spot Test Analysis: Clinical, Environmental, Forensic and Geochemical Applications**. By Ervin Jungreis
Vol. 76 **Receptor Modeling in Environmental Chemistry**. By Philip K. Hopke

Vol. 77 **Molecular Luminescence Spectroscopy: Methods and Applications** (*in three parts*). Edited by Stephen G. Schulman
Vol. 78 **Inorganic Chromatographic Analysis**. Edited by John C. MacDonald
Vol. 79 **Analytical Solution Calorimetry**. Edited by J. K. Grime
Vol. 80 **Selected Methods of Trace Metal Analysis: Biological and Environmental Samples**. By Jon C. VanLoon
Vol. 81 **The Analysis of Extraterrestrial Materials**. By Isidore Adler
Vol. 82 **Chemometrics**. By Muhammad A. Sharaf, Deborah L. Illman, and Bruce R. Kowalski
Vol. 83 **Fourier Transform Infrared Spectrometry**. By Peter R. Griffiths and James A. de Haseth
Vol. 84 **Trace Analysis: Spectroscopic Methods for Molecules**. Edited by Gary Christian and James B. Callis
Vol. 85 **Ultratrace Analysis of Pharmaceuticals and Other Compounds of Interest**. Edited by S. Ahuja
Vol. 86 **Secondary Ion Mass Spectrometry: Basic Concepts, Instrumental Aspects, Applications and Trends**. By A. Benninghoven, F. G. Rüenauer, and H. W. Werner
Vol. 87 **Analytical Applications of Lasers**. Edited by Edward H. Piepmeier
Vol. 88 **Applied Geochemical Analysis**. By C. O. Ingamells and F. F. Pitard
Vol. 89 **Detectors for Liquid Chromatography**. Edited by Edward S. Yeung
Vol. 90 **Inductively Coupled Plasma Emission Spectroscopy: Part 1: Methodology, Instrumentation, and Performance; Part II: Applications and Fundamentals**. Edited by J. M. Boumans
Vol. 91 **Applications of New Mass Spectrometry Techniques in Pesticide Chemistry**. Edited by Joseph Rosen
Vol. 92 **X-Ray Absorption: Principles, Applications, Techniques of EXAFS, SEXAFS, and XANES**. Edited by D. C. Konnigsberger
Vol. 93 **Quantitative Structure-Chromatographic Retention Relationships**. By Roman Kaliszan
Vol. 94 **Laser Remote Chemical Analysis**. Edited by Raymond M. Measures
Vol. 95 **Inorganic Mass Spectrometry**. Edited by F. Adams, R. Gijbels, and R. Van Grieken
Vol. 96 **Kinetic Aspects of Analytical Chemistry**. By Horacio A. Mottola
Vol. 97 **Two-Dimensional NMR Spectroscopy**. By Jan Schraml and Jon M. Bellama
Vol. 98 **High Performance Liquid Chromatography**. Edited by Phyllis R. Brown and Richard A. Hartwick
Vol. 99 **X-Ray Fluorescence Spectrometry**. By Ron Jenkins
Vol. 100 **Analytical Aspects of Drug Testing**. Edited by Dale G. Deustch
Vol. 101 **Chemical Analysis of Polycyclic Aromatic Compounds**. Edited by Tuan Vo-Dinh
Vol. 102 **Quadrupole Storage Mass Spectrometry**. By Raymond E. March and Richard J. Hughes (*out of print: see Vol. 165*)
Vol. 103 **Determination of Molecular Weight**. Edited by Anthony R. Cooper
Vol. 104 **Selectivity and Detectability Optimization in HPLC**. By Satinder Ahuja
Vol. 105 **Laser Microanalysis**. By Lieselotte Moenke-Blankenburg
Vol. 106 **Clinical Chemistry**. Edited by E. Howard Taylor
Vol. 107 **Multielement Detection Systems for Spectrochemical Analysis**. By Kenneth W. Busch and Marianna A. Busch
Vol. 108 **Planar Chromatography in the Life Sciences**. Edited by Joseph C. Touchstone
Vol. 109 **Fluorometric Analysis in Biomedical Chemistry: Trends and Techniques Including HPLC Applications**. By Norio Ichinose, George Schwedt, Frank Michael Schnepel, and Kyoko Adochi
Vol. 110 **An Introduction to Laboratory Automation**. By Victor Cerdá and Guillermo Ramis
Vol. 111 **Gas Chromatography: Biochemical, Biomedical, and Clinical Applications**. Edited by Ray E. Clement
Vol. 112 **The Analytical Chemistry of Silicones**. Edited by A. Lee Smith
Vol. 113 **Modern Methods of Polymer Characterization**. Edited by Howard G. Barth and Jimmy W. Mays
Vol. 114 **Analytical Raman Spectroscopy**. Edited by Jeanette Graselli and Bernard J. Bulkin
Vol. 115 **Trace and Ultratrace Analysis by HPLC**. By Satinder Ahuja
Vol. 116 **Radiochemistry and Nuclear Methods of Analysis**. By William D. Ehmann and Diane E. Vance

Vol. 117 **Applications of Fluorescence in Immunoassays**. By Ilkka Hemmila
Vol. 118 **Principles and Practice of Spectroscopic Calibration**. By Howard Mark
Vol. 119 **Activation Spectrometry in Chemical Analysis**. By S. J. Parry
Vol. 120 **Remote Sensing by Fourier Transform Spectrometry**. By Reinhard Beer
Vol. 121 **Detectors for Capillary Chromatography**. Edited by Herbert H. Hill and Dennis McMinn
Vol. 122 **Photochemical Vapor Deposition**. By J. G. Eden
Vol. 123 **Statistical Methods in Analytical Chemistry**. By Peter C. Meier and Richard Züd
Vol. 124 **Laser Ionization Mass Analysis**. Edited by Akos Vertes, Renaat Gijbels, and Fred Adams
Vol. 125 **Physics and Chemistry of Solid State Sensor Devices**. By Andreas Mandelis and Constantinos Christofides
Vol. 126 **Electroanalytical Stripping Methods**. By Khjena Z. Brainina and E. Neyman
Vol. 127 **Air Monitoring by Spectroscopic Techniques**. Edited by Markus W. Sigrist
Vol. 128 **Information Theory in Analytical Chemistry**. By Karel Eckschlager and Klaus Danzer
Vol. 129 **Flame Chemiluminescence Analysis by Molecular Emission Cavity Detection**. Edited by David Stiles, Anthony Calokerinos, and Alan Townshend
Vol. 130 **Hydride Generation Atomic Absorption Spectrometry**. Edited by Jiri Dedina and Dimiter L. Tsalev
Vol. 131 **Selective Detectors: Environmental, Industrial, and Biomedical Applications**. Edited by Robert E. Sievers
Vol. 132 **High-Speed Countercurrent Chromatography**. Edited by Yoichiro Ito and Walter D. Conway
Vol. 133 **Particle-Induced X-Ray Emission Spectrometry**. By Sven A. E. Johansson, John L. Campbell, and Klas G. Malmqvist
Vol. 134 **Photothermal Spectroscopy Methods for Chemical Analysis**. By Stephen E. Bialkowski
Vol. 135 **Element Speciation in Bioinorganic Chemistry**. Edited by Sergio Caroli
Vol. 136 **Laser-Enhanced Ionization Spectrometry**. Edited by John C. Travis and Gregory C. Turk
Vol. 137 **Fluorescence Imaging Spectroscopy and Microscopy**. Edited by Xue Feng Wang and Brian Herman
Vol. 138 **Introduction to X-Ray Powder Diffractometry**. By Ron Jenkins and Robert L. Snyder
Vol. 139 **Modern Techniques in Electroanalysis**. Edited by Petr Vanýek
Vol. 140 **Total-Reflction X-Ray Fluorescence Analysis**. By Reinhold Klockenkamper
Vol. 141 **Spot Test Analysis: Clinical, Environmental, Forensic, and Geochemical Applications**. *Second Edition.* By Ervin Jungreis
Vol. 142 **The Impact of Stereochemistry on Drug Development and Use**. Edited by Hassan Y. Aboul-Enein and Irving W. Wainer
Vol. 143 **Macrocyclic Compounds in Analytical Chemistry**. Edited by Yury A. Zolotov
Vol. 144 **Surface-Launched Acoustic Wave Sensors: Chemical Sensing and Thin-Film Characterization**. By Michael Thompson and David Stone
Vol. 145 **Modern Isotope Ratio Mass Spectrometry**. Edited by T. J. Platzner
Vol. 146 **High Performance Capillary Electrophoresis: Theory, Techniques, and Applications**. Edited by Morteza G. Khaledi
Vol. 147 **Solid Phase Extraction: Principles and Practice**. By E. M. Thurman
Vol. 148 **Commercial Biosensors: Applications to Clinical, Bioprocess and Environmental Samples**. Edited by Graham Ramsay
Vol. 149 **A Practical Guide to Graphite Furnace Atomic Absorption Spectrometry**. By David J. Butcher and Joseph Sneddon
Vol. 150 **Principles of Chemical and Biological Sensors**. Edited by Dermot Diamond
Vol. 151 **Pesticide Residue in Foods: Methods, Technologies, and Regulations**. By W. George Fong, H. Anson Moye, James N. Seiber, and John P. Toth
Vol. 152 **X-Ray Fluorescence Spectrometry**. *Second Edition.* By Ron Jenkins
Vol. 153 **Statistical Methods in Analytical Chemistry**. *Second Edition.* By Peter C. Meier and Richard E. Züd
Vol. 154 **Modern Analytical Methodologies in Fat- and Water-Soluble Vitamins**. Edited by Won O. Song, Gary R. Beecher, and Ronald R. Eitenmiller
Vol. 155 **Modern Analytical Methods in Art and Archaeology**. Edited by Enrico Ciliberto and Guiseppe Spoto

Vol. 156 **Shpol'skii Spectroscopy and Other Site Selection Methods: Applications in Environmental Analysis, Bioanalytical Chemistry and Chemical Physics**. Edited by C. Gooijer, F. Ariese and J. W. Hofstraat

Vol. 157 **Raman Spectroscopy for Chemical Analysis**. By Richard L. McCreery

Vol. 158 **Large (C> = 24) Polycyclic Aromatic Hydrocarbons: Chemistry and Analysis**. By John C. Fetzer

Vol. 159 **Handbook of Petroleum Analysis**. By James G. Speight

Vol. 160 **Handbook of Petroleum Product Analysis**. By James G. Speight

Vol. 161 **Photoacoustic Infrared Spectroscopy**. By Kirk H. Michaelian

Vol. 162 **Sample Preparation Techniques in Analytical Chemistry**. Edited by Somenath Mitra

Vol. 163 **Analysis and Purification Methods in Combination Chemistry**. Edited by Bing Yan

Vol. 164 **Chemometrics: From Basics to Wavelet Transform**. By Foo-tim Chau, Yi-Zeng Liang, Junbin Gao, and Xue-guang Shao

Vol. 165 **Quadrupole Ion Trap Mass Spectrometry**. *Second Edition.* By Raymond E. March and John F. J. Todd

Vol. 166 **Handbook of Coal Analysis**. By James G. Speight

Vol. 167 **Introduction to Soil Chemistry: Analysis and Instrumentation**. By Alfred R. Conklin, Jr.

Vol. 168 **Environmental Analysis and Technology for the Refining Industry**. By James G. Speight

Vol. 169 **Identification of Microorganisms by Mass Spectrometry**. Edited by Charles L. Wilkins and Jackson O. Lay, Jr.

Vol. 170 **Archaeological Chemistry**. *Second Edition.* By Zvi Goffer

Vol. 171 **Fourier Transform Infrared Spectrometry**. *Second Edition.* By Peter R. Griffiths and James A. de Haseth

Vol. 172 **New Frontiers in Ultrasensitive Bioanalysis: Advanced Analytical Chemistry Applications in Nanobiotechnology, Single Molecule Detection, and Single Cell Analysis**. Edited by Xiao-Hong Nancy Xu

Vol. 173 **Liquid Chromatography Time-of-Flight Mass Spectrometry: Principles, Tools, and Applications for Accurate Mass Analysis.** Edited by Imma Ferrer and E. Michael Thurman

Vol. 174 **In Vivo Glucose Sensing.** Edited by David O. Cunningham and Julie A. Stenken

Vol. 175 **MALDI Mass Spectrometry for Synthetic Polymer Analysis.** By Liang Li

Vol. 176 **Internal Reflection and ATR Spectroscopy.** By Milan Milosevic